BOOKS PRODUCED FOR THE MEYER PROJECT

The Facsimile

An exact reproduction of Meyer's 1570 treatise (using all the same materials and construction) based primarily on the painted copy in the Leipzig University Library, supplemented by pages from several other copies containing annotations and other use marks.

The Prestige Edition

A copy of this translation constructed using the same materials and methods as the facsimile, with restored artwork and the English text laid out and formatted to match the original book as closely as possible, including historical typefaces.

The Reference Edition

A two-volume set: volume 1 contains this translation laid out side-by-side with MICHAEL CHIDESTER's transcription of the 1570 German, with an introduction by CHRIS VANSLAMBROUCK and an appendix summarizing the Meyer Census; volume 2 contains a complete set of painted Figures at 150% size, black and white Figures at 100% size, indexes of all references to each Figure, and appendixes containing the illustrations from the München, Lund, and Rostock manuscripts.

The Reading Edition

A copy of this translation with the Figures separated into individual pairs of fencers and placed inline in the text whenever they are referenced, with an introduction by ROGER NORLING.

READING EDITION

FOUNDATIONAL DESCRIPTION OF THE ART OF FENCING

The 1570 Treatise of Joachim Meyer

Written by Joachim Meyer
Translated by REBECCA L. R. GARBER

with introduction by ROGER NORLING

HEMA Bookshelf

Published by HEMA Bookshelf, LLC.
47 High St #433
Medford, MA 02155
www.hemabookshelf.com

All illustrations in this volume were created by MICHAEL CHIDESTER, MARIANA LOPEZ RODRIGUEZ, and BRYCE and SYDNEY JOHNS based on media from the public domain.

HEMA Bookshelf endeavors to respect copyright in a manner consistent with its educational mission. If you believe any material has been included in this publication improperly, please contact us so we can make revisions to future editions.

Designed and edited by MICHAEL CHIDESTER.

Version 1.2 (2024)

ISBN 978-1-953683-34-2 (hardcover)
ISBN 978-1-953683-35-9 (softcover)
ISBN 978-1-953683-36-6 (e-book)
ISBN 978-1-953683-43-4 (large-format spiralbound)

Typeset in Libertinus Serif Display and Libertinus Sans, which are used under the Open Font License
http://libertine-fonts.org/
German text typeset in Humboldt Fraktur, used under the 1001Fonts Free For Commercial Use License
https://www.1001fonts.com/licenses/ffc.html

Printed by Ingram.

TABLE OF CONTENTS

For MIKE CARTIER *(1966–2019),*

*first Captain of the Meyer Frei Fechter Guild, whose example
continues to drive our generational pursuit of martial knowledge
and research, with sword in one hand and book in the other.*

And for everyone who loves Meyer in their heart.

Acknowledgements

This project has stretched over the better part of two years, and as with any project spanning so much time, a large cast of supporters came together to make it possible.

First and foremost, my partner KENDRA BROWN not only not only reminded me to cook food and go outside periodically during this project, but also reviewed drafts of various parts of these books and offered tons of feedback that improved the finished products immensely.

The reviewers of Dr. GARBER's draft translation were CAMERON ATKINSON, JACK BAKER, JAYSON BARRONS, CHRISTOPH BUSCHE, LIAM CLARK, CAIUS EHMKE, ADAM FRANTI, SCOTT MACDONALD, LYNDA PRESTON, JAMES REILLY, AURELIA SEDLMAIR, DAVIS SIMS, JOHN TSE, JAY VAIL, CHRISTOPHER VANSLAMBROUCK, and GINDI WAUCHOPE. Once she finalized her translation, the first round of editing was done by SHAWN M. ALLBEE, ANDY WILLIAM HILLIARD, and JEFFREY WILL. The first round of editing of ROGER NORLING's introduction was likewise done by MEG PIEKARSKA.

The gorgeous color illustrations in the Prestige and Study Editions are based on the copy of Meyer housed in the Albertina Library of the University of Leipzig, which scans were graciously placed online and deeded to the public domain. The tooling design on the cover of the Prestige Edition was created by ELLIOTT RAZO based on this Leipzig copy. TRACY SEYFERT and the artisans at *Grimm Book Bindery* produced the beautiful Prestige Edition itself.

The black and white artwork for the Reference and Reading Editions was prepared by MARIANA LOPEZ RODRIGUEZ of *Metropolitan Historical Fencing Academy* (MHFA) and BRYCE and SYDNEY JOHNS of *Ox and Plow*, based on scans of the copy housed in the Arms & Armor Collection of the Metropolitan Museum which were likewise deeded to the public domain.

CHARLOTTE MAULER HAYES of *Shutterbug's Creations* prepared the illustrations used on the covers of the Reference Edition.

Staff and volunteers at the Library of Congress in Washington, D.C.; the *Metropolitan Museum* in New York City, NY; the *Museo delle Arti Marziali* in Botticino, Italy; the *New York Public Library* in New York City, NY; and the *Worcester Art Museum* in Worcester, MA were very accommodating when I examined their original copies of Meyer to learn more about the construction and physical properties of the book itself.

In addition, ELLEN ALM, CHRISTOFF AMBERGER, WILLIAM BUSCHUR, CHRISTIAN BÜTTNER, TROELS ROSTED CHRISTENSEN, HENRIETTA DANKER, LONDON DARCE, KAT FANNING, MALCOLM FARE, JEFFREY FORGENG, HÅKAN HÅKANSSON, DIETRICH HAKELBERG, CHRISTIAN HERRMANN, NICOLIEN KARSKENS, TEA KEW, PETER KLINGER, RAINER KÖBELIN, HEDVIKA KUCHAŘOVÁ, HANNAH LÜSCHER, JO MADDOCKS, RALPH MOFFAT, REINIER VAN NOORT, PATRICK PAGLEN, HENRIK PERSSON, JESSICA ROZEK, OTMAR SINGER, MARTINA STEDEN-PAPKE, ESTHER STURM, and MIRIAM VOGLAAR all examined copies that were out of my reach and sent me notes and photographs of their findings.

JEAN HENRI CHANDLER, OLIVIER DUPUIS, KEVIN MAURER, ROGER NORLING, MANUEL VALLE ORTIZ, CHRISTOPHER VANSLAMBROUCK, ONDŘEJ VODIČKA, and ADELHEID ZIMMERMAN all contributed their expertise at various points in the project.

Wiktenauer, a project of the 501(c)(3) *HEMA Alliance*, made a generous financial contribution to this project in support of our goal to produce a free translation for the community.

Finally, I thank the many supporters who took a leap of faith and contributed money to this project with no guarantee that it would produce anything worthwhile.

MEYER PROJECT BACKERS

KASPER AASE • VLADIMIR ADAMEC • SCOTT ALDINGER • SHAWN ALLBEE • DAVID AL-LEN • JONATHAN ALLEN • NICHOLAS ALLEN • MICHAEL ALLENSON • JOSE ANTONIO ALVES DA CUNHA COUTINHO • JAMES ANDERSON III • ADAM ANDREWS • MIGUEL BONILLA ARCAUTE • KRISTEN ARGYLE • JOHN L. ARMITAGE • EUGENE ARNOLD • GRANT H. ARNOLD • GABRIEL ASATRYAN • CAMERON ATKINSON • ERIC AVILA • SCOTT BAI • JACK BAKER • MICHAEL BAKER • ANTHONY BARAJAS • SETH BARCH • PAUL BAROLET • DUSTIN BARRETT • SARAH AND MICHAEL BARSNESS • JOHN C. BAUSCHATZ • JIM BA-ZAR • SAMUEL BEARDSLEY • IAN BELL • THOMAS BELLOMA • MANUELA BELTRAMI • MULI BEN-YEHUDA • SIMON BERG • JACK BERGGREN ELERS • JARON BERNSTEIN • JOSEPH BERRY • CHRISTOPHER BERTELL • NICHOLAS BEVAN • TYLER BEXTON • THOMAS BIE-BAUW • HILDO BIERSMA • DAVID BIGGS • THOMAS ALLEN BILITER • FLORIAN BINDER • ERWAN BINEAU • NJAL BJORN • JON BOEGER • LEAH BONSER • DEVON BOORMAN • DOUGLAS BOSTIC • RONALD W. BOYD • ARMELLE BRABELLEC • MATTIAS BRÄNNSTRÖM • LORENZO BRASCHI • PHILIP BREWER • TOMAS BROSNAN • KENDRA BROWN • SHANE L. BROWN • ADAM J. BRUCE • FREIRAUM BUCH • CHRISTIAN J. BUETTNER • ANTHONY BUO-NOMO • DANIEL BURGER • NICHOLAS BURGESS • SARAH BURRIDGE • CHRISTOPH BUSCHE • CHARLES BUSCHMANN • WILLIAM BUSCHUR • RAYMOND GLENN CABREDO • ALEXIS-LOUP CAMPEAU VALLÉE • AARON CARTER • DAVID CASSERLY • CHUCK CASTEL-LANOS • RUNE CASTILLO • CHRISTOPHER CASTLE • PIERRE-ALEXANDRE CHAIZE • JEAN CHANDLER • ANNE CHAUVAT • PIERRE CHAUVAT • STEPHEN CHENEY • SEAN M. CHIN • ERIC CHRISTENSEN • DONALD CHURCH • COSMAS SEBASTIAN CIEPLAK • SAMUEL CIORTEA • DANIEL BARTHOLOMÄUS CIUPKA • LIAM CLARK • NATHAN CLARK • GORDON CLAYTON • DAVID CLEGG • DAVID COBLENTZ • MARK COGAN • BRETT COKER • RYAN COKER • DAN COLLINS • BENJAMIN CONAN • PETER CONCANNON • CHRISTY CONLEY • KATHERINE CONNOR • SHAUN CONRARDY • BEN CORFIELD • STEVEN CORWIN • CONNER CRAIG • ALYSSA CRAWFORD • PETER CROOK • AARON CUNEO • MYLES CUPP • MATTHEW J. CURREY • DAVID D'ANTONIO • JAMES R. DARLING • TAYLOR C. DARNELL • HARRY M. DAVID • HUGO DAY • CURTIS DEEM • JOSEPH DEVINCENZO • AURELIE DEYME • JOHN DICKENS • BENJAMIN D. DIETRICH • MIMI DIONNE • MILOSH DJORDJEVICH • JASON DRAGO • KATINA DUNHAM • KEAGAN DUNHAM • ERIC DUSSEL • COURT DYKE • AGNES VALYI-NAGY EATON • ANDREW EDWARDS • CAIUS EHMKE • JOHAN ELDER • BERUNI ENRIQUEZ • ANNA ERMAKOVA • ROMAIN ESSERTEL • HEIKKI EUROPAEUS • COLIN FARA-BEE • HOLLY FARLESS • TOM FARMER • MATTHIAS FASSNACHT • STEFAN FEICHTINGER • BRADLEY FELL • JOHN FENZ • ASHLIN FERGUSON • STEAPHEN FICK • JOHAN FIRMENICH • WILLIAM FISHER • MAX FISHMAN • DEAN FLAGG • SETH FLORES • ETHAN FORD • PHILIP L. FRANK • SEAN FRANKLIN • OLA JOHN ERIK FRANZÉN • KEVIN FROST • CAROLINE FRY • TIM FUKE • JOSHUA FURRATE • CHRISTINE GAMACHE • NICOLAS GAMBERA • JOSHUA GARDNER • NATHAN GEDE • ROLF GEISSBÜHLER • SHANE GIBSON • ANTHONY GILJUM • JOSEPH D. GIULIANO • REBECCA GLASS • RUPAL GOKHALE • NATHAN GOREE • WILLIAM GRANDY • ANDREW GREENE • JAMES GREENING • ERRO GREENWOOD • NATHAN GRE-PARES • CHUCK GROSS • KIMBERLY GRUNDVIG • PATRICK GSTYR • YUNQI GU • DIERK HAGEDORN • BEN HALLIWELL • DANIEL HAMBRAEUS • JEPPE H. HANSEN • ERIC HARDE-MAN • JOHN HARMSTON • GEOFFREY HARRINGTON • LARA HART • ELISA HARTMAN • CHAD HEALEY • MORGAN HEDSTRÖM • MICHAEL HEIN • GERHARD HELLEMANN • JAY HELWAGEN • MILES HENDERSON • ARTHUR HENRY • THOMAS HEYDENREICH • M. B.

HILDRETH • JOSEPH MICHAEL HILLEN • ANDY HILLIARD • AARON HIMMLER • JACQUES HINKLEY • NICKOLYS HINTON • MARK HOLGATE • ANNIE HOLMES • STEVE HRADSKY • TING HSU • EDWARD HUGHES • WILLIAM HUGHES • ZACHARY THOMAS HUNT • HARRY HUPMAN • JOHN HURWOOD • SAMULI HYTTINEN • JU HYUM • TYLER JACHETTA • JEFF JACOBSON • LAURA JAKOBSSON • JASON JAMES • OLIVER JANSEPS • DANIEL JEFFERY • LIEW JIA HUI • KENNETH JOHANSEN • BRYCE JOHNS • EMIL JUHL • FRANZISKA JUST • PETER KAHLE • ALEXANDER KALYWAS • TOM KARNUTA • ROBERT KAY • BEN KEILLER • HEIKKI KERKKÄNEN • STEFAN KERN • BEN KERR • THOMAS KESLER • TEA KEW • JAMIE KIKILIDIS • ADRIAN KIM • LEELUND KIM • DON L. KINDSVATTER • JASON KIRK • MICHAEL DAVID KITE • MATTHEW KLEIN • NOEL KLING • JAMES KLOCK • ANTHONY KLON • DYLAN KNOWLES • RYAN KOHLER • MIKKO KOLI • TIES KOOL • CLAIRE KOPENHAFER • JESSE KORHONEN • ALEX KOTARAKOS • DIANA KOVACHEVA • DANIEL KRAUSS • JUSKA KUHANEN • YOUVAL KUIPERS • JORDAN KUNEYL • LELA LAM • JASON LAPRADE • FEDERICO LASAGNI MANGHI • PATRIK LASOTA • HÉLÈNE LEBLANC • CECILIO LEDESMA • JACOB LEE • KIM LEE • JESSE LEFRANC • UNNI LEINO • QUOC M. LIEU • JOSEPH LILLY • CHARLES H. LIN • JOAKIM LINDE • ANDREAS LINDEGREN • JOSH LITTLE • PHILIP LIU • ADAM LONGLEY • CHALIN LUKAS • MAGNUS LUNDBORG • JULIAN MADDOX • ŁUKASZ MAJEWSKI • TIMOTHY MAKOFSKE • TUOMAS MALKKI • HAROLD MARION • GAVIN MARKEE • SOPHIE MARSHALL • ROBERT MARTIN • GILLES MARTINEZ • JUSTIN MASTERS • ERIC MAUER • ROBERT FRANK MAULER HAYES • JAY MAXWELL • MICHELLE MAY • RYAN MAY • CLEMENS MAYER • ROBERT MAYO • ALEXANDRA MAZZEO • DAVID MCDARBY • STUART MCDERMID • ARTHUR MCGRAW • KEN MCKENZIE • VERONIQUE MCMILLAN • JOHN MCPHERSON • LOUIS MELANCON • GREGORY D. MELE • TRACY A. MELLOW • ANDREW MENDEZ • JÖRG MENHORN • KIRSTEN MEREDITH • MICHAEL-FORREST MESERVY • CAMERON METCALF • EMANUEL MEYER • PARKER MIDDENDORP • MANUEL MIGONE • CONNIE MILLER • NIKOLAS MILLER • MARK MILLMAN • ANDREW J. MINEO • JUSTIN MINGO • MATTHEW MOLE • ELIOT MOOK • TREVOR MORA • ALICIA MOREIRA MAYO • SEAN MORGAN • DIEGO AVENDANO MORINEAU • JAMES MORRIS • PETER MORROW • MATTHEW MOST • KEVIN MURAKOSHI • JASON MURDEY • CHARLES MURDOCK • MIKKO MUSTAJÄRVI • KEITH MYERS • JENNIFER NAGLE • GRYPHON M. NAYMAN • KEITH NELSON • CHAD NEWCOMER • MARCUS NEWTON • JIM NG • BRANDON NGUYEN • LUCAS NGUYEN • DANIEL NIEDERAU • RICHARD NOLAN • JOSEPH NORTH • JACOB NORWOOD • MAXIM NOSSEVITCH • LYDIA NTAFA • ERIC NUNLEY • MARTIN OBRIEN • YANCY ORCHARD • SCARLETT ORD • ANDREAS ORPHANIDES • CHRISTIAN OUELLET • GEORGQ PALMER • JAMES PARK • GRANT S. PARRINELLO • JOHN PARTIKA • JULIUS ALEXANDER PEDERSEN • PATRICK PERREAULT • JENNIFER PERRY • STEFAN PETERSON • VESSELIN PETKOV • DAVID CHRISTOPHER PETREE • LOUKASMÄKI PETRI • AMANDA PITCHER • DANIEL POPE • GREGORY POULOS • JOAKIM PRAMANIK-JONSSON • LYNDA PRESTON • BRIAN R. PRICE • ANDREW PRYDE • HAEDEN PURYEAR • DEAN PYE • ROBIN RABECHAULT • DOMENIK RADKE • DAVID RAWLINGS • BETH RAWSON • DUSTIN REAGAN • JAMES REILLY • NANCY REIMERS • ANDREW IAIN RENSHAW • JEFF RICHARDSON • ANTHONY RISCHARD • JEA RITCHEY • ADAM RITZ • ALEX ROBERTS • JAMES ROBERTS • BRAD ROBINSON • JASON ROGGENBUCK • GUDET ROMAIN • JOE ROMEO • EETU RÖPELINEN • JOHN D. ROTHE • JARYD RULE • BJÖRN RÜTHER • DYLAN SAFFORD • EDWARD SAGRITALO • JAMES DAVID SAMUEL • DIMITRI SANBORN • JULIA SANCHEZ SIMON • PHILIPP SATTLER • ANDRÉ SAUL • ALEXANDER SCHEWTSCHENKO • BERRY SCHMAAL • JOHN SCHMIDT • ANTHONY SCHMITT • NICHOLAS SCHNEIDER • THOMAS CHRISTIAN SCHRATWIESER • TILL

EDITOR'S PREFACE

"...About the Meyer Symposium, in May. We would love to have you there as our Guest. You don't need to give a lecture, or class, just come and hang with us. I wish we could help you with the Airfare, but we got some good folks from Europe we are flying over. Man, I hope you consider this. It's a fun time, and a very different HEMA event. Look forward to seeing you there?"

I got this message from KEVIN MAURER, senior researcher of the *Meyer Freifechter Guild* (MFFG), in January 2014. I had interacted with members of guild online quite a bit, and I also knew the guild founder and captain from a decade earlier when we were both members of the old *Association for Renaissance Martial Arts* (ARMA), though I had never run into any of them in real life. In fact, my only experience with the teachings of Joachim Meyer was some workshops that JAKE NORWOOD taught in the mid '00s.

But decided to make it happen. In May, I rolled into Chicago O'Hare Airport and was picked up by CHRIS VANSLAMBROUCK and HEIDI ZIMMERMAN, who carted me off to Lincoln, IL for my first Meyer Symposium. It was a crazy weekend, with everyone packing into a gym during the day and stacked up in only a couple hotel rooms at night punk-rock style. But it also ended up being a turning point in my path through HEMA.

KEVIN ended up not even being able to attend Meyer Symposium that year, and instead I met CURT DUNHAM, GREG MELE, ROGER NORLING, JAY VAIL, and many other teachers for the first time, establishing friendships that last until the present day. I also got a cool new nickname. And I finally met MIKE CARTIER (𝕲𝖔𝖙𝖙 𝖘𝖊𝖎 𝖎𝖍𝖒 𝖌𝖓𝖆̈𝖉𝖎𝖌), the founder and captain of the guild and a true phenomenon.

MIKE was an old-school punk anarchist who still had the tattoos and crazy hair to show it (never the same florescent color twice in all the times I ran into him). He also had the ideals to prove it, and his reluctant leadership (he tried to resign multiple times that very weekend, or propose restructuring the guild to eliminate the position of captain, but was lovingly shouted down every time) was founded in the idea that people don't need leaders, all they need is a shared vision and the willingness to help each other. You don't join the guild, he'd say, because everyone who loves Meyer in their heart is already a member—all you have to do is realize you belong there.

That weekend, he and I chatted at length about how we could grow the Meyer community, and my thought was that the biggest need was a free translation of Meyer's 1570 treatise (far and away his longest piece of writing). My friend JEFFREY FORGENG's 2006 translation had been widely pirated, of course, primarily in the form of a 2003 draft that circulated in PDF form. But something that was owned by the community and could be shared and remixed and customized and used in video and book projects without any fear of copyright could usher in a whole new era of Meyer study.

Years and Meyer Symposia came and went, but it wasn't until 2019 that I decided that Meyer's book was probably too large for anyone to take on as a free project. I needed to make it happen, and that meant hiring a professional translator and raising money to pay for it. I started casting about for a translator with the necessary skills: not only freelance, but with a strong background in Medieval and Renaissance languages and some familiarity with fencing—but ideally not familiarity with Meyer in particular, because I knew that someone weaned on JEFFREY's translation would be hard-pressed to produce an objective work free from that influence. I also started transcribing the text of the book on Wiktenauer so the Joachim Meyer page would be ready when the time came.

The obvious candidate for this job was REBECCA L. R. GARBER. She and I had both been members of the Cambridge HEMA Society (CHEMAS) for many years at that point and knew each other well. CHEMAS met once a week on MIT campus to transcribe, translate, and interpret whatever fencing treatise seemed most interesting at a given time (usually linked to the public historical fencing demonstrations that some of the group members would then present at libraries, museums, and other such venues); REBECCA lived a thousand miles away in Michigan, but would attend via video call. With KENDRA BROWN's Latin skills, AMY WEST's German skills, and REBECCA's ability to switch between both languages with ease, we slowly chewed through sources including *Jörg Wilhalm–München #1* (JWM1), *Fiore de'i Liberi–Paris* (FLP), and many sections of the sprawling manuscripts of Paul Hektor Mayr (PMD, PMM, and PMW).

She was the obvious candidate, but usually had a full roster of German corporations occupying her professional time. In that sense, the arrival of COVID in 2020 was almost beneficial, bringing with it corporate shutdowns around the world. Noting her sudden underemployment, it seemed like the ideal time to approach REBECCA with this project: a new translation of Meyer's 1570 treatise, which would be released in print by HEMA Bookshelf and then made free for use in perpetuity through Wiktenauer. To pay for her time, we'd not only sell preorders of the new translation, but also produce a facsimile of the book. I also decided to launch a new idea I'd been turning over in my head for a while: recreating the book in every detail but in English rather than German. "Dammit", she replied, and in May of 2021, she began her work. She estimated it would take about six months.

Meyer was HEMA Bookshelf's first facsimile of a printed book. I had resisted the idea for the first few years because printed treatises are often boring: "clean" copies that have only survived the centuries because no one ever used them and they were preserved, unopened, on a library shelf somewhere—time capsules that could give a limited report of the moment of time in which they were created, but tell us nothing about the centuries of history that they experienced.

Manuscripts are the result of a long process of individual, manual work by craftsmen of another age—parchmenters or papermakers, scribes, rubricators, illuminators, tanners, smiths, metalworkers, and binders, each applying their arts to contribute to something designed to last for centuries. They tend to be old and worn and full of stories and lessons. What those craftsmen wrote and drew in it tells us what the person who designed or ordered the book thought was important. We can learn about how it was intended to be used based on its size and shape and the materials used to create it, and we can learn how it was actually used based on the damage it suffered and the marks left by readers. Notes and doodles, pages pasted in or torn out, even the crayon scribbles of children—these tell us about the readers, and their relationship to the book.

But what if there were copies of Meyer that held these stories in them too? The scans of the book that circulated online at the time tended to be these "clean" copies, but no one even knew how many copies existed, let alone what was inside them. SARAH BARSNESS started to change that when she compiled a catalog of 38 copies of Meyer in 2019. I worked on expanding that catalog and grew it to its present size of 71 (which will be published in *Acta Periodica Duellatorum* later this year) and started sending friends and associates to actually look at them in the libraries and museums where they live. After the COVID lockdowns ended library visits for a few years, I ordered scans of many more copies of text and recruited librarians and curators to look at copies that couldn't be scanned.

In 2019, we already knew about two painted copies of Meyer (including what is still the most beautiful known copy, in Leipzig), but through this project we discovered six more. My hopes of finding a copy with significant hand-written 16th- or 17th-century notes haven't come true yet, but we did find several copies with minor annotations. We've also found many other use marks in the form of doodles, smears, fingerprints, food and drink stains, the odd rust imprint from a paperclip, pages added in or torn out, and so on.

Once visits started again, I realized I needed hands-on experience, so I traveled to examine three of the four copies of the 1570 known to be in the United States (and later looked at a fourth when I visited ROBERTO GOTTI's collection during the IFHEMA Symposium in Botticino, Italy in 2022), and also saw five 1600s along the way.

With scans of many of these altered pages in hand and having an idea of the material properties of these surviving copies, a facsimile suddenly seemed less like producing just another book and more like recreating a manuscript—something that bears the scars of 450 years of history.

I find Meyer interesting and want to see our understanding of his art expanded, but I am not a Meyer expert. Furthermore, I was long-since familiar with REBECCA's stance that writing introductions was agony and the worst part of trying to publish anything. So to improve this book, I approached some of my friends who are real Meyer experts. I contacted ROGER NORLING and CHRIS VANSLAMBROUCK, both of whom I'd remained friends with ever since that first Symposium, and I asked if they'd be interested in coauthoring an introduction to the eventual book.

I'd attended a lecture that ROGER gave at a subsequent Meyer Symposium in which he discoursed for two hours about the world and society that Meyer lived in (and didn't even get through all his slides). CHRIS, meanwhile, had taken a different tack in his research and spent the better part of a decade diving deep into the history of the fencing guilds and the life stories of the historical figures associated with Meyer's career, and especially his students and patrons. Between the two of them, I knew they could put together the most comprehensive study of the life and times of Meyer that had ever been written.

Unfortunately, due to complications of life and work and living on different continents, the level of collaboration the three of us were hoping for never happened. Instead, after many months of missed connections they agreed that they would each write parts of the introduction focused on their expertise, and I would stitch them together into one document.

They then proceeded to each deliver a document that was twice as long as I requested their combined introduction be.

Rather than try to cut out any of the fascinating history that they were offering, I decided that since we had two editions of the book planned—the Reading Edition that you're holding and the two-volume, dual-language Reference Edition—I could give one edition to each of them. This book contains ROGER's.

Two years later, here we are.

(Six months ended up being wildly optimistic, and REBECCA didn't deliver her final draft until December of 2022.)

I've spent many hundreds of hours over the past six months formatting and cleaning up pictures, designing layouts, and editing and polishing both the translation and the introductions by CHRIS and ROGER. My objective with my books is not simply to put out new translations, but to produce the kinds of translation *books* that I've always wanted to have. The introductory material represents the advances in Meyer research made over the past decade, and the translation threads a careful line of being faithful to the words and intent of the original text while also being clear and readable English.

REBECCA included a host of footnotes in her translation discussing the language and her translation choices, and I added many more that associate the text with Meyer's other works and with other fencing treatises that Meyer had access to. Any errors in those footnotes are mine and mine alone.

One problem with using Meyer's book has always been the Figures. They are complex and intricate, illustrating several plays at once, and they are often cluttered with background elements that are artistic but distracting to someone trying to understand the fencing. For this book, we've had artists cut out the individual fencers and separate them from the full pictures. These are then inserted into the translation inline, below or alongside the text that describes them. (If you want the big picture, both literally and figuratively, consider also getting the Reference Edition Volume 2.) Since the figures often overlap each other, the illustrations have frequent gaps where someone else's weapon or body partially covered them up (see the example of Sword Figure M). Drawing in the missing parts in your copy of the book is not required, but is encouraged.

I also added illustrations from the *Joachim Meyer–Lund* (JML) sword section to the Third Part of the Sword Section in this book, since the text is largely identical between the two and some plays were illustrated there which weren't illustrated in the 1570. I likewise added in two illustrations of rapier guards from *Joachim Meyer–München* (JMM) which weren't illustrated in the 1570. I justify these inclusions by pointing out that those manuscripts are still Meyer's work, and might provide useful visual reference for readers.

The mission of HEMA Bookshelf from the beginning has been to add something beautiful to the historical fencing community, and with these books I think we've succeeded.

And once all the books are shipped to backers of this project, the translation will go up on Wiktenauer for all to use. We did it, MIKE.

MICHAEL CHIDESTER
Project Editor

THE BOOK

Foundational Description of the Free, Knightly, and Noble Art of Fencing, with all commonly used weapons, decorated with many beautiful and useful figures was printed in Strasbourg in 1570 at the shop of Thiebolt Berger, in the Treubel house in the Wine Market district; its introduction is dated 24 February. It is printed in landscape format and the uncut original was approximately 200×250 mm. It is bound in quarto in half sheets and numbered in folio in three separate units plus several unnumbered leaves ([8], I–LXIIII, [1], I–CVII, I–XLVII, [1]), for a total of 456 pages. The paper varies in weight, and sometimes a single signature will use two different paper stocks (indicating that the signatures were cut and stored as individual sheets before binding).

78 prints (from 64 individual woodblocks) are scattered through the text which were produced in the workshop of Hans Christoph Stimmer. Most illustrations occupy the entire page and include between two and six pairs of figures engaged in fencing activities in the midst of elaborate architecture, often with spectators in the background. There is also a decorative title page print, a frontispiece containing the Wittelsbach heraldry (for its dedicatee, Johan Kasimir, future Count Palatine of Simmern), two prints that only show a single figure and occupy less than a full page, and two very small geometric designs.

The first edition shows signs of having been corrected during printing. The Figure on page I.17 originally had two small solid rectangular marks in the frame, one between the arms of the small middle tenants and one between the legs of the small upper left tenants, but these were excised from the block and are only present in some copies. Furthermore, some copies of the book have pages II.97–98 (which would properly be numbered XCVII and XCVIII) mis-numbered XCXVII and XCXIX (which would be something like 'ninety seventeen' and 'ninety nineteen'), while others correct II.97 but change II.98 to 'XCXVIII' ('ninety eighteen'). A few copies also have page II.96 numbered 'XCVII' rather than XCVI, but this was successfully corrected in most. The differences on I.17 and II.96–98 do not seem to be linked with each other and there are extant copies which have all possible combinations of the two (perhaps due to the pages being stored and assigned randomly during binding).

A second edition was printed posthumously in Augsburg in 1600 by Michael Manger (for Elias Willer). It was originally approximately 170×230 mm. It used all of the original woodblocks (which show occasional damage, most notably the loss of about a centimeter from the bottom of the title page) apart from the frontispiece, and the reset text is unchanged apart from spelling variations and implementing the errata (and introducing a few new typographical errors). This includes the dedication to Johan Kasimir, even though he was deceased at that point. An attempt was made to replicate the layout of the first edition and all of the illustrations are on the same pages in both editions, but the typeface that Manger used was wider than Berger's so the text breaks across pages in slightly different places.

The total size of either of these two print runs is unknown, but at least 40 copies of the 1570 and 31 copies of the 1600 are known to have survived until recent years (including partial copies). At least ten copies of the 1570 had some or all of their prints painted, and many copies contain annotations and other signs of use by early readers. They are now distributed across a large swathe of the globe, having traveled from Strasbourg as far east as Moskva, Russia, and as far west as Austin, Texas, United States.

A fencing master from Sword Figure A of the 1570 treatise of Joachim Meyer, which may be a representation of Meyer himself. By Hans Christoph Stimmer, 1560s.

Introduction: Joachim Meyer and His World

Roger Norling[1]

In his (too) short, 34-year life, Fechtmeister Joachim Meyer appears to have had quite a success-ful career as a fencing master, instructing burghers and nobility both while also authoring at least four different fencing treatises and possibly more. This while, at least for a time, also working as a Messerschmidt—a cutler. His works became renowned far outside of his own country, known even in Italy, and were copied into other works well over 100 years after his death.[2]

Today we are blessed with a fantastic body of work for study—a treasure trove we still, after some 15–20 years of work, have only just begun scratching the surface of. We are not the first on the stage of recreating these arts, though. With the first resurrection of the historical Euro-pean martial arts starting in the late 1800s, a wealth of material on the many aspects of the fencing culture which Fechtmeister Joachim Meyer lived in, was collected, presented, and tested by pioneers and fencers like Cpt. Alfred Hutton, Egerton Castle, Cpt. Carl A. Thimm, Gustav Hergsell, Karl Wassmannsdorff, Luigi Barbasetti, Georges Dubois, Arsène Vigeant, Cpt. Emil Fick, Archibald Corble, and many others. Several of these men would also be involved in what would soon evolve into the modern Olympic Games, in the hopes of restoring the declining art of fencing.

However, the actual *biographical* data we have on Meyer has been rather limited, and what little we have is to a large degree due to the efforts of Meyer scholar Olivier Dupuis, who for the past two decades tirelessly has sifted through the large mass of handwritten material in the city archives of the hometown he shares with Meyer: *Strasbourg*.[3] Likewise, members of the Meyer Freifechter Guild, not least my fellow Meyer scholars Kevin Maurer and Chris VanSlambrouck, have invested countless hours into researching Meyer and his times. And of course, the Meyer Freifechter Guild would never even have existed without its dear, de-parted captain and founding member Mike Cartier. I should also thank Jean Chandler for the many wonderful and fascinating conversations over the years! His knowledge of the medieval and Renaissance cities is incredible and there are few researchers I respect as much. Together, we are now well into our second decade of attempting to paint a richer and more colourful picture of Meyer's life and work. My hope is that the following chapters, will help in that goal.

In order to do this, I have chosen to lay a puzzle that I hope will give a reasonably cohesive image of the multi-faceted world a man 450-years-dead lived in—a puzzle depicting numerous small scenes which includes, among other things: Renaissance humanism and its relation to Antiquity; contemporary views on civic duties versus the "noble" employment of amoral mer-cenaries of war; social aspects of city structure in the Holy Roman Empire; martial culture in 16th century German cities; views on athleticism, astrology, and personalities of men; as well as pieces on oral vs. written culture, a brief linguistical analysis of 15th and 16th century

[1] I'd like to thank Jean Chandler, Olivier Dupuis, Kevin Maurer, and Chris VanSlambrouck for reviewing this introduction, and my wife and family for supporting me as I wrote it.
[2] See the works of Jakob Sutor von Baden (1612) and Theodori Verolini (1679).
[3] See Dupuis 2006 and Dupuis 2016.

instruction manuals, pedagogical inventions of the Renaissance versus scholasticism, and finally a short piece on the particulars of Meyer's art of fencing.

I. THE LIFE OF JOACHIM MEYER

HISTORICAL BACKDROP

When beginning our studies of a treatise on the art of fencing such as this one, we should perhaps start with asking ourselves a very fundamental but important question: *why was it written?* But with 450 years having passed, how do we even begin to understand what motivated a man to write it—beyond the more obvious aspects of financial and social gain? Perhaps one answer can be found in the world he lived in and which shaped him, presenting him with a great many threats, opportunities, worries, temptations, and obstacles.

Looking at the culture of Meyer's time, its roots, and the perceptions people then living had of it, we appear to be able to trace a cohesive thread starting with the fall of Rome and ancient astrological and social concepts surrounding virtue, and especially *civic virtues* and duties, related to thoughts on athleticism, health, and the development and corruption of character. This thread continues through the humanists of the Renaissance to men like Machiavelli, Schwendi[4] and Fronsperger,[5] and into arguments over the advantages of militias of men trained in the art of fencing versus armies of unreliable mercenaries who often engaged in murder, rape, thievery, and looting. This is a topic we will discuss more soon.

For most of us, the world we live in today is in many ways radically different from Meyer's reality, which was more comparable to life in parts of modern Africa or the Middle East, with a much closer proximity to death caused by violence or natural disasters; with frequent and deep religious conflicts, extremism, and persecution; with war in the fields and the city streets; and with mass murders rooted in religious conflict. However, this was also a world of great technical, philosophical, and social developments; of frequent travel; continents newly discovered; exotic and fantastic beasts; and exploration of both the world and the supernatural otherworld. It was a world of increasing industrialization via the harnessing of waterpower and the invention of progressively more advanced mechanisms; of growing trade; of powerful guilds and careful protection of precious skill and knowledge; of developing international economies with banking, stock markets,[6] and insurance businesses;[7] and of a spreading

---

[4] Lazarus von Schwendi was an imperial commander and military strategist who worked with Leonhard Fronsperger in the military campaign of 1566, and who also wrote a number of books on war and had an extensive private library, including a heavily annotated copy of Machiavelli's *Discorsi*. Like Machiavelli, he favoured a militia over mercenaries.

[5] Not to be confused with Georg von Frundsperg, the father of the *Landsknechte*. Leonhard Fronsperger served the Habsburgs in a number of military roles, and took part of the Siege of Marseille in 1535, the Siege of Buda & Pest in 1541–42, in the Schmalkaldic War in 1546, as well as imperial campaigns from 1553–63 (against the Ottomans) and from 1568–73, before dying due to an exploding cannon in 1575. He also wrote four different treatises on war: 1555 *Fünff Bücher von Kriegß Regiment und Ordnung*, 1557 *Von Geschütz und Fewerwerck*, 1565 *Geistliche Kriegßordnung*, and 1573 *Kriegsbuch*.

[6] See the Belgian Van der Beurze family and the Brugse Beurse.

[7] Invented in Genoa in 1347.

curiosity and enlightenment, especially in the cities—the hubs of most progress and change. In short, this was a world of constant change, new discoveries, and many unknowns.

In addition to the terrible and deadly conflicts between Catholics and Protestants and the shocking Sack of Papal Rome by Karl V's Imperial (and Catholic) troops in 1527, the Ottoman Wars in Europe would also have been in constant awareness, with the Battle of Mohacs in 1526, the Siege of Vienna—the resident city of the Imperial House of Habsburg and the heart of the Holy Roman Empire—in 1529, the Siege of Buda—which left Hungary in Ottoman control for 150 years—in 1541, and finally, the Battle of Lepanto in 1571 (the same year Meyer died). Straßburg, where Meyer lived, was rife with religious and socio-political conflict, and in 1576 (just five years after Meyer's death), we see Johann Kasimir, Count Palatine[8] of Simmern (to whom Meyer had dedicated this fencing treatise), leading Protestant troops outside of the city, heading for Saint-Nicolas to support persecuted French Huguenots.

But I am getting ahead of myself. Let us start with looking at some key dates of the history of the Alsatian region where Meyer lived.

KEY DATES

While there is not enough space to delve deeper into them, and before continuing to the cities associated with Meyer, I will now list a few important key dates that provide a little more context for the world Meyer lived in:

476 – The fall of the Western Roman Empire. The Eastern Roman Empire continues based in Constantinople.

800 – Charlemagne is crowned emperor of the Frankish (Holy) Roman Empire, followed by Otto I in 962.

1272 – The Battle of Hausbergern, where the burghers of Straßburg defeat the army of the prince-bishop of the city, liberating them from episcopal rule and making Straßburg a *Free City of the Holy Roman Empire.*

1438 – Albrecht V von Habsburg is crowned Emperor of the German Holy Roman Empire. The empire would remain in the hands of the House of Habsburg until 1740.

1439 – Gutenberg invents the movable type printing press in Straßburg. The same year, the city's Münster cathedral is completed after nearly 300 years of construction.

1444 – The Battle of St. Jakob an der Birs, a war between the Swiss Confederacy and French mercenaries (mostly Armagnacs) taking place just 1 km from Basel within the St. Alban district. During this conflict, Strasbourg houses 7,000 peasants from the surrounding area. The battle between the Swiss pikemen and Armagnac cavalry is unusually fierce, with the 1,500 pikemen refusing to surrender to the French army of 30,000, ripping arrows from their bodies and continuing to fight even after having been stabbed by lances or having lost hands. Because of

---

[8] "Count Palatine" (or *Pfalzgraf*) originally a Roman title, was used in the Holy Roman Empire for representatives of the Holy Roman Emperors who were granted quasi-monarchical powers while residing in, and maintaining, one of the dozens of palaces which the emperors had at their disposal throughout the empire.

this, the *Dauphin* decides to have his forces turn back and sign a peace treaty with Basel and the Confederacy.

1453 – The fall of Constantinople to Sultan Mehmet II, *Kayser-i-Rum*, and the final end of the Eastern Roman Empire. Greek scholars flee to Western Europe, boosting the Renaissance and a spreading fascination for Antiquity. They bring "new" texts for study with them and teach the world to read Greek. As a result of efforts to find Greek texts in monastic libraries, Roman texts by Tacitus, Vitruvius, and others are rediscovered.

1455 – The Latin Gutenberg Bible is printed in Mainz, starting the so-called 'print revolution'. By the end of the 1400s, 270 cities in Europe have their own Gutenberg printing presses and more than 20 million copies of texts will have been printed.[9]

1460 – The first printed Latin Bible in Straßburg is published by Johann Mentelin.

1466 – Mentelin's German Bible is printed. This is the first Bible printed in any language other than Latin, and for the first time, people who didn't know Latin could read the Bible. Sixteen more editions in German would follow prior to Luther's translation. In 1471, the first Bible in Italian is printed, and Bibles in French and Spanish are printed in 1476. The English would not receive a Bible in their own language until 1545, four years after the Swedes.

1474 – The Marxbrüder fencing guild is officially formed.

1477 – The Battle of Nancy, the last and deciding battle of the Burgundian Wars, takes place. Charles "the Bold", Duke of Burgundy, fights René II, Duke of Lorraine, and the Swiss Confederacy, and he ends up dead alongside of most of his army, with casualties counted in the thousands.

1486–88 – Maximilian I, Holy Roman Emperor, begins hiring Landsknechte (mercenaries), to organize a 'pike and shot' army modelled on the Swiss Reißläufer, first giving a unit to Eitel Friedrich II, Count of Hohenzollern, who employed Swiss instructors in Bruges. The Landsknechte, alongside the Reißläufer and the Italian *condottieri*, would grow to become a hugely important source of power (not least under commander Georg von Frundsberg), but also a problematic and unreliable weapon, both within and outside of the Holy Roman Empire, and one which could just as easily strike back at the empire itself.

1487 – The Marxbrüder receive their privileges from Emperor Frederich III.

1492 – Ships commanded by Genoese Cristoffa Corombo encounter the American continent, later explored by Amerigo Vespucci. Due to the popularity of fictionalized accounts of Vespucci's travels, 'America' becomes the accepted name of the new continent within 40 years.

1517 – Martin Luther's *95 Theses* are posted. In 1518, they are translated from Latin to German and within two weeks, they would spread throughout Germany, reaching France, England, and Italy within a year.

1517 – A cutler who may have been Joachim Meyer's father is listed as a burgher of Basel.

[9] See FEBVRE & MARTIN 1976.

1518 – The Straßburg Dancing Plague, in which some 40–50 people dance for weeks. The causes remain unclear, but proposed theories include food poisoning and stress-induced mass hysteria.

1520 – Pope Leo X orders Martin Luther to renounce his works in the *Exsurge Domine.*

1521 – The Diet of Worms. Emperor Karl V orders Martin Luther to renounce his works in response to the pope's *Exsurge Domine*, but Luther refuses. He is excommunicated by the pope and declared an outlaw by the emperor.

1522 and 1534 – Luther's New Testament and Bible are printed in German.

1526 – The Battle of Mohács, Hungary and the Holy Roman Empire vs. the Ottoman Empire lead by Suleiman the Magnificent. The Ottomans capture Buda and Pest.

1526 – The Imperial Diet of Speyer.

1527 – The Sack of Rome by the Imperial troops of Karl V. Ten months later, its population has dropped from 55,000 to 10,000. The split between Catholics and Protestants is made permanent and the Italian High Renaissance ends.[10]

1528 – Imperial Commander Georg von Frundsberg, the "father of the Landsknechte" **dies** at the age of 54, after over 35 years serving as a military commander for the empire.

1529 – The First War of Kappel between the Protestant and Catholic Cantons of the Swiss Confederacy.

1529 – The Imperial Diet of Speyer.

1529 – The Siege of Vienna. For two weeks, Suleiman the Magnificent besieges the city with 100,000 men, including many *Sipahi* and *Janissary* elite troops.

1530 – The Augsburg Confession, seeking to restore religious and political unity in the face of the Ottoman threat.

1530 – Jehan Cauvin, the father of Calvinism, breaks away from the Roman Catholic Church.

1531 – The Schmalkaldic League, a Lutheran defensive military alliance similar in some respects to today's NATO, is formed by Johan Friedrich I, Elector of Saxony, and Philipp I, Landgrave of Hesse, father-in-law of Heinrich, Count of Eberstein (who Meyer mentioned in his last, unfinished treatise).

1532 – Suleiman the Magnificent's second campaign against Vienna is halted.

1532 – The Second War of Kappel, between the Protestant and Catholic Cantons of the Swiss Confederacy.

1532 – Straßburg becomes a member of the Schmalkaldic League.

1537 – Joachim Meyer is born in Basel.

1541 – Siege of Buda. Ottoman vassal Johann I of Hungary dies and as a result, the Habsburgs besiege Buda. Suleiman the Magnificent himself leads a relief army and defeats the Imperial Habsburg army, killing 20,000 men. As a result, central Hungary is under Ottoman control for 150 years.

---

[10] The Swiss Guard's bravery in defending Pope Clement VII during the Sack of Rome is still commemorated by having new recruits to the Swiss Guard being sworn in on 6 May every year.

1546–47 – The two-year Schmalkaldic War between Karl V's Imperial forces and the Schmalkaldic League starts in Füssen, about 150 km east of both Basel and Straßburg. It ends with the League defeated, several leaders imprisoned, and 28 cities stripped of their independence. However, Luther's ideas had already spread so much by then that they could no longer be contained.

1552 – The Second Schmalkaldic War is fought and leads to...

1555 – The Peace of Augsburg, a treaty between Emperor Karl V and the Schmalkaldic League. Lutheranism is officially recognized in the Empire. Calvinism would not be recognized until 1648.

1558 – Meyer's father and mother rent a house with a garden and a vineyard in St. Albanvorstadt, Basel.

1560 – Joachim Meyer marries Appolonia Ruhlmann, a baker's widow, in Krutenau, Straßburg.

1560 – The Grindelwald Fluctuation, an important date for the *Little Ice Age*, begins.

1561 – Joachim Meyer seeks and receives permission to arrange a Fechtschule in Straßburg.

1561 – Meyer signs his fencing treatise (JMM) dedicated to Georg Johann, Count Palatine of Veldenz, using the title Freifechter.

1562 – The Massacre of Vassy. 50 Protestant Huguenots are murdered and a hundred injured, including women and children, by Catholic troops led by François, Duke of Guise. The same year, massacres of Huguenots also take place in Toulouse, Sens, and Tours.

1562 – The French Wars of Religion start, killing an estimated 2–4 million people before their end in 1598.

1562–63 – The First French War
1563–67 – The Second French War
1568–70 – The Third French War

1563 – 4 Sep, Joachim Meyer petitions to arrange a Fechtschule in Straßburg

1563–64 – The Plague hits Basel and Straßburg, killing roughly 5,000 people in each city.

1566 – 15 June, Joachim Meyer petitions to arrange a Fechtschule in Straßburg.

1566 – Emperor Maximilian II elevates Johannes Sturm's *Gymnasium* by granting privileges as an Academy.

1567 – 1 Feb, Joachim Meyer petitions to arrange a Fechtschule in Straßburg

1568 – 28 June, Joachim Meyer petitions to arrange a Fechtschule in Straßburg and is described as a Fechtmeister for the first time.

1568 – Otto, Count of Solms-Sonnewalde, visits the *Gymnasium* in Straßburg and Meyer likely delivers a fencing treatise (JML) to him.

1570 – Joachim Meyer becomes the treasurer of the Smiths' Guild of Straßburg.

1570 – Joachim Meyer publishes the book translated here, dedicated to Johann Kasimir.

1570 – Joachim Meyer has in his possession a compilation of several older fencing treatises (JMR) and has added in his own treatise on the rapier. This also has a design for a frontmatter, with teachings attributed to Heinrich, Count of Eberstein.

1570 – The Imperial Diet of Speyer, with Catholics and Protestants seeking unification in the face of the Ottoman threat. Johann Albrecht I, Duke of Mecklenburg-Schwerin, attends this event, and Meyer possibly meets him there.

1571 – Joachim Meyer accepts employment as fencing master for Johann Albrecht in Schwerin.

1571 – Joachim Meyer dies two weeks after arriving at the court in Schwerin.

1571 – The Battle of Lepanto, a naval battle between the fleet of rowed ships of the Holy League of Catholic states (mostly Spain and Italy) defeat the fleet of the Ottoman Empire in the Gulf of Patras. In total, more than 400 warships were involved. This was a turning-point for Ottoman military expansion into the Mediterranean.

1572 – Joachim Meyer's widow marries for a third time, again to a cutler, again granting citizenship to him.

1572 – The St. Bartholomew's Day massacre. Charles IX, King of France, orders the execution of leaders of the Protestant Huguenots and triggers slaughtering of Huguenots in Paris, which spreads into the countryside and to other urban centres. An estimated 5,000–30,000 people are killed.

With those key dates out of the way, let us zoom in on some particulars of Meyer's world, starting with the city he spent his young years in, and where his parents lived.

THE CITY OF BASEL

Our recorded history is the history of cities, although it only provides us with a small, yet important slice of the lives of the past, this, since up until the middle of the 1800s, about 95% of all people lived *outside* of the cities, not *in* them, and most historical material revolves only around a few percent of the remaining urban population.

In Meyer's case, the city where he was born and spent his youth, was Basel. In the 16th century, Basel was a very small town by *modern* standards with about 12,000 men and women,[11] and in fact, as late as 1850 it still only consisted of some 28,000 inhabitants.[12] However, by *contemporary* standards it was still a very large city, and an *important* city for reasons we will return to later. Like so many other cities before the *Industrial Age*, most of it existed within its city walls and the number of inhabitants fluctuated, sometimes quite dramatically with the frequently returning plagues. Basel is said to have been founded by Roman general Lucius Munatius Plancus in 44 BCE as a military settlement called *Augusta Raurica* or *Colonia Raurica*—today two towns called Augst and Kaiseraugst—which was located 10–20 km upstream of the Rhein, east of Basel. The place takes its name from Roman Emperor Augustus

---

[11] "Basel aus der Vogelschau". *HMB Magazin, Das Magazin des Historischen Museum Basel* 4.
[12] See "Basel: Historical Population". *Wikipedia*. Retrieved 7 Aug 2022. <http://en.wikipedia.org/wiki/Basel#Historical_population>

and a local Celtic tribe called the Rauraci which was living in the area. It is the oldest known Roman colony on the Rhein and at its peak in the 2ⁿᵈ century CE it was an important trading centre with some 20,000 inhabitants and a Roman theatre with near 10,000 seats. The settlement is believed to have included a fortress on the hill where the Münster Cathedral is placed today, overlooking the Rhein, but remains of an even older Celtic fortification called *Basel Oppidum* have also been found on the place. Already in 374 CE, a Roman historian named Ammanius Marcellinus names the city *Basilia*, which would later become Basel.

Following a series of wars, the bishop's seat was transferred from Augusta to Basel in the 7ᵗʰ century. Then in 1006, by decree of Emperor Heinrich II of the Ottonian House, the civil power of the bishop was established in Basel. 74 years later in 1080, the first city walls were built, following what today are called, from west to east, Petersgraben, Leonhardsgraben, Kohlenberg, Steinenberg and St. Albangraben. In 1083, the first monastery of Basel, the St. Albankloster, was founded, and in 1270, the church St. Albankirche was built.

Skipping ahead, the biggest earthquake in the recorded history of Central Europe takes place in Basel in 1356. It is estimated at 7.1 on the Richter scale and took place at about 10 p.m. on 18 October, 𝔖𝔞𝔫𝔨𝔱 𝔏𝔲𝔨𝔞𝔰 𝔗𝔞𝔤 (St. Luke's Day). The earthquake and the resulting fires in the city killed about 300 people, ruining most of the city and its largely wooden houses as well as all major churches and castles within a 30 km radius of the city. The town was soon rebuilt, though, and the building of the Third Town Wall began in 1362, with its completion in 1398. This made the city a well-defended one, but also distinctly separated the burghers from the peasant population.

In 1431, Basel had the attention of the Christian world with the Great Council of Basel, where the council sought to reform the Church in the context of the Hussite Wars and the Ottoman threat. The council also declared itself superior to the pope, and eventually, after an internal split—and the majority transferring first to Ferrara in 1437, and then Florence in 1439—the few remaining councilmen of Basel elected Amadeus VIII, Duke of Savoy, as an anti-pope.

The Rhein River connects Basel to many cities, and the city naturally built important relationships with both Zürich and Straßburg. These cities were situated a fair distance away, but connected via the Rhein, and through the river still so close that it used to be said you could put soup on a boat in Basel and serve it still hot in Straßburg upon arrival to the city.[13] It was also in Straßburg that Gutenberg had invented the printing press with movable type in 1439; this labour-saving device for scribes to keep up for the increasing demand for manuscripts would eventually revolutionize the world by allowing the rapid spread of ideas. This came to be hugely important, not just for the growing Protestantism, but also for knowledge transfer in general including that of the, up until then, secret art of fencing. With a quickly growing desire for books and pamphlets, Basel soon became a centre for the printing of books and also for paper making,[14] with the first "Swiss" printing press established in Basel in 1468. Various

---

[13] Likewise, Johann Fischart, in his poem *Das glückhafft Schiff von Zürich*, recounts how 54 citizens of Zürich sailed to Straßburg for a shooting match in 1579, doing so in a mere 19 hours, bringing with them a large pot of porridge which they managed to keep warm until their arrival in Straßburg. This was in fact the second porridge run from Zürich to Straßburg, with the first taking place already in 1456, after a bet, then taking 22 hours. Since 1946, this trip has again been repeated every 10 years, performed in historical costume to commemorate the event. However, today the trip takes 2½–5 days, due to the many new hinderances on the way.

[14] Paper mills were also established in Basel in 1448, 1449, 1453, 1472, 1476, 1482, 1489, 1525, and 1634.

associated professions, like carvers of printing blocks and illustrators, quickly became fast expanding groups of workforces in response to this, and by Meyer's time Basel was thus a near-two millennia-old city that had grown into a centre of education, spreading of information and religious reformation, protected by strong city walls and tightly connected to many other cities—not least the Protestant and, like Basel, Free Imperial City of Straßburg.

It was also during this period, in 1444, that the Battle of St. Jakob an der Birs takes place. This was a war between the Swiss Confederacy and French mercenaries, mostly Armagnacs, which took place just 1 km from Basel within the district of St. Alban. The battle between the Swiss pikemen and Armagnac cavalry was unusually fierce, with 1,500 pikemen refusing to surrender to the 30,000-man strong French army, reportedly ripping arrows from their bodies and continuing to fight after having been stabbed by lances or even after having lost hands. Meanwhile, the city of Straßburg chose to protect and give safe harbour to peasants from the surrounding region by housing more than 7,000 of them. As a result of the incredible resistance and determination of the Swiss infantry, the French *Dauphin* decided to have his forces turn back and signed a peace treaty with the Confederacy and with the city of Basel.

Twenty-seven years later, the University of Basel was formed by Pope Pius II, and in 1471 Emperor Friedrich III conferred rights for the city to hold two fairs every year, further enhancing its importance. Then in 1501 Basel was accepted into the Swiss Confederacy, the precursor of Switzerland.

In 1529, Johannes Oekolampad, the German Reformer, introduced the Reformation into Basel, after which the bishop left the city. This is important, since it was likely part of the foundation of Meyer's religious views and connections, as indicated by the Calvinist dukes and princes he was associated with. With that in mind, we will briefly look at how this came to be.

Oekolampad, born in Weinsberg, Germany, was already a preacher in Basel in 1515 and a vicar in St. Martinskirche in 1522. He also read and lectured in the Holy Scripture at the University of Basel. He worked together with other famous early Protestants like Desiderius Erasmus and Huldrych Zwingli and even translated works by Erasmus. He came to be a strong inspiration for later generation protestants like Heinrich Bullinger and Jehan Cauvin. Cauvin, in turn, came to Basel in 1536 to escape the violent uprisings against Protestants in France. Later the same year, he would be invited to the already-Protestant city Straßburg by Martin Bucer, ending up preaching in several churches in Krutenau, an area that as then in the outer parts of Straßburg where many new citizens and inhabitants of the city lived (including French Huguenot refugees).

By Meyer's time (and as we will discuss in more depth later), the fascination with Antiquity and the Greeks and Romans was also at its peak, with the Holy Roman Empire claiming heritage straight from the ancient Roman emperors, and its elected 'King of the Romans' crowned emperor by the pope in Rome. In honour of its purported founder, Basel raised a statue of Roman general Lucius Munatius Plancus in 1580 (a statue which Meyer thus never had a chance to see), which still stands in the Rathaus of Basel. We can also clearly see this fascination expressed in Meyer's treatise of 1570, and Meyer refers to Greco-Roman culture in several ways throughout this book. While we don't know for certain, it is entirely possible that he had seen first-hand parts of the ruins of a Roman wall later excavated in the region.[15]

---

[15] At least in Straßburg, parts of the Roman wall appear to have remained standing as late as 1444, when they were described by Enea Silvio Piccolomini on a visit to the city.

Now, let us turn to what we believe we know about Meyer and his associations with the city of Basel.

JOACHIM MEYER AND BASEL

The 1560 marriage certificate of Joachim Meyer and Appolonia Rulmennin

The firſt piece of proper data we have on Joachim Meyer is a record from Straßburg, not Basel, but which ties him to Basel. On 4 June 1560, Joachim Meyer of Basel married Appolonia Rulmennin in the Church of St. Guillaume. At the time, Appolonia was the widow of baker Jacob Wickgaw. Meyer then officially became a proper burgher of Straßburg, as a member of the Schmíden Guild (the smiths), which the cutlers sorted under.

From this we have a few intereſting pieces of data to ſtart building our Basel puzzle with: Joachim Meyer was a cutler, as noted in this and other later Straßburg records, and as also indicated by the cutler coat of arms (the crown with three swords) on both the title page of his 1570 book as well as in the backgrounds of five of the Figures. We also know he comes from Basel. These two pieces are seemingly quite important when put beside what we find next.

The baptismal records of Joachim Meyer, 1537

Looking to the birth records and liſts of burghers of Basel there are very few people in the city by the name of Joachim Meyer in the period of intereſt with data fitting "our" Meyer. However, on 16 Auguſt 1537, it is noted[16] that paper maker Jacob Meyer and his wife Anna Freund has a son baptized as Joachim.[17] They live in the area of St. Alban, then a very sparsely-populated semi-rural area outside of the original innermoſt city walls but inside of the new walls protecting the city. Moſt likely they had their son baptized in the St. Albankirche, the only church in St. Alban, with a congregation consiſting mainly of people who worked with the water-wheels: the smiths (including cutlers), the millers, the sword sharpeners and polishers, and the armour makers.[18] Intereſting to note here is that Meyer's father is noted as a paper maker, which fits well with the area he lived in but less so with his son's trade.

St. Alban and Dalbdych/Dalbeloch

Craftsmen of different trades commonly lived in different parts of cities. In Basel, moſt of the fishermen lived in St. Johanns-Vorſtadt, the weavers in the Steinenvorſtadt, and the sawmill workers likely in Kleinbasel.[19] The area of St. Alban and Dalbedych in Basel was known, among other things, precisely for the paper making business, and that is also where moſt of the paper makers lived—although not juſt people of that trade, as we shall return to shortly. Named after the St. Alban Monaſtery located here, St. Alban was a semi-rural area all through the 1500s, with only about 30 residential buildings; as late as 1815, only about 2,000 men and women

---

[16] Document from the Parish of St. Guillaume, Archives de Strasbourg, microfilm 5Mi482/12.
[17] CAMES 1995: 2641.
[18] SIMON-MUSCHEID 1988: 78.
[19] Basel aus der Vogelschau, HMB Magazin, Das Magazin des Historischen Museum Basel /4

living there. In Meyer's time, it was an unusual mix of industrial and residential buildings, with craftsmen, patricians, and nobility living together, right next to vineyards and fruit trees. The St. Alban-Tor was a nearby gate which provided easy passage in and out of the city.

At the peak of its business life, there used to be 7 + 9 water wheels at work in St Alban, arranged in sets of 2–3 wheels attached to 6 + 6 buildings, where paper makers, smiths, and others worked. Another two wheels where in use at the sawmills just outside of the north-eastern gate to Kleinbasel. Mills powered by water or wind were commonly placed at the out-skirts of the cities, likely due to the risk of dust explosion and possibly also due to the noise associated with the powerful wooden mechanics. The wheels seem to have been powered by the Birs River coming from the south and ending east of St. Alban. The Birs and the Birsig River filled the moat, and the latter also functioned as the city sewers. The mills were used by the professions organized into so called 𝖂𝖆𝖘𝖘𝖊𝖗𝖋𝖚̈𝖓𝖋 ("Water Fives"): the millers, the armour makers, the sword sharpeners, and the cutlers, with colleges of five elected representatives, who alongside of water masters, arranged for fair use of the canals and water wheels for the different trades, households, and workhouses, as well as for maintenance and the fining of unapproved usage.[20] The water wheels were split up between the trades, with the wheels clos-est to the city walls seemingly having been dedicated to paper making. The water wheels sit-uated the closest to the St. Albankirche however seem to have been used for other purposes,

The quartier of St Alban, showing the water wheels associated with various crafts. On the right we can also see St. Albankirche, the church where Meyer most likely was baptized. The image also clearly shows the old, covered entrance to the church which was removed in 1845. By Matthäus Merian d.Ä, 1615.

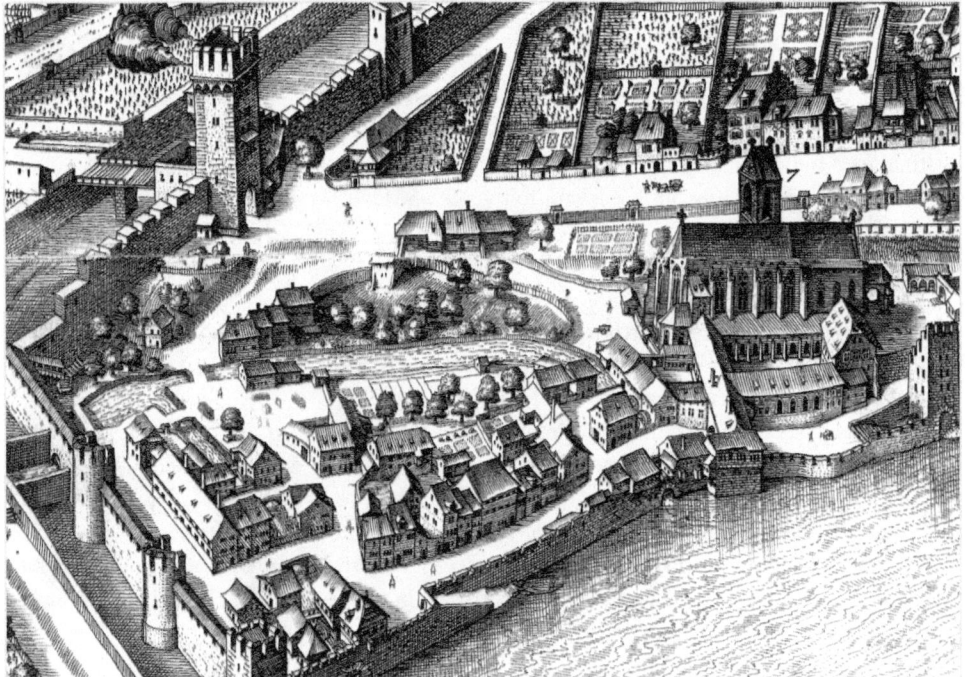

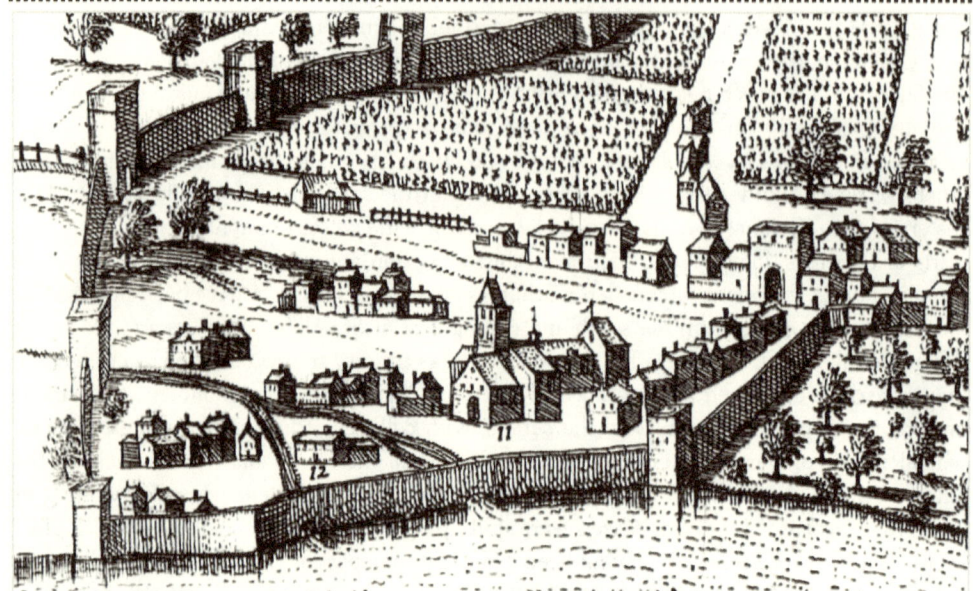

St Alban, the neighbourhood in Basel where Fechtmeister Joachim Meyer lived and played as a child. Residential area at the bottom, and vineyards in the upper half of the overview. St. Albantor is the third wall tower from the bottom left. The overview is from about four years after his death in 1571. By Georg Braun and Frans Hogenberg, ca 1575

like knife making, sharpening, and polishing swords and armour. Two of these buildings still remain, as do several of the paper making buildings. Unfortunately, only one of the wheels remains today, part of the very active Basler Papiermühle Museum, and still in use for its original purpose.

Right next to this, west of the water wheels, we also find the St. Albankirche, the only church in this neighbourhood and quite relevant to the life of Meyer as we shall soon see.

Jacob Meyer the 𝕸𝖊𝖘𝖘𝖊𝖗𝖘𝖈𝖍𝖒𝖎𝖉, 1517

As in the case of the name 'Joachim Meyer', there are few suitable listed candidates for the father of Joachim in other records. However, there does seem to have been a "Jakob Meyer, 𝕸𝖊𝖘𝖘𝖊𝖗𝖘𝖈𝖍𝖒𝖎𝖉" who becomes a burgher of Basel in 1517.[21]

This Jacob Meyer is currently not known to be associated with any of the larger Meyer families of Basel, but the fact that he is listed as a 𝕸𝖊𝖘𝖘𝖊𝖗𝖘𝖈𝖍𝖒𝖎𝖉—a cutler—like Joachim Meyer is interesting. Up until learning about the city architecture of Basel and St Alban, it was quite puzzling having two men by the name of Jacob Meyer, one a cutler and the other a paper maker, that potentially could be the father of Joachim Meyer. However, realizing that the two trades worked in the very same place in St. Alban, using the water wheels of the workhouses

---

[21] LUTZ 1819: 210. Note that only burghers—citizens—would be listed here. Non-citizens like *Einwohner* (inhabitants), and travelling *Kauffleute* (merchants), who merely lived in the city, weren't listed and also did not have the rights, or obligations, of a citizen.

in the area to manufacture their goods, it seems plausible that the two are one and the same. With the still expanding paper and book industries, switching trade somewhere between 1517 and 1537 would likely have been fairly easy to do, thus making it quite possible that Jacob Meyer arrived as a cutler but switched to paper making. He may then have transferred some of his knowledge of knife making onto his son, who picked up on the trade again and became a master cutler in Straßburg. This, of course, is speculation.

Yet one more note for Meyer's family's life in Basel exists, this time relating to their rental of a house and part of a vineyard, on November 9, 1558.

> "Jakob Meyger, the papermaker and citizen of Basel, and his wife Anna Fründt confess with the above-mentioned guarantors that they shall pay an annual rent of 2 florins to master Michel Gernnler, bailiff of Frau Mergelin Höfflin, widow of Ludy Gernler, for their house and garden in St. Albanvorstadt as well as a 1/2 Jucharten of vines in front of St. Albantor for the sum of 40 florins."[22]

Unfortunately, we do not know more about Meyer's family or relatives, but several Meyer families are known in Basel, and we will now examine some of them.

The Meyer families of Basel

While we currently do not know for certain which family Joachim Meyer's father belonged to, seven larger families are known and there were also families of unknown roots, as well as unlisted men named Meyer staying in the city.

The seven families are:

- Meyer zum Hirzen/Hirschen ("of the Deer")
- Meyer von Hüningen
- Meyer von Balderstorf
- Meyer mit dem Pfeil ("with the Arrow")
- Meyer zum Haasen/Hasen ("of the Hare")
- Meyer zum Schlüssel ("of the Key")
- Meyer zum Sternen ("of the Star")

So, how high up on the social ladder was Meyer's family? Despite his common roots, Joachim Meyer appears to at times have moved around in the higher echelons of society and he seems to have had a fair amount of education and a respectable standing, possibly indicating that he came from a fairly well-off family. His second treatise (JML) is dedicated to Otto, Count of Solms, and was given to the count after having taught him in the fencing arts. Some of the topics of the 1570 treatise indicate that he was well-read on history and had received some form of higher education. In the introduction to the book, he also speaks directly to Johann

---

[22] *Jakob Meyger, der Papierer und Bürger zu Basel und seine Frau Anna Fründt bekennen mit genannten Bürgen, dass sie dem Meister Michel Gernnler, als dem Vogt der Frau Mergelin Höfflin, Wittwe des Ludy Gernlers, einen jährlichen Zins von 2 fl. von ihrem Hause und Garten in der St. Albanvorstadt sowie von einer 1/2 Jucharten Reben vor dem St. Albantor um die Summe von 40 fl. verkauft haben.* Staatsarchivs Basel-Stadt, K 7, f 107. Trans. by OLIVIER DUPUIS.

A *Juchart* was defined as "what a man can plow in a day". So, half a *Juchart* equals ca. 0.44 acres or 1800 m².

Kasimir, stating that the count palatine personally, and numerous times, had asked him to write a book, seemingly pointing to a conversation that had extended over some time. Likewise, his claims of having studied under famous fencing masters may point to the same.

Zum Hirzen and Jacob Meyer the Prediger of St. Alban

Looking to the known Meyer families of Basel, the zum Hirzen family is interesting for several reasons. One of the things they are known for is Jakob Meyer (1437–1541) who served as mayor of Basel between 1530–41, and who was a friend of Calvinist reformer Wolfgang Capito.[23]

Second, in 1555 Jacob Meyer zum Hirzen (1526–1605) was the pastor of St. Albankirche, the church where Joachim Meyer was baptized eighteen years earlier, and thus likely to have preached to Joachim's parents (who still lived in the area). This seems more of a coincidence though, but certainly deserves investigating more. On 5 September 1555, Pastor Jacob Meyer was married to Agnes Capito, the daughter of Wolfgang Capito and Wibrandis Rosenblatt, and her mother was in turn a several-time widow, also married to renowned Protestant reformers Johannes Oekolampad and Martin Bucer both.

Zum Haasen

Jacob Meyger zum Haasen is another interesting possible relative to Joachim Meyer. He was born in 1482 in Basel. His father was a shopkeeper, also named Jakob, and like Joachim Meyer's father, his grandfather was a paper maker. Jacob Meyger took the 'Haasen' name from the house he bought—a house which had a hare (Haase) as its symbol. He had a rich and interesting life, working as a money changer and trader, and served as a landsknecht for Emperor Maximilian I between 1507–15. By 1508, he had gathered enough wealth to be able to purchase the Castle of Gross Gundeldingen in the suburbs of Basel. In 1516, he became the first mayor of Basel to be a tradesman and member of a guild, rather than an aristocrat or patrician. However, already in 1521, he fell from grace and was accused of receiving pensions from France and was arrested alongside of six other council members. After being released, he first joined the Grand Council of Basel, and then retired to Freiburg im Breisgau in 1529, two years before his death in 1531.

***

Now, some of these similarities and possible connections, if not all, are bound to be no more than coincidences, and the use of biblical and Roman names was common. However, looking through family trees we can see how it was, and still is, common with groups and combinations of names running through the family lines, sometimes skipping a generation or two. *Jacob* for instance, is common in some Meyer families, especially the zum Hirzen, but not at all in others. Likewise, professions commonly transferred naturally within the same family, between the generations, due to easy access to learning of otherwise carefully guarded trade secrets. However, more research is needed. Looking through the family trees of contemporary zum Haasen and zum Hirzen, no mentioning of any Joachim Meyer has so far been found. Likewise, the name is not listed in the *Baslerisches Bürgerbuch*, indicating that he himself never was a

---

[23] HEINZE 2006.

burgher and full member of a guild of the city—which isn't surprising considering both his fairly young age, still in his very early 20s, and that he must have been preparing to relocate already around the last few years of the 1550s.

With this, we leave Basel for the next chapter in the life of Meyer; the Free Imperial City of Straßburg.

THE FREE IMPERIAL CITY OF STRAßBURG

The city of Straßburg has its roots in a Roman outpost belonging to the Roman province *Germania Superior* and was at first called *Argentoratum*, or in medieval Latin, *Argentina*. It was first settled in 12 BCE, and after several fires and rebuilds, it gained its properly Roman fortified form after 97 CE. It was thus made the permanent station of *Legio VIII Augusta*, which included cavalry, and later also made temporary station for the *Legio XIV Gemina* and the *Legio XXI Rapax*. The centre for the camp was located near today's *Rue des Hallebardes* ("Halberd Street"), close to the cathedral that would be built almost 1,500 years later. From 300 CE onwards, Straßburg was the seat of the Bishopric of Straßburg, and excavations at the current *Église Saint-Étienne* have unearthed remnants of a church going back to late fourth or early fifth century, which today is considered to be the oldest known church in all of Alsace. The

Argentoratum–Strasbourg, from Civitates Orbis Terrarum. *By Braun & Hogenberg, 1572. Krutenau, where Joachim Meyer married and likely lived is at the bottom right.*

place is considered to be the first seat of the Roman Catholic Diocese of Straßburg.

The region was of course filled with conflicts between the native Germanic tribes and the Romans, and in 357, the Battle of Argentoratum was fought. Future Roman Emperor Julian defeated the Alemanni and King Chonodomarius was taken prisoner. Nine years later, the Alemanni again tried to invade the Roman Empire, and early in the fifth century, the Alemanni appear to finally have been successful in crossing the Rhine, conquering and settling in today's Alsace, and also large parts of what is today called Switzerland. In the ninth century, the city was commonly known as *Strazburg*, as documented in 842 by the *Oaths of Strasbourg*. Already a major commercial centre, in 923 the town came under the control of the Holy Roman Empire. After a long period of conflicts between the citizens and the bishop, and after the Battle of Oberhausbergen in 1262, Philip, King of Swabia, granted the city the status of Free Imperial City, thus the city was no longer under bishopric rule.

After a revolt in 1332 when the government of the city was redefined with close participation of the burgher artisan guilds, Straßburg declared itself a free republic. The city ruled with three councils, but the **Ammestre**[24] would not be properly defined until 1482.

As was common at the time, Strasbourg's local Jewish residents faced a number of re-strictions under city laws. As they were forbidden from bearing arms under specific Imperial ordinances, they were regarded by other citizens as dishonourable to attack, akin to clergy, women, children, or the elderly. Jews were also required to pay the city for protection.

Tragically, following major sociopolitical changes within the city and an outbreak of bubonic plague in 1348, over a thousand Jews were publicly burnt to death on 14 February 1349 during a six-day massacre. This massacre followed similar events in Basel and Freiburg and stemmed heavily from power shifting in the city from the nobility and wealthy burgher patricians to the craftsmen guilds.

Within a few generations, Straßburg would come to be known as a centre for humanist thought and scholarship, and not least for Johannes Gutenberg's invention of the printing press, or rather *moveable type* printing, in this city in 1439, thus providing the revolutionary tools for the humanists and the spread of the Reformation in the early 1500s. Gutenberg himself was originally a blacksmith and goldsmith, and like most burghers, part of the Straßburg militia (as was the customary obligation of the households of the empire).

THE MÜNSTER CATHEDRAL

Construction of the dominant Münster Cathedral had begun already in the 12th century, and it wasn't fully completed until about 300 years later, in 1439—the same year that Gutenberg invented his printing press. However, only one of the originally planned two towers was actually built, likely due to financial concerns, and remains in this peculiar state even today.

At the time the cathedral was the world's tallest building, surpassing even the Great Pyramid of Giza in Egypt. It would have been a dominant feature of the landscape, easily seen for miles, not to mention expressing an overpowering presence over the other low buildings surrounding it in the city. By Meyer's time, the cathedral would already be a 400-year-old landmark,[25] and

---

[24] For more reading on the *Ammestre* and the governing of Straßburg, see DUPUIS 2013.
[25] Although only a completed one for about 120 years.

The Münster Cathedral in Straßburg. By Daniel Specklin, 1566.

we can imagine Meyer might have walked its aisles more than once as it was one of the most important, public buildings of his time.

In the 1520s, during the Protestant Reformation, the city officially embraced the teachings of Martin Luther under the political guidance of Jacob Sturm and the spiritual guidance of Martin Bucer. As a result, many existing Catholic churches were destroyed or vandalized as Protestants sought to erase the extravagant and luxurious decorations of the Catholic Church, desiring a more minimalistic and simplistic expression of worship (closer to God, without distractions). Interestingly, in 1532 Straßburg legates 𝔄mmeíſter Jakob Meyer and 𝔅urgermeíſter Jacob Sturm, met Albrecht of Brandenburg and Count Palatine Ludwig V in Schweinfurt to discuss Straßburg's joining of the Schmalkaldic League and managed to convince Bucer to subscribe to the Augsburg confession, which allowed the city to join the league.[26]

Four years later in 1536, in the same year the theologian and pastor Jehan Cauvin published his greatly important *Institutes of the Christian Religion*, Bucer invited him to Straßburg, and in September 1538, Cauvin became the minister of a church of French refugees. A few months later he applied for, and was granted, rights as a burgher of Straßburg. He held office successively in the Saint-Nicholas Church, the Saint Madeleine Church and the Temple Neuf, churches which all still stand, although none of them in the shape they had in his time. It is believed that Meyer was a follower of Cauvin's teachings, both based on the fact that most of the patrons he was associated with were known Calvinist princes and dukes, and on his close proximity to Calvinist and Huguenot communities and Cauvin's own churches.

Nearby, the town of Speyer held three (of many) Imperial Diets of Speyer in 1526, 1529, and 1570, which were hugely important in the shaping of the Protestant Reformation. It has been suggested by both DUPUIS and FORGENG that Meyer attended the latter. That is likely where he met Johann Albrecht I, Duke of Mecklenburg, and procured his employment as fencing master at the duke's court. At the Diet of 1570, imperial commander Lazarus Schwendi attended and proposed a standing Imperial army to Emperor Maximilian II, made up of Catholics and Protestants, for the unification and protection of the empire against the Ottomans. This army would also stand in opposition to the unreliable and problematic 𝔏andskneḑte, who were loyal only to their own personal greed. However, his efforts were not successful.

By Meyer's time, only some 15–20,000 burghers lived in Straßburg,[27] though this was sufficient for it to remain one of the largest cities of Germany. As already described, the river Rhein connected Straßburg to Basel, as well as Zurich, Heidelberg, Frankfurt-am-Main, Köln, and even Rotterdam on the Dutch coast, and from there, access to the world by sea.

KRUTENAU, A NEIGHBOURHOOD FOR CALVINISTS, FISHERMEN, AND BOATSMEN— AND MEYER

We do not know for certain where Meyer lived, but considering the time when he arrived at the city—probably sometime between 1558–1560—and considering where he got married, he most likely lived in Krutenau.

---

[26] See RUMMEL 2015.

[27] KLIPFEL, MONIQUE. "L'importance démographique de la ville". *Académie de Strasbourg*. Retrieved 9 Aug 2022. <http://www.crdp-strasbourg.fr/data/histoire/strasbourg_1400/ville_opulente.php>

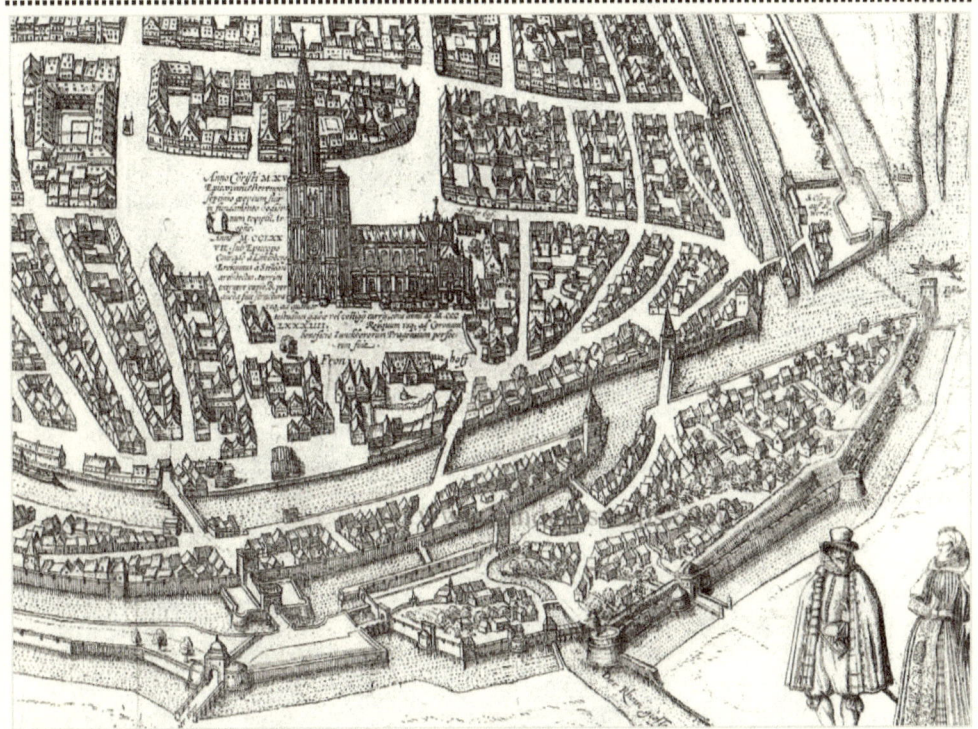

Krutenau is the area in the bottom right quarter. St. Guillaume is in the centre of Krutenau.

The whole area in the northwestern districts were built from the 1400s through the 1500s, and in the early 16th century, several parts of the area still remained fields and agricultural area, with much of the population of Krutenau consisting of fishermen and boatsmen of various stripes. This is also the area where the churches Jehan Cauvin worked and preached in are located, and as previously mentioned, thus directly or indirectly served as influences on Meyer's worldview.

MEYER MARRIES IN SAINT GUILLAUME CHURCH

On 4 June 1560, Meyer married Appolonia Rulmennin in the Church of St. Guillaume, with the marriage confirmed and blessed by its priest, Mattheus Hägelin. As part of the city's requirements, Meyer would here have been expected to prove ownership of arms and armour, which was a standard for all married men and their households. Officially Joachim then became a proper burgher and member of the Cutler's Guild. The church records of St. Guillaume state:

"Joachim Meyer of Basel, cutler and Apolonia Rülmann, widow of Jacob Wickgaw, celebration of nuptials, 4 June, 1560."[28]

---

[28] *Joachim meÿer v Basel Messerschmid / Apoloni Rülmannin Jakob Wickgaws nachgelassen witwe / Celebrar nt nuptiae / 4. Junÿ .An. 1560.* Document from the Parish of St. Guillaume, Archives de Strasbourg, microfilm 5Mi482/12.

Furthermore, the Straßburg book of burghers then mentions:

"Joachim Meyer of Basel has received citizenship from Appolonia Rülmann, widow of the late Jakob Wickgaw, baker, and will serve with the smiths. Enacted 10 June 1560."[29]

This was possibly—perhaps even likely—a marriage of convenience, as many people suffered premature deaths due to illness, plagues, lack of medical care, dangerous working conditions, wars, and simply high rates of violent crime, with many citizens having been marrying three or four times before reaching old age. Meyer in turn would be seeking citizenship as a burgher and entry into the smiths' guild of the city, and this may have been a practical solution for all parties. However, it may well also be that Meyer just moved in the same social circles and had met and fallen in love with her. Naturally, the one does not have to exclude the other.

As noted before, the church is located in the strongly Calvinist district of Krutenau, just a short walk from churches where Jehan Cauvin preached. In fact, the street near the church is today called *Rue Calvin.*

Schematic overview showing St Guillaume on the top left, the fisherman's gate and tower at the bottom left. By Martin Weiss (1711–51)

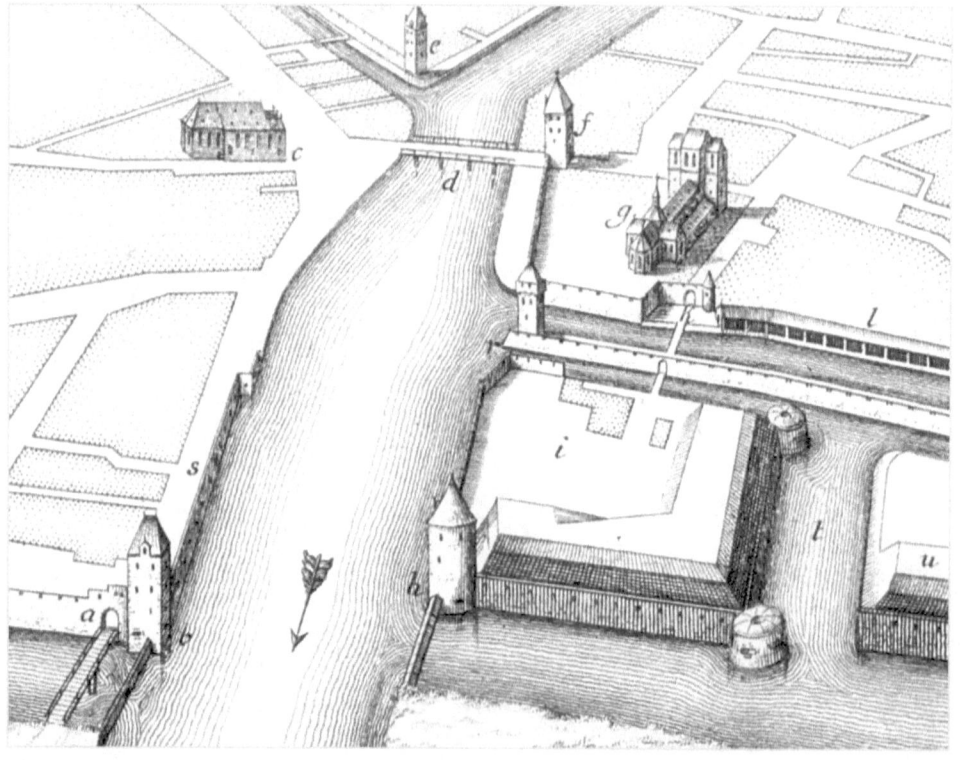

The Church of St Wilhelm in the 16th century. By Alfred Touchemolin (1829–1907), date unknown.

The summer and winter of 1560–1561 were unusually harsh, beginning the Grindelwald Fluctuation (1560–1630), a period which is part of the Little Ice Age of Europe. This period started with the Grindelwald glacier expanding, but soon more serious and unusual occurrences followed: severe draughts, storms, floods, frosts, and blizzards, which caused crop failures and famines, eventually leading to mass starvation at a global scale. As a consequence, antisemitism mounted, as well as scapegoating of and violence against minorities who were blamed and murdered for these troubles. Times were difficult indeed.

Let us now turn our attention to the known fencing activities of Meyer, starting with the question of how he at all learned the art of fencing.

WHO TAUGHT MEYER TO FENCE?

"(Panthino) Some to the warres, to try their fortune there;
Some, to discouer Islands farre away:
Some, to the studious Vniuersities;
For any, or for all these exercises,
He said, that Protheus, your sonne, was meet;
And did request me, to importune you
To let him spend his time no more at home;
Which would be great impeachment to his age,
In hauing knowne no trauaile in his youth.

"(Antonio) Nor need'st thou much importune me to that
Whereon, this month I haue bin hamering.
I haue consider'd well, his losse of time,
And how he cannot be a perfect man,
Not being tryed, and tutord in the world:
Experience is by industry atchieu'd,
And perfected by the swift course of time."[30]

So, how did Meyer learn fencing? While we don't know much with certainty, based on his own words he appears to have learned in a number of different ways. We can also speculate a little. Considering how he was already well-established as a skilled fencer in his early twenties, we can assume that he would have started learning fencing quite young, as a boy. Given how common it was for young boys to learn to grapple and fence with wooden dusacks, this would not have been unusual.

Meyer also leaves us with several tantalizing hints, but unfortunately never mentions any names of the men he learned from. We find several examples in this treatise, where he states, "I (without any fame to speak of) not only learned the noble knightly art of fencing from skillful and famous Masters, but also have now practiced [this art] for many years and have instructed some young princes, counts, lords, and those of noble descent in it".[31] He again alludes to foreign masters of rapier in a third passage, saying: "However, because such foreign customs increase daily in many locations, it has also become more necessary for us that such foreign and alien customs of those people become known and familiar to us—indeed, that we practice and equip ourselves not less than they (as much is proper for necessary defense) so that we can shield ourselves from them (when it becomes necessary), to properly oppose them and be victorious. Therefore, I will present and describe rapier fencing in an orderly fashion—as I have learned it from well-considered people and through daily practice of the same..."[32] A fourth passage does not make it clear if the fencing master he learned from was alive, or if Meyer just learned from a treatise of his: "So that you shall also understand the Zedelfechter who laid out the saying for me..."[33]

[30] Shakespeare, William. *The Two Gentlemen of Verona.* Act I. Scene iii. 1589–1593.
[31] Meyer (1570): a IIIᵛ. All quotations from Meyer (1570) are from the present translation by GARBER.
[32] Meyer (1570): II.50ᵛ–51.
[33] JMM: 20ᵛ; trans. from MAURER 2022.

Brothers of the 𝔐𝔞𝔯𝔵𝔟𝔯ü𝔡𝔢𝔯 Guild

Qualified and licensed instruction in his time was primarily given by the 𝔐𝔞𝔯𝔵𝔟𝔯ü𝔡𝔢𝔯 fencing guild ("Brothers of St. Mark"),[34] and part of his early training is likely to have occurred there. The whole disposition of this treatise also appears to follow the structure of fencing guild teaching, as far as we understand it today.

Particularly noteworthy here is also that we find rhyming verses on fencing with dusack in this treatise,[35] which are then also published 19 years later, word for word, in a 1589 publication by 𝔐𝔞𝔯𝔵𝔟𝔯𝔲𝔡𝔢𝔯 𝔐𝔢𝔦𝔰𝔱𝔢𝔯 𝔡𝔢𝔰 𝔩𝔞𝔫𝔤𝔢𝔫 𝔖𝔠𝔥𝔴𝔢𝔯𝔱 Christoff Rösener.[36] We can only speculate whether Rösener borrowed it from Meyer or whether Meyer in turn borrowed it from a 𝔐𝔞𝔯𝔵𝔟𝔯ü𝔡𝔢𝔯 tradition that preceded his work, though if Meyer were associated with the budding 𝔉𝔯𝔢𝔦𝔣𝔢𝔠𝔥𝔱𝔢𝔯 guild and had written this poem, it would probably make the 𝔐𝔞𝔯𝔵𝔟𝔯ü𝔡𝔢𝔯 less inclined to use it.

Another noteworthy curiosity is Meyer's inclusion of the rarely mentioned 𝔉𝔩ü𝔤𝔢𝔩𝔥𝔞𝔲𝔴, the "wing cut" mentioned in an anonymous poem called 𝔉𝔢𝔠𝔥𝔱𝔩𝔢𝔯𝔢 ("fencing teaching"), partially included in *Hans Talhoffer-København* (HTK) in 1459, and also in *Hans von Speyer* (HS) in 1491. The 𝔉𝔩ü𝔤𝔢𝔩𝔥𝔞𝔲𝔴 is also listed by 𝔐𝔢𝔦𝔰𝔱𝔢𝔯𝔰𝔦𝔫𝔤𝔢𝔯 ("master singer") and 𝔐𝔞𝔯𝔵𝔟𝔯𝔲𝔡𝔢𝔯 Hans Sachs,[37] and later by Paul Hektor Mayr in the 1540s,[38] and then by Rösener as one of the main cuts a 𝔐𝔞𝔯𝔵𝔟𝔯ü𝔡𝔢𝔯 should learn. It is also included in a fencing poem by Martin Syber[39] as well as in the *Kölner Fechtregeln* (K). Finally, a contemporary of Meyer's, fellow Straßburg citizen Johann Fischart, mentioned the 𝔉𝔩ü𝔤𝔢𝔩𝔥𝔞𝔲𝔴 in his German translation of *Gargantua* in 1575, when describing fencing with the 𝔐𝔞𝔯𝔵𝔟𝔯ü𝔡𝔢𝔯.[40]

Furthermore, in his *von Veldenz fight book* (JMM), Meyer writes, "the rhyme stands in the old and the Frankfurt 𝔷𝔢𝔡𝔢𝔩𝔰...".[41] Frankfurt am Main was then the city of residence of the 𝔐𝔞𝔯𝔵𝔟𝔯ü𝔡𝔢𝔯 guild, as well as where they licensed the 𝔐𝔢𝔦𝔰𝔱𝔢𝔯 𝔡𝔢𝔰 𝔩𝔞𝔫𝔤𝔢𝔫 𝔖𝔠𝔥𝔴𝔢𝔯𝔱.

---

[34] The full name of the brotherhood was *Bruderschafft vnser lieben Frawen der Raine junckfrawe Maria vnd des hailigen vnnd gwaltsamen himelfürsten Sannct Marxen* ("Brotherhood of our beloved Lady the Pure Virgin Mary and the holy and mighty prince of heaven Saint Mark").

[35] Meyer (1570): unnumbered page between I.64 and II.1.

[36] Rösener (1587): F iii–F iii^V.

[37] *Er sprach, der kunst zu eym eingang / lert man ober unnd unterhaw, / Mittel und flügel haw genaw.* Sachs (1558): CCCCX.

[38] See, e.g., PMD: 25^V, 39^R, 41^R, and 48^V.

[39] Meyer's treatises share a certain set of unusual terminology with the works of two masters in particular: Martin Syber and Hans Lecküchner. Here we find a whole range of terms that are quite specific to them and masters who came after, with terms like *Schneller*, *Bogen*, and *Blindhauw* in Syber, and *Entrüsthauw*, *Bastey*, *Stier*, *Eber*, *Bogen*, *Wacht*, and *Luginsland* in Lecküchner. Furthermore, both Syber (1491) and Talhoffer (1459) share some material which may both come from the same anonymous poem entitled *Fechtlere*, a poem which mentions *Rosen* and *Rädlin*, *Wechselhauw*, *Stürtzhauw*, *Wecker*, *Treiben*, and *Flügelhauw*. Like Syber, Talhoffer also uses terms like *Wechselhauw* and *Stürtzhauw*, and also *Wecker* and *Eisenport* in his version of the *Fechtlere*. Meyer had a copy of Martin Syber's poem in JMR. Talhoffer has been suggested to be a founding member of the *Marxbrüder*, and is also depicted with a pendant of the winged lion of St. Mark in HTG and HTK.

[40] *den ober vnd vnderhau, mittel vn͟d flügelhau, im tritt, mit kurzer vnd langer schneid.* Fischart (1575): T v^V.

[41] JMM: 15^V; trans. from MAURER 2022.

University fencers

Although we have no records of Meyer's studies, his works and his words certainly indicate that he was an educated man who would have studied a certain amount at universities, academies, or gymnasia, perhaps in his hometown Basel or at Johannes Sturm's *Schola Argentoratensis*, the gymnasium of Straßburg, since both were important centres of learning. As already discussed, this would also have been a good opportunity to learn in an environment where many young men were more dedicated to practicing fencing than to the studies they were expected to do at the university.

From other fencing treatises and masters

As is probably familiar to most readers, Meyer was part of a centuries old fencing tradition, a lineage of fencing masters extending back to "grandmaster" Johannes Liechtenauer and perhaps beyond. Like so many other fencing masters before him, Meyer frequently refers to Liechtenauer and his Zedel.[42] Again, from JMM, we read, "in the word Indes stands all the art of fencing, then it decides all things as Liechtenauer's old Zedell and other Zedell say of it, as can be seen in the rhymes,"[43] and "Liechtenauer says in his secret words, Guard yourself against parrying, if need befalls you it will hurt you."[44] Likewise, in 1570 he wrote, "Liechtenauer states correctly in his enigmatic verses: The Crosswise Cut takes away / that which comes from above. Item, Crosswise Cut with the strong, / take note of your work with it".[45]

Looking at his treatises, there may also be some influence and inspiration from Austrian Freifechter and Trabant ("bodyguard") Andre Paurenfeyndt. Compositionally, Paurenfeyndt's 1516 treatise[46] bears a distinct resemblance to Meyer's treatises, although it is also possible that they both simply followed Marxbrüder teaching structure. Also interesting to note here is how Meyer in JMM uses the term "Zedelfechter", a term which otherwise only appears in a verse in Frankfurt printer Christian Egenolff's 1530s reprint of Paurenfeyndt's work:[47] "A Zedelfechter I proclaim myself, Indomitable in Sword and Messer".[48] This particular verse simply doesn't exist in Paurenfeyndt's original, which here seems significant.[49] Furthermore, also in JMM, Meyer mentions the "Twelve Rules" and quotes one of them (the fourth), which is almost certainly a reference to Paurenfeyndt/Egenolff.[50] In this source, we also find a unique technique

---

[42] A didactic poem consisting of rhyming verses on fencing.

[43] JMM: 14^V; trans. from Maurer 2022.

[44] JMM: 8^V–9^R; trans. from Maurer 2022.

[45] Meyer (1570): I.55^V.

[46] Paurñfeyndt's original treatise one had on edition in 1516, though a French translation was later printed in 1538 (excluding Liechtenauer's verses). At least one later manuscript copy created by *Freifechter* Lienhart Sollinger also exists (JWA).

[47] Between the early 1530s and 1558, four editions were printed in total, with the last coming after Egenolff's death in 1555 as his heirs continued to operate his business.

[48] Egenolff (1530s): 4^V; trans. by Christian Trosclair.

[49] This may also have been a term used by the Marxbrüder specifically.

[50] JMM: 21^V; Meyer (1570): I.23^V. Compare Paurenfeyndt (1516): A 2, and Egenolff (1530s): 2^V.

called **Türcken zug** ("Turkish pull/draw") which Meyer includes in JMM, again indicating a link between it and Meyer.[51]

Looking to other sources instead, the dagger section of the 1570 treatise bears a striking resemblance to the dagger illustrations of Achille Marozzo (1536), with a statistically significant overlap between the two.[52]

Finally, Meyer's last, unfinished fencing treatise (JMR) contained copies of fencing treatises by the aforementioned Martin Syber, as well as by Sigmund Ainringck, Martin Huntsfeld, Lew, and Andre Lignitzer, all of which he therefore must have been familiar with.

While not naming them, he also refers to other masters: "Dear Reader to whom I have granted my didactic fencing poem which, based on the Authorities, I have composed, improved, and correctly organized...",[53] and earlier in JMM: "So that you shall also understand the **Zedelfechter** who laid out the saying for me...".[54]

Furthermore, he doesn't just refer to the Liechtenauer tradition, but in his own rapier section of JMR, he also mentions having studied foreign sources: "Fencing with the rapier brought together from the Italians, Spanish, Neapolitans, French, and Germans and whereupon whose proper foundation it stands."[55] Who these Italian, Spanish, Neapolitan, and French and German rapier sources and masters are remains to be discovered, and it is of course quite possible that he either met foreign masters locally, or if he himself travelled Europe,[56] perhaps as part of his studies. Still, all this remains speculative for now and we know nothing for certain about his studies or potential travels. We can only make educated guesses based on the social and professional circuits he appears to have moved in, and on his use of the title **Freifechter** when signing treatises and various petitions.

The cutlers' guild, or family tradition

Other possibilities which merit further research are the customs and practices of the Cutlers Guild. We know Meyer was a cutler by trade, and possibly also his father before him. And although certain simpler trades like the furriers and hammersmiths were more common in the **Marxbrüder** fencing guild, the cutlers, alongside the watchmakers and other more skilled trades, were more frequent among the **Freifechter von der Feder**. Cutlers specifically are very often depicted performing sword dances combined with fencing displays. As the 1597 fencing

---

[51] A technique called *dürkisch haw* is also taught in the pseudo-Peter von Danzig gloss (see PD and G) but is rather different: a cut to the neck in mounted fencing. The technique also appears in several other treatises, although not named after the Turks.

[52] At least 11 individual pairs of fencers were reproduced in Meyer's dagger Figures, out of 22 in Marozzo and 24 in Meyer.

[53] Meyer (1570): I.45^V.

[54] JMM: 20^V; trans. from MAURER 2022.

[55] *Fechten im Rapier Zusamt gethragen aus dem Italianisch, Spanischen Neapolitanischen, Francösischen und Theutschen und worauf dess selbigen grund und Rechte Fundament stande.* JMR: 123^R.

56 This is also reminiscent of how 15th century fencing master Martin Syber, whose *New Zettel* (first attested in HS, 1491) also appears in JMR, describes being a well-travelled master who had learned from masters in Hungary, Bohemia, Italy, France, England, Alamannia, Russia, Preussia, Greece, Holland, Brabant and Swabia.

Sword dance of the Nürnberg cutlers, including fencing on platforms made of swords. Anonymous, 1600.

ordinances of Prague show, the Cutlers' Guild had a prominent position in contemporary fencing practices:[57]

"2. Organising fencing tournaments has always been in power of the cutlers' guild in order to train bravery and manly fencing.

"3. The fencing tournament was established so that the cutlers, who live an honourable and humble life, may improve their fencing skills."

"9. Anyone wanting to organise a fencing tournament must ask the elder cutlers in advance, so that they can accompany him and put in a good word for him before the town council."

"12. A fencing tournament requires the usual wooden weapons, such as dusacks, staffs and halberds, as well as the steel weapons, such as swords and rapiers. These weapons must be held ready in sufficient numbers by the cutlers' guild and they are obliged to lend them to the organiser of a fencing tournament. If a weapon is damaged or broken during the tournament, the cutlers will make a new one.

"13. For this reason, anyone organising a fencing tournament should pay the cutlers' guild the appointed fee of threescore of Meissner groschen. In case a sword or rapier breaks during the tournament, the organiser pays them another thirty kreuzer for making a new sword and twenty for a rapier, however there is no such payment for breaking a wooden weapon."

[57] See VODIČKA 2019.

There are also records in Straßburg about cutlers who were fencing maſters, travelling from Basel to arrange fencing tournaments in the city. In 1558, an unnamed cutler and "weapons maſter" from Basel applies to arrange a tournament. Six years later in 1564, another Basel cutler and fencing maſter named Hans Jacob Meyer came to Straßburg and petitioned to arrange a Ƒechtſchule. The former may well be Joachim Meyer himself, and it is quite possible that Ƒechtmeiſter Hans Jacob Meyer is a relative, pointing to the possibility of a family tradition.

As a cutler, Meyer would also have spent several years as a geſell ("apprentice"), and then become a wandergeſell ("journeyman") who, after finishing his apprenticeship,[58] travelled for several years to other cities and even countries to work under and learn from other maſters of the trade, and through this, gain enough experience to pass as a maſter of their trade. This, too, would have been an opportunity for Meyer to learn from both German and foreign fencing maſters.

***

However Meyer learned, we do know that he was a Ƒreifechter and Ƒechtmeiſter of Straßburg, and that he associated with several nobles who were all tied to each other by blood or marriage, which will be the next topic of discussion.

MEYER AND NOBILITY

Outside of the three princes mentioned earlier, we don't know for certain who among the nobility Meyer had direſt contaſt with. However, when we trace the family trees of the princes who are associated with Meyer and his life as a Ƒreyfechter, one name in particular ſtands out, tying them all together, and that is Anna von Mecklenburg-Schwerin (1485–1525), mother of Philipp I (1509–1567), Landgrave of Hesse and founder of the Schmalkaldic League.

Placing these in chronological order we find the following:

- In 1561, Meyer dedicated his first known treatise to Georg Johann ("Jerrihans") I, Count Palatine of Veldenz. He was the son of Ludwig II and cousin of Wolfgang von Zweibrucken (via his uncle Rupert von Veldenz). Wolfgang married Anna von Hesse, daughter of Philipp I, in 1544. Jerrihans, in turn, married princess Anna of Sweden, daughter of Gustav Vasa I, King of Sweden, in 1563.
- In 1568, Meyer dedicated and gave his second treatise to Otto, Count of Solms. Otto was the grandson of Anna von Mecklenburg-Schwerin and nephew of Phillip I.
- In 1570, Meyer dedicated his third, printed treatise to Johann Kasimir, Count Palatine of Simmern. Johann Kasimir was the grandson of Kasimir von Brandenburg-Kulm-back, whose aunt Barbara Jagiellon was the mother of Christina von Sachsen, who married Philipp I in 1523.
- Also in 1570, Meyer's unfinished fourth treatise was based on teachings from Stephan Heinrich, Count of Eberstein. Heinrich married Margarethe von Dietz in 1577, daughter of Phillip I and Christina von Sachsen, after her first husband, Johann Bernhard von Neu-Eberstein, died in 1574.

---

[58] For comparison, statutes from the cutlers' guild in Sheffield, England in 1563 set the minimum length of service as an apprentice to seven years.

- Still in 1570, Meyer received an employment as 𝔉𝔢𝔠𝔥𝔱𝔪𝔢𝔦𝔰𝔱𝔢𝔯 at the court of Johann Albrecht I, Duke of Mecklenburg-Schwerin. Johann Albrecht I was the nephew of Anna von Mecklenburg-Schwerin.

Considering this, investigating Philipp I and his mother Anna von Mecklenburg-Schwerin to seek further links to Meyer and his activities in the 1560s may prove fruitful in the future.

Moving on from the noble families, we will now take a look at the tournaments and students Meyer is known to have been associated with.

𝔉𝔢𝔠𝔥𝔱𝔰𝔠𝔥𝔲𝔩𝔢𝔫

"Or our Chyrogargantua visits the fencing schools and fencing floors / when he did his school right / hew upwards / walked in with the dusacken / in which lead was poured / in the arch / in closed and simple downthrow / seated himself in the bulwark / shewed himself in all knightly arms / how they were laid / the sword / long knife / spear / staff / rod / daggers / poleaxe / rapier / parade sword / leather dusacken to let the parting crack / resist the Mark's brothers / the Frankfurt Master of the long sword / wrote with ink / as fought with blood / the feather must hover above / and it should steady his young life and he dared it God's might / struck it that crashes at the skin / challenged the highest blood wound / that crest / the school / a glass of wine / as the crafts companion desires / dry or wet / edged or blunt / naked or bare.[59]

In the city records of Strasbourg, OLIVIER DUPUIS has discovered 1,545 petitions to arrange 𝔉𝔢𝔠𝔥𝔱𝔰𝔠𝔥𝔲𝔩𝔢𝔫[60] in the city in the years 1540–1670, with the city council approving 1,342 of them.[61] Among the 𝔉𝔢𝔠𝔥𝔱𝔪𝔢𝔦𝔰𝔱𝔢𝔯 petitioners for these tournaments, 613 came from other cities, with 31 from Straßburg. Twenty-five percent of these 𝔉𝔢𝔠𝔥𝔱𝔪𝔢𝔦𝔰𝔱𝔢𝔯 were also furriers by trade. Among these petitions we also find several examples of Joachim Meyer seeking to arrange fencing tournaments.

In February 1561, Meyer sought and received permission to arrange a 𝔉𝔢𝔠𝔥𝔱𝔰𝔠𝔥𝔲𝔩𝔢 in Straß-burg, alongside of Christoff von Elias.[62] The 𝔉𝔢𝔠𝔥𝔱𝔰𝔠𝔥𝔲𝔩𝔢 was approved for a window of eight

---

[59] *Oder unser Chyrogargantua besucht die Fechtschulen unnd Fechtböden/ da that er sein Schulrecht/ hub auff/ gieng ein mit Dusacken/ darinn Bley gegossen war/ im Bogen/ in geschlossenen und einfachen sturtz/ lägert sich in die Pastey/ erzeigt sich in allen Ritterlichen Wehren/ wie sie vor Augen lagen/ im Schwert/ Messer/ Spieß/ Stangen/ Stängeln (Stänglin)/ Dolchen/ Hellenbart/ Rapier/ Paratschwert/ Le-dern Dusacken zum Platzmachen/ sträußt sich wider die Marxbrüder/ die Franckfortische Meister deß Langen Schwerts/ schrieb mit Dinten/ so ficht wie Blut/ die Feder must ihm oben schweben/ un[d] solt es festen sein junges Leben/ er wagts in Gottes macht/ schlug drauf das der Peltz kracht/ focht umb die höchst Blutruhr/ vmb daß Krantzlein/ vmb die Schul/ ein Glaß mit Wein/ wie es der Gesell an ihn begert/ trocken oder naß/ scharff oder stumpff/ nackend oder bloß.* Fischart (1575). Trans. from KLEINAU 2011.
[60] For a contemporary description of a *Fechtschule*, see EDELBECK 1574: 81–82. For further reading on the *Fechtschulen*, see WASSMANNSDORFF 1870.
[61] See MONDSCHEIN & DUPUIS 2019.
[62] *Joachim Meier burgher alhie bittet jme zuuergonnen ein fechttschul zuhalt des gleichen bittet Christoff Elias derso von jme Joachim meiern gelert jme vber acht tag hernach gleich gestalt ein schul zuhalten. Erkant jnen beiden iren geberen zulassen doch sagern dz sye fursehung thun dz ordenlich gefochten und*

days, ending on 22 February. On 7 March that same year, he signed his fencing treatise dedicated to Jerrihans, using the title 𝔉𝔯𝔢𝔦𝔣𝔢𝔠𝔥𝔱𝔢𝔯 for the first time.

DUPUIS has also discovered another four petitions from Meyer to arrange fencing tournaments on 4 September 1563,[63] on 15 June 1566,[64] on 1 February 1567,[65] and on 28 June 1568.[66] A second note regarding the 1568 𝔉𝔢𝔠𝔥𝔱𝔰𝔠𝔥𝔲𝔩𝔢 also exists, describing how they had trouble finding a suitable location since they were forbidden to use a canon's residence[67], but Kuno, Count von Manderscheid, offered them a place. They thus sought the council's permission, despite the restrictions regarding canons' residences. This was approved, but not appreciated.[68]

Also noteworthy here is that he is called a 𝔉𝔯𝔢𝔦𝔣𝔢𝔠𝔥𝔱𝔢𝔯 in the 1566 petition, while in the notes of 1568, Meyer is recorded as a 𝔉𝔢𝔠𝔥𝔱𝔪𝔢𝔦𝔰𝔱𝔢𝔯 for the first time, even though he had already used the term as early as 1561.

The students of Meyer

While we only have a few actual names of students of Meyer, he states that he taught soldiers and civilians, boys and men, of both high and low social standing. A few examples here follow:

kein Unlust furgehe. Minutes of the Council and the Twenty-One, Series R, 1R24, f 57[R], Saturday 15 Feb, 1561.

[63] *Joachim meyer burgher und messerschmid pitt Jme ein fechtschül zühalten, Züerlaübenn. Erkandt Jme Zülassen.* 1R26, f 352[V], Saturday 4 Sep 1563.

[64] *Joachim meyger d freyfechter bitt meyne hn wöllen jme vergonnen vnd zulassen ds er bis Montag nechst künfftig eyn fechttschul halten möge. Erkanndt mann solls jme zulass doch ds er nuhr wid pson j [pfenning] nemme, sagen jme h Schenkber h Kniebß.* 1R29, f 226[R], Saturday 5 June 1566

[65] *Wygand Brack der fechttMeister bitt meyne herren wöllen jme erlauben bis Montag vber achtt tag eyn offene freye fechtschul zuhalten. Also bitt auch Joachim meyger der Messerschmidt jme eyn fechtschule zu erlauben, so wöll er Samuel Shilling zu so auch zugegen freyen, derselbig bitt jme allß dann auch eyn fechtschul zu zulassen. Erkannt mann solls Wygand Bracken bis montag vber achtt tag, Joachim Meyger bis Montag über vier=zehen tag eyn fechttschul zuhalten zulassen, daby sagen ds sy es dahin rechtten, Das Es gesellisch freündlich vnd bescheidenlich abgang, mann werd geburlich auffsehens haben. Den dritten betreffendt soll manns zu disem mal gestellen, jst jme hernach was angelegen mag er wider ansuch.* 1R30, f 44[R], 1 Feb 1567

66 *Weygandt Brack der fechttMeister bitt jme zuerlauben biß Montag vber acht tag eyn freye offne fechttschul zuhalt Allso bitt auch Joachim Meyger der fechttMeister jme zuzulassen, biß kunfftigen Montags eyn freye offne fechtschul zuhalten. Erkanndt weyl sy bede burg soll manns jnen zulassen, doch dß sy von eyner person nit mer dan eyn pfenning nemmen vnd sich fridlich vnnd bescheidenlich halten. So by eyns Rhats... sagen jnen h Jörg Schoner vnnd heynrich Sypell.* 1R32, f 253[R], 23 June 1568.

[67] A canon is a member of an organization existing subjected to church rule.

[68] *Joachm Meyger der Messerschmidt vnnd Weygandt Brack die beden FechttMeister supplicieren an meyne heren, dennach jnen beden zwo vndeschiedliche fechttschuolen zuhalten erlaubt daby aber vndesagt warden de jme keyns Thumbheren hoff zuhalt. Sy abe keyne zunfftstuben bekhommen konnen, hab graff Chuono von Mondersch jnen bewilligt. Weyl den sy der school deselbig zuhalt schon auff geschlag vnndt mann meyne heren diener gern zulassen, denhalb keyn mangel erscheynen werde. So bitten sy jnen zuuergonnen, ds sy zu disen mal die schuol jne den gemelten Thumbherenhoff halten mugen. Erkanndt Dweyl sy sonst keynen platz bekhommen mugen soll manns jnen jezmal zulassen jnn des Thumbherenhoff zuhalten / doch daweben sagen meyne heren haben Mißfallen ds sy trage frage die fechtschuol an dem one angestelle, sagen jnen h Sixt Baltner vnd Sebastian Scheschtt.* 1R32m ff 261[R]–261[V], 28 June 1568

"I (without any fame to speak of) not only learned the noble knightly art of fencing from skillful and famous Maſters, but also have now praĉticed [this art] for many years and have inſtruĉted some young princes, counts, lords, and those of noble descent in it. And it has been requeſted multiple times, graciously and sympathetically, by you [and other] princes, counts, gracious lords, and noble [men], that I would draft the abovementioned praisewor-thy art of fencing in a certain order and then publish it for public consumption as a printed text, so that the same [art] is revealed for the use of many people of our nation. Therefore, as I ought not set myself to resiſt such gracious and sympathetic entreaties..."[69]

"...that it should thus be good and quite serviceable to many persons of high and low ſtand-ing, who have love, joy, and desire for the art of fencing..."[70]

"Therefore, this [book] can provide no minor service and benefit for them—and especially for young lords and for others of noble birth..."[71]

"...this aforementioned cause is claimed to be sufficient and occurs among knowledgeable fighting men..."[72]

"...it is necessary for a wise, knowledgeable military leader to be beſt equipped and provided with good equipment for war, including all accessories..."[73]

"...those youth, as they grow older and after they have learned from a true maſter—yet do not have the [maſter] by them at all times—can be reminded from this [book] and praĉtice each day at the time they have scheduled..."[74]

Some of his ſtudents are known from records, like Chriſtoff Elias, who is liſted as a ſtudent of Meyer in 1561, and Samuel Schilling, who asked to be allowed to arrange a Sechtschule under Meyer's authority. Weygand Pranck/Brack was a fellow fencing maſter who petitioned to ar-range fencing tournaments in Straßburg alongside Meyer and would naturally have been a familiar face, given the comparatively small size of cities in their time. In JML, he also says he taught the young Otto von Solms-Sonnewalde:

"...and your Grace... has summoned me to it as an honeſt combat-maſter to inſtruĉt your Grace in this art; and I have undertaken this in all humility with willing spirit."[75]

According to research by DUPUIS, Meyer also trained and prepared four fencers for the examination to become fencing maſter, this over a period of four years.[76]

---

[69] Meyer (1570): a iii[V].
[70] Meyer (1570): a iv.
[71] Meyer (1570): b i.
[72] Meyer (1570): a ii.
[73] Meyer (1570): b iii.
[74] Meyer (1570): b i.
[75] FORGENG 2016: 38.
[76] See DUPUIS 2022.

Johann Kasimir, Count Palatine of Simmern. By Tobias Stimmer, 1576.

Whether he ever instructed Johann Kasimir remains to be confirmed, but it is a possibility. This is also true for Jerrihans and Heinrich, both of whom he dedicated treatises to.

Of course, right before his end, he also took up employment with Johann Albrecht I, Duke of Mecklenburg, to teach both him and his sons fencing, although he likely never got the opportunity to do so (outside of perhaps some small, informal lessons).

***

We know Meyer was the most active with teaching fencing in the middle of the decade. Both Straßburg and Basel were struck extremely hard by the plague during the years of 1563–64, and a nearly third of the population in each city died, leaving both streets and churches empty.[77] In Basel, thirteen members of its council and the rector of the university died. So did Wibrandis Rosenblatt, mother-in-law of pastor Jakob Meyer of St. Alban and widow of Oekolampad and Bucer as well as two other husbands. One can only imagine how badly afflicted Meyer's society would have been, how profoundly struck by fear and sadness, having lost mainly young, pregnant wives, new mothers still recovering from childbirth, and older relatives. At the time of writing this, the recent COVID-19 pandemic is but a bleak shadow, but may have given us slight sliver of understanding of what these people went through. Yet they persevered, carrying on with their lives, as people tend to do.

Four years later, in 1568, Meyer likely met his former student Otto, who revisited the *Schola Argentoratensis* academy where he did an important part of his studies, and Meyer handed over the handwritten treatise on fencing he had dedicated to von Solms.[78]

***

At this time, we do not have specific locations for Fechtschulen arranged by Meyer, but certain places in Straßburg are known to have been used for such purposes.

The courtyard at Rue des Dantelles no. 11 appears to have been one such location, and coincidentally, the street just outside of it is today called Rue du Bouclier or Schildtgass—Buckler Street.[79] Nearby, you also find the Church of the Buckler. Likewise, Roßmarkt (the horse market) was used also for this purpose. The more common purpose for the Roßmarkt was selling horses, but fencing tournaments and even jousting were occasionally performed there as it was the most open place of the city and thus the most suitable for safely entertaining a crowd of spectators.

Meyer had greater ambitions though and decided to accept the request of Johann Kasimir von Simmern to publish a treatise on the art of fencing, which brings us to the next important person connected to this book, its printer Thiebolt Berger.

---

[77] For more information about the plagues in Switzerland, see ECKERT 1978: 49–80. See also Rebmann, Jean Roger. "Die Pest von 1563/64". Altbasel.ch Gesichte, 2011. Retrieved 9 Aug. 2022. <http://altbasel.ch/zeittafel/pest_basel_1563.html>

[78] See NORLING 2012A.

[79] A buckler is a small, dinnerplate-sized shield.

View along the Münstergasse to Roßmarkt in Straßburg. By Hans Baldung Grien, ca. 1517–1545.

MAKING THE ART OF FENCING PUBLIC

After having been repeatedly asked to do so by Johann Kasimir, Meyer had some time in the 1560s and decided to spread his understanding of the art of fencing to as many as possible, getting his skills and knowledge in print.[80] Meyer selected book printer Thiebolt Berger and the workshop of Hans Christoph Stimmer, brother and partner of one of the finest artists of his time, Tobias Stimmer, to aid him in making it happen. Berger had had his shop at Barfußerplaz since 1551, but moved it in 1566 to Alte Weinmärkt. This is where most of the work on preparing Meyer's treatise would have been done, in the Haus zur Traube. [81]

The book was largely based on his earlier manuscripts but contained even more material and expanded on his earlier teachings. This would be only the second original, printed German fencing treatise to be published, and the first in several generations[82] to be designed to teach the broader span of contemporary single combat, including both wrestling and the typical weapons of war of the period. In fact, it was the first printed treatise to teach a number of aspects of single combat which had never been published in German before: for example, teaching the use of the rapier, rapier and dagger, halberd, and pike, as well as the use of tactics and psychology in fencing. This makes his efforts rare and very significant.

A BUSY 1570 AND RISING IN THE RANKS

The year of 1570 must have been one of the most intensely busy years of Meyer's life. He became the treasurer of the smiths' guild of Straßburg,[83] which might indicate that he by then was considered a senior and reliable master in the guild. By this time, he would have spent six to eight years as a gesell, perhaps several years as a wandergesell, and ten years in Straßburg as a master cutler. At the same time, he was finalizing the preparations for the sale of his newly-printed book and finally seeing a financial return on the 300 kronen he had borrowed[84] to invest in creating this lavishly illustrated treatise. He possibly also visited the Diet of Speyer, meeting Johann Albrecht I and being offered a position as fencing master at his court in Schwerin, which he also accepted in this year.

---

[80] See Meyer: a Iii^v.

[81] Berger would here remain until 1579, when he moved to Gewerbslauben no. 83, near the Rossmarkt. See BENZING 1963: 420 and RESKE 2007.

[82] Meyer's 1570 printed treatise, *Gründtliche Beschreibung der freyen Ritterlichen unnd Adelichen kunst des Fechtens*, is only the second treatise on fencing printed in German, with Hans Wurm publishing on grappling in the 1490s, Andres Paurñfeyndt on fencing in 1516, and Fabian von Auerswald, again on grappling, in 1537. (Christian Egenolff also rearranged and republished Andre Paurñfeyndt's text in the 1530s). In other languages, we also see Pedro Monte's two very early printed fencing treatises of 1492 and 1509, written in Latin, and in Italian, Antonio Manciolino in 1523, Achille Marozzo in 1536, and Marc'Antonio Pagano and Camillo Agrippa in 1553. Most of these were printed in small runs and would have been difficult to come by.

[83] *Rechner ins gericht: ... Joachim Meÿger, Messerschmidt.* Archives of the Guilds, XI138: Register of the Officers of the Company of Smiths, 1570.

[84] This sum is mentioned in the letter which Meyer's brother-in-law, Anthon Ruhlmann, sends to the duke of Mecklenburg. Brief von den Stadtsrat zu dem Herzog zu Mecklenburg, Stadtarchivs zu Strasburg, V14 110, ff 1–6.

Johann Albrecht I, Duke of Mecklenburg, with his wife Anna Sophia. Artist and date unknown.

Johann Albrecht I is particularly interesting for several reasons. To begin with, he was the grandson of Magnus II of Mecklenburg—the father of Anna von Mecklenburg-Schwerin, who married Otto von Solms-Laubach and was mother of Philipp I von Hesse and grandmother of

Otto von Solms-Sonnewalde (Meyer's student). He was also the duke who gave the Freifechter von der Feder their first official recognition, their charters, and a small kleinod ("privilege") in 1570 (which may be significant to Meyer's story).

At this time, Meyer also had a compilation fencing manuscript in his possession (JMR) which included a number of older fencing treatises and a "book" of his own writing on the use of the rapier. This also included a draft for a foreword mentioning teachings from Heinrich von Eberstein. He could conceivably have had manuscripts of any of the further books that he mentions in the text of his 1570 treatise.

SCHWERIN, DEATH, AND BEYOND

Having secured the position as fencing master at Johann Albrecht's court in Schwerin, Meyer left Straßburg for Schwerin on 4 January, but became ill while travelling the 500 miles to the court. He arrived on 10 February, having travelled for just over a month and died soon thereafter. On 24 February 1571, his death commented upon by Heinrich Husanus, chancellor and councillor of the ducal council, who lamented that he could have been saved had they had a physician available at the court.[85]

His brother-in-law, Anthon Ruhlmann, subsequently wrote a letter to the city council (on behalf of his sister Appolonia) seeking assistance from the council in requesting the court of Schwerin to have Meyer's personal effects, including a box of his printed fencing treatise, returned to the widow in Straßburg. A formal letter was sent from the council, but the duke replied that the books had been ruined and he offered to send some compensation alongside the remaining personal effects to the widow. For unknown reasons, JMR appears to have not been sent back with his belongings, as it ended up in the Ducal library in Schwerin before its eventual transfer to the Rostock University Library.

The woodblock engravings and original drawings are believed to have been sold to pay off debts resulting from the book printing and later surfaced in Augsburg.

---

[85] This was discovered by KEVIN MAURER: Merkel, Johannes. *Heinrich Husanus. 1536 Bis 1587.: Eine Lebensschilderung.* L. Horstmann, 1898. p 188.

Husanus himself had an interesting life. Born in Eisenach, Germany, he worked as an apprentice for the Hanseatic League in Bergen, Norway, but due to bad experiences returned to Germany, before studying at the university of Jena and Wittenberg in 1551 and 1553. Three years later he attended the university of Ingolastadt in 1556, continued to the university of Bourges in France in 1557, and then the university of Padua, Italy, in 1559. He returned to Jena and became a professor at law. The next year he received a position as councillor at the ducal court of Saxon, but falling out of favour with his duke in 1566, he was forced to flee, leaving all his possessions behind. In 1567 then, Johann Albrecht I, Duke of Mecklenburg appointed him as chancellor and councillor. Meyer arrives, and dies in 1571, and two years later, unhappy with his position, Husanus leaves for Lüneburg, where he spends the rest of his days.

Schwerin town and castle as it looked like in 1653. During Meyer's brief visit, the Dutch star-shaped fortifications wouldn't yet have been present. By Johann Stridbeck I (after Matthäus Merian).

DID MEYER HAVE CHILDREN?

The idea of Meyer having children and the possibility of his bloodline continuing to the present day is tantalizing. However, no children are for certain known to have come from the marriage between Meyer and Appolonia. That said, there is an interesting note in the margins of the letter to Johann Albrecht mentioned previously which mentions "Joachim Meyer's wife and child".[86]

Still, considering the fact that Appolonia had already been married before she became Meyer's wife, it is quite possible that this child mentioned was the result of that previous marriage and not Meyer's child by blood.

APPOLONIA MOVES ON WITH LIFE

Meyer's widow would soon marry another cutler, Hans Küele, on 14 April 1572—her third known marriage. This is likely to be part of the guild tradition of providing a new husband for the widows of its members, as a means both to care for the guild's widows and fatherless children and to grant membership to prospective guild members, just like occurred with her marriage to Meyer. Hans Küele thus became a member of the smiths' guild as a 𝔐eſſerſchmið in the same manner that Meyer did.

"Hans Küele, the cutler, has received citizenship through widow Appolonia Ruollmännin Weylands, wife of Joachim Meyer (also cutler), and serves with the smiths."[87]

Three years later in 1575, as discovered by CHRIS VANSLAMBROUCK, Hans Küele was noted building a house at No. 14 of des heiligen Liehts gässel (today Rue des Chandelles) with his initials carved into a beam.[88]

It is believed that Appolonia and either Hans or Anthon were involved with the second, Augsburg printing of Meyer's book in 1600. According to some sources, it was reprinted yet again in 1610 and in 1660, although no copies of these editions are known to have survived and they probably never existed.

II. THE SOCIAL AND MARTIAL CONTEXT OF JOACHIM MEYER

We now return to the question asked at the beginning of this text: what motivated Meyer to write a treatise on fencing? To understand this, we need to explore the culture and history which shaped him. Looking through the key dates listed at the beginning may also be helpful before continuing.

---

[86] *Joachim Meygers / Wittwe und Kind / Johann Albrecht / Herzog zuo / Meclenburg / Fürschrift.* Dated to Monday 14 may, 1571. Strasbourg city archive, 1R38, f 442[R].

[87] *Hans Küele der Messerschmidt hatt das Burgkrecht Empfang von Appolonia Ruollmännin in weÿlands Joachim Meÿers auch Messerschmidt sellig wittwen seiner ehefrawen vnd dient zu den Schmieden.* Dated to April 14, 1572. Add reference to Str. Archive.

[88] "Das Haus N° 14 zeigt am Holzbalken im ersten Stock das Datum 1575 und HK, dh Hans Kuele, ein Messerschmidt, der das Haus gebaut hat." p75, *Das alte Strassburg, vom 13. Jahrhundert bis zum Jahre 1870,* Seyboth Adolph, J. H. Ed. Heitz, Heitz & Mündel, 1890

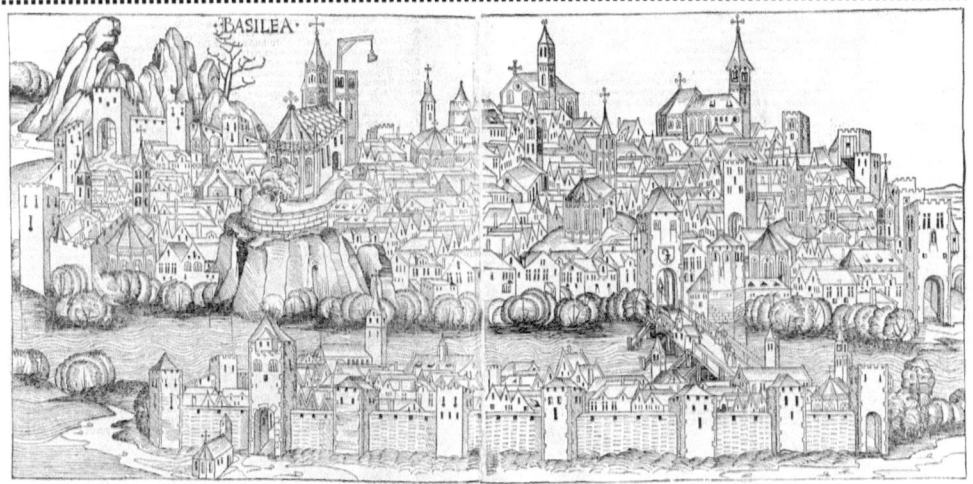

Basel 1493, from the Nuremberg Chronicle. *By Michael Wolgemut and Wilhelm Pleydenwurff, 1493.*

FORTIFIED CITIES AND THE MILITARISTIC CULTURE OF RENAISSANCE GERMANY

The introduction of gun powder and cannons to Europe in the mid- to late-14th century would, over time, drive the development of more and more refined fortifications, with whole cities being fortified in quite complex manners using shapes and designs intended to deflect balls or otherwise protect against increasingly more powerful cannons. Medieval towers were changed from square to round shapes, triangular and pentagonal forms were introduced, and by the late 16th and early 17th century, cities and fortifications had taken on very complex star-shaped forms[89] in the "Dutch" style—a shape which provided carefully planned overlapping lines of fire from the walls and bastions, using several defence lines consisting of walls, bulwarks, canals, waterways, and moats to hinder the advance of the besiegers. This went hand in hand with military tactics for infantry and cavalry, forming complex formations of men armed with pike, halberd, and harquebusier.

To better understand the power of these cities and fortifications, a decent comparison to today would perhaps be a modern warship, with its powerful defences making it extremely difficult and highly dangerous to physically approach for an enemy force. Its mere sight would have been both impressive and quite intimidating. Basel, the city where Meyer was born, went through this architectural process, and in contemporary artwork we can clearly see its development. This is even more true for Straßburg, the city where Meyer spent most of his professional life as a cutler and fencing master, and which changed radically between the 15th and 17th centuries, as can be seen in the following series of images.

[89] This had its origin in the *trace italienne* style, which had begun developing already in the mid-1400s with fortifications quite literally shaped like fully symmetric stars or snowflakes. This came to be the ideal city shape of the Renaissance, spreading outside of Italy from around the 1530s. These designs would often remain until the mid-1800s, after which city fortifications commonly were torn down, in order to create space and light.

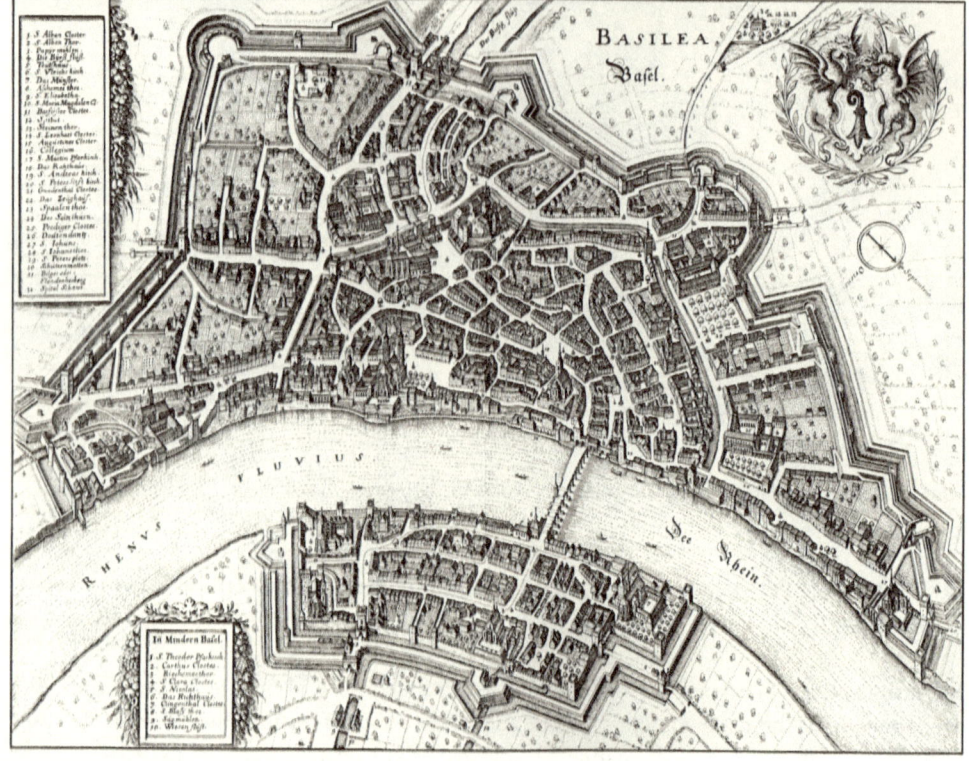

Basel. By Matthäus Merian d.Ä, 1642. The older walls can be seen clearly.

Strong walls and obstructing moats are insufficient for extended sieges, and preparations for such emergencies required the storing of necessities, while actually defending a city demanded a trained militia, both of which were common in the German cities, as Machiavelli describes in The Prince:

"German cities are completely independent, don't have much territory around them and obey the emperor only when it suits. They are not afraid of him, nor any other powerful rulers in the area. This is because these towns are so well fortified that everyone realizes what an arduous wearisome business it would be to attack them. They all have properly sized moats and walls; they have the necessary artillery; they have public warehouses with food, drink and firewood for a year; what's more, to keep people well fed without draining the public purse, they stock materials for a year's worth of work in whatever trades are the lifeblood of the city and whatever jobs the common folk earn their keep with. They hold military exercises in high regard and make all kinds of arrangements to make sure they are routinely practiced."[90]

In this place and time, all able-bodied, adult men who were citizens had to serve in the town militia, and every household had to provide a man who would also serve as town guard,

---

[90] PARKS 2009.

keeping the city streets in order. [91] And as all men, apart from the clergy and Jews, were expected to own and keep arms and armour as well as to bear arms in public, and since drunken brawls (not least among soldiers) and even the occasional riot were common, this was no small task. Conflicts between university students, who often lived under church law, and burghers, who commonly lived under town law—and, unlike the students, served as town guards—were also frequent.

However, burghers who were citizens as opposed to mere inhabitants of a city [92] were at all times expected to stand ready to act in defence of the city, serving not only as town guards, protecting against natural disasters and invading armies, but also against internal threats to public order, ready to stop any violence, disorderly conduct, misbehaviour, or otherwise suspicious behaviour taking place inside of the city walls. As such, every burgher would represent the power of the Burgermeister and the whole community. This also in part explains the requirement for all men to keep and bear arms.

Another important burgher duty required of all able-bodied men was to serve as a fire guard and as honorary guard for occasions where dignitaries visited the city. Naturally, these obligations placed the training of fencing in a position of relative importance, and men of all social stations practiced regularly, with Meyer himself mentioning practicing every day. Public and bloody fencing tournaments were therefore also quite common and popular. In modern eyes, these were fairly brutal but heavily regulated affairs, fought for honour and money both, with the fencers who caused the highest bleeding wounds crowned as winners.

However, by Meyer's time, the art of fencing was perceived as having declined due to the gun taking the place of bladed weapons. This was to the detriment of society, as a man with no skill or honour could easily kill a man trained and highly skilled in the art of fencing. Meyer—and Paul Hektor Mayr before him—lamented this change, as would Leonhard Fronsperger in his *Kriegsbuch* of 1573.

Still, especially compared to today, it should be recognized that fundamentally Meyer's society was heavily militaristic, and some degree of martial training was common for most men, with a sizeable percentage having personal experience of combat and war. This said, others like French medievalist FRANCIS RAPP, were less impressed with the military skills of the Alsatian militias and the infantry of Straßburg. [93]

---

[91] If the household had no man who could serve, one could also be hired on their behalf. See TLUSTY 2011.

[92] As a citizen a man had both rights as well as responsibilities which non-citizens like women, children, visiting foreigners, and others did not. This included the possibility of serving in the city council, but also the duty to protect the security of the city and its citizens, among many other things.

[93] See DUPUIS 2021.

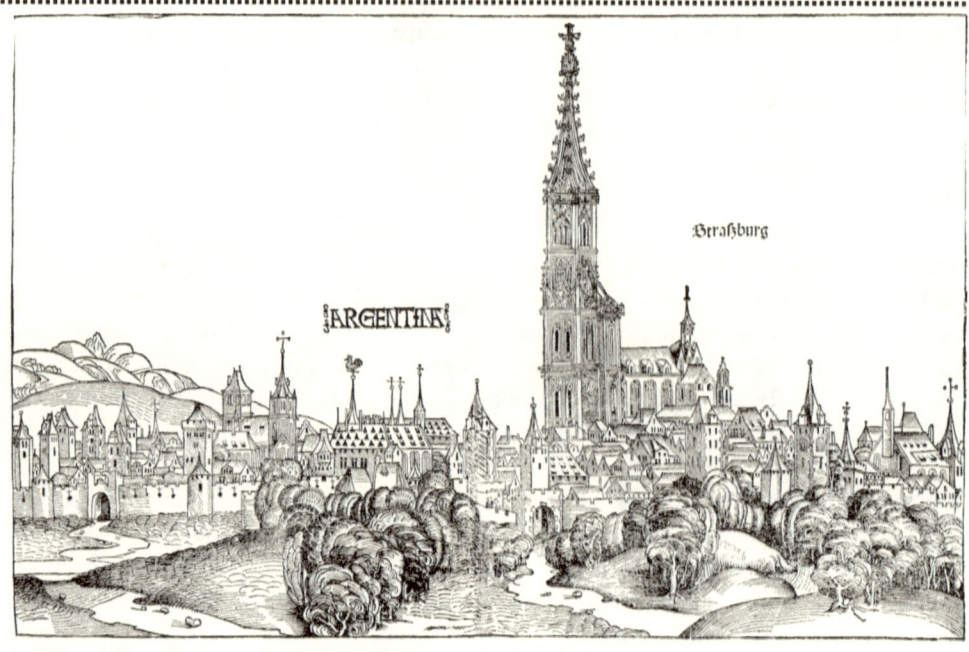

Straßburg, from the Nuremberg Chronicle. *By Michael Wolgemut and Wilhelm Pleydenwurff, 1493.*

Straßburg. By Braun & Hogenberg, 1572.

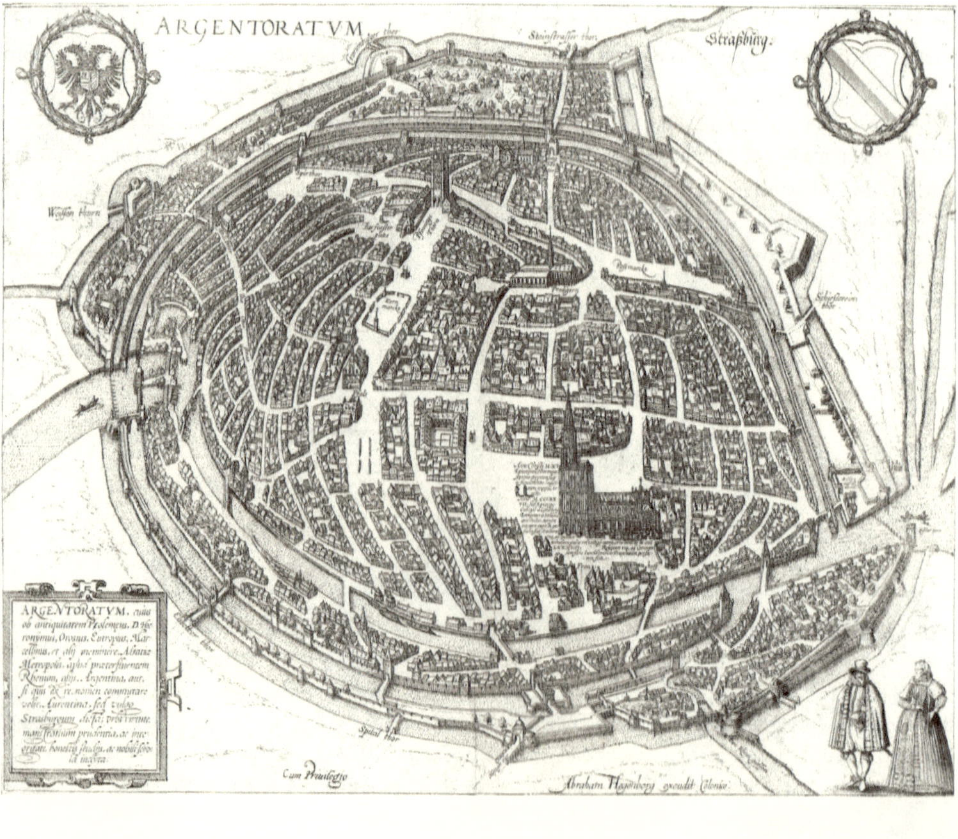

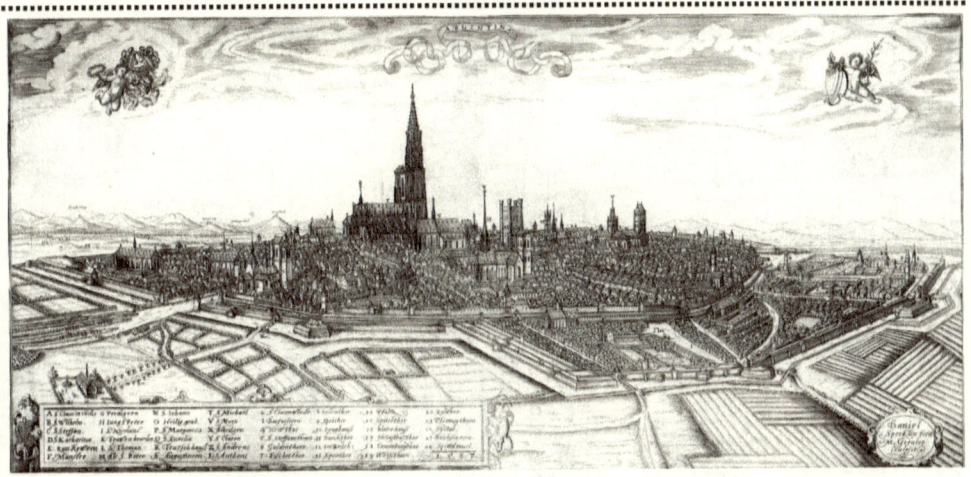

Straßburg. by Daniel Specklin, 1587.

Straßburg. by Matthäus Merian d. Ä., 1643. (Rotated to match the orientation from 1572.)

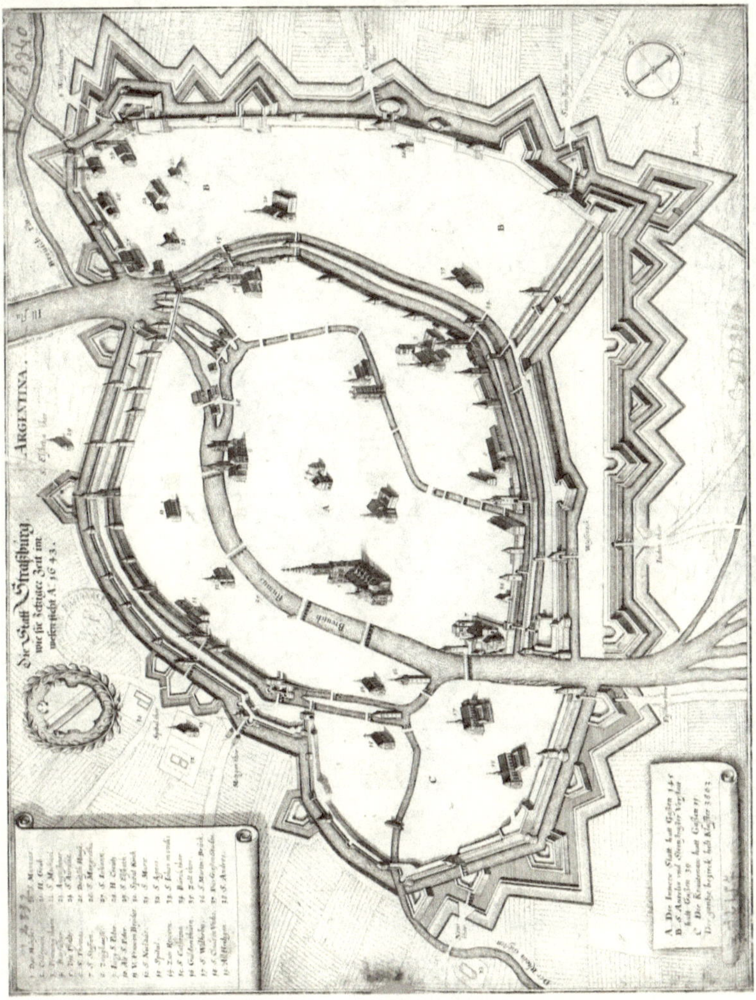

Brawling students and burghers in Erfurt. By Justus Jonas d. Ä, 1506.

Students training fencing at the university of Tübingen in Württemberg, in one of thirteen engravings of university life in Germany, designed by Johann Christoph Neyffer and produced by Ludwig Ditzinger sometime between 1589 and 1600. All stances are recognizable as those taught by Meyer.

INSTRUCTION IN THE ART OF FENCING

Given the importance of the town militia and guard duties, receiving some form of instruction in the art of fencing was part of most urban young men's education in this time, especially those who were expected to have some form of military responsibilities. At the universities, fencing was so popular among the young men that university faculties felt concerns about it having a negative effect on their studies, causing "all sorts of nonsense", and even though the faculties at the universities regarded the practice of the art of fencing as useful, specific orders were written requiring fencing masters to ask permission to arrange Fechtschulen, not only from the Burghermeister ("mayor"), but also from the rector so they together could decide a convenient date which would least hinder the youth in their studies.[94]

Even the mind-bogglingly rich patrician Fugger family's young men received training, and the family specifically chose to include fencing scenes with little putti to decorate the frames of the portraits of their family members *in Das Ehrenbuch der Fugger* from 1545–48.[95] Likewise, dusacks were included among the objects symbolizing the education of its family members in the portraits of their *Fuggerorum et Fuggerarum* of 1618. Their accountant, Matthäus

---

[94] See, e.g., *Leges Academia Witebergensis de studiis et moribus auditorium.* Wittemberg: 1557.

[95] These bear a strong resemblance to the artwork of the fencing treatises of Paul Hektor Mair, especially PMM, which is perhaps not so surprising since the artist who illustrated both was Jörg Breu d. J. See NORLING 2011.

Schwarz, chose to have himself depicted with a 𝔉𝔢𝔠𝔥𝔱𝔰𝔠𝔥𝔴𝔢𝔯𝔱,[96] a blunt training sword, in hand in his *Trachtenbuch*, a book on his clothes created between 1520–60.[97] His son Veit Konrad was also depicted receiving fencing instructions under a master in 1561, with the following description:

"On 13 March 1561, I was taught together with Jörg Ulstat, Connratt Mair, Victor Vöhlin, Christoff Stamler, David Sultzer, and others by Kirschgen Eiser von Köln, an oathbound-master of the long sword, an armorer, and citizen here at Augsburg, and the sword fighting was done in Hanns Behams', innkeeper, dance master outside Our Lady's Gate. We went to him for two hours every day."[98]

In the *Weisskunig* (1506),[99] we even see Emperor Maximilian I himself learning fencing under the supervision of a fencing master, with a 𝔉𝔢𝔠𝔥𝔱𝔰𝔠𝔥𝔴𝔢𝔯𝔱[100] in hand and training gloves, Messers,[101] and staffs on the floor around him.

Similarly, in Hans Burgkmair's *Triumphzug der Kaiser Maximilians* (1517), we see several groups of fencers carrying training swords, quarterstaffs, and even flails in Emperor Maximilian's triumphal parade.[102]

Finally, in the Swedish National Archive, letters from the young Swedish Erik Oxenstierna, son of Lord High Chancellor of Sweden, Count Axel Oxenstierna, are still preserved. These were written years before he himself would become count and Lord High Chancellor of Sweden, and they provide us with a small insight into the life and international connections of young men in the higher echelons of society. Reading through these letters, we find that in the just over two years which he spent travelling from Sweden to Germany, Holland, France, Italy, and Switzerland, studying Dutch, Italian, history, literature, and architecture, at different universities, he also enjoyed learning fencing in Amsterdam, Leiden, and Rome and visited two military camps along the way. Oxenstierna also paid visits to a number of courts to establish important connections. In Rome he also described meeting many foreigners, including Poles, Frenchmen, Dutchmen, and a few Danes, who were themselves on similar journeys.

While these examples are mostly from the upper social classes of society, we should not, however, think that the art of fencing in this time was reserved for those only. They are simply chosen to exemplify that even the upper classes learned largely the same art of fencing as the regular burghers of a city did.

---

[96] See also JASER 2014: 207–23.

[97] See RUBLACK & HAYWARD 2015: 87, 253–254.

[98] See RUBLACK & HAYWARD 2015: 224, 371–371.

[99] See p92a Weißkunig, 1775, <https://digi.ub.uni-heidelberg.de/diglit/maximilian1775/0146 > Retrieved Aug 4, 2022.

[100] A blunt training sword with flanges on the part of the blade closest to the cross. The use of these started in Germany sometime in the first half of the 1400s, and they were used for unarmoured fencing practice.

[101] German for "knife", essentially single-edged one-hand swords, but with the grip constructed of two wooden plates peened to the sides of the tang.

[102] For flail, see also Image M under the polearms section of this treatise.

Maximillian I practicing fencing, from Weißkunig *(produced in 1514–1516, but not published until 1775).*

CONTEMPORARY WEAPONS OF WAR

What weapons were in use in Meyer's time, in the period of mid 1530s to early 1570s? Firearms (arquebuses, muskets, and pistols) were the most important weapons of Meyer's time, but as Meyer himself expresses it in the foreword of this treatise:

"Whereas the knightly and noble art of fencing has devolved in current times due to many and various reasons, yet there is no doubt that the greatest and most eminent cause, among others, is namely the recent emergence of the harmful gun, which has taken the upper hand to such an extent that the most virile and courageous heroes are robbed and deprived of their lives by the same, sometimes even by the least and most timid, and at times both friends and enemies are unintentionally injured and aggrieved.

Therefore, it is no surprise that this practice of free knights has not merely declined, but has even become somewhat contemptible and not lacking in prejudice toward the age-old, praiseworthy custom. Elsewhere, this aforementioned cause is claimed to be sufficient and occurs among knowledgeable fighting men, principally because, in the case of guns, nothing functions correctly without other equipment, weapons, and arms,[103] as occasionally the entire battle must be stopped due to such fragile weapons and handguns (because the guns cannot be used due to some reason), as knowledgeable fighting men attest."[104]

For these reasons, common edged and bladed weapons of war at this time were the very popular and fashionable rapier, a cut-and-thrust sidearm used for both civilian and military purposes, and when given a heftier blade and carried as a riding sword simply called a Reitschwert. The rapier was already carried on the hip of young and old burghers alike by the 1540s, but was still regarded as foreign in Meyer's time. The manner of its use was also considered to be in conflict with contemporary ethics regarding thrusting at one's fellow countrymen, a serious issue—not least with mercenaries commonly employed by foreign powers to attack their own countrymen—which Meyer brought up in this treatise, although he also recognized that it was important to learn the use of the thrust.

The dusack, an early sabre equivalent to the Italian storta and the Ottoman kilij, named with a Czech word for "fang" or "claw", also saw frequent use, especially in the late 1500s and early 1600s. This weapon was common in large parts of the Holy Roman Empire, and Meyer himself expressed that after the (two-handed) sword, it was the weapon most used by the Germans. It also became an important weapon in Scandinavia, and as early as 1539, Sweden was already importing them alongside Schlachtschwerter, with customs records listing "(Two) dozen battleswords... (two) dozen dusacks",[105] while fifty years later in 1589, Norwegian King Christian IV began importing at least 8,000 dusacks from Switzerland. He ordered all his peasants to arm themselves, and thus required them to buy dusacks from him at their own expense. This great number, combined with the fact that only about 350,000 people lived in Norway at the time, would mean that about one in ten Norwegian men were armed with a dusack, and peasants in particular.[106]

While still in occasional use, the two-handed sword[107] saw less and less use in this period but can still be seen in artwork depicting contemporary battles and conflicts. It was also

---

[103] Meyer appears to be using *gewehr* in its collective meaning of weapons or arms that are not projectile weapons. He is further distinguishing between traditional weapons = *waffen* = swords (and polearms, etc.) and *geschütze* = guns. *Rüstung* can mean armor, but can also mean all of the equipment required for a fighting knight or man at arms.

[104] Meyer (1570): a ii.

[105] *(Två) dussin Slagsuerdh .. (två) dussin tesshakar.* From *Tullbok från Stockholm. Lokala tullräkenskaper.* Manuscript.

[106] See NORLING 2012B.

[107] Pedagogically, the sword, together with the quarterstaff, was also used by Meyer, I would suggest, to teach the use of the *Schlachtschwert*, the great two-hand sword of war, just like the staff was used to teach the halberd and pike and the dusack to teach all one-hand swords.

sometimes carried by bodyguards, sometimes with complex hilts, well into the 17th century. Of course, greying and battle-proven gentlemen in their late forties and beyond would have been more comfortable carrying such swords, having learned to use them twenty or more years earlier, and are thus commonly depicted with them in Meyer's time.[108] Some of these elder men would have served as quarter or lane captains in the militias of their cities and towns, as the majority of such were of a mature age between forty and sixty.[109]

The Schlachtschwert (or Zwei- or Bidenhänder),[110] the *spada di due mani* (or *spadone*), and the *montante*—terms for great two-hand war swords in German,[111] Italian, and Spanish and Portuguese, respectively—saw battlefield use well into the second half of the 1600s, although they were consigned specialized functions: they were commonly assigned to the defence of cannons, banners, and high-ranking officers, coming into play when the lines were broken or in various outnumbered scenarios. The great two-hand swords were also carried by captains on ships and town guards for similar situations. This is described in several fencing treatises; for example, Giacomo di Grassi (1570):

"The two hand Sword, as it is used nowadays being four handfuls in the handle, or more, having also the great cross, was found out, to the end it should be handled one to one at an equal match, as other weapons, of which I have entreated. But because one may with it (as a galleon among many galleys) resist many swords, or other weapons: therefore in the wars, it is used to be place near unto the Ensign or Ancient, for the defence thereof, because being of itself able to contend with many, it may the better safeguard the same. And it is accustomed to be carried in the City, as well by night as by day, when it so chances that a few are constrained to withstand a great many. And because his weight and bigness, requires great strength, therefore those only are allotted to the handling thereof, which are mighty and big to behold, great and strong in body, of stout and valiant courage. Who (for as much as they are to encounter many, and to the end they may strike the more safely, and amaze them with the fury of the sword) do altogether use to deliver great edge blows, downright and reversed, fetching a full circle or compass therein, staying themselves sometimes upon one foot, sometimes on the other, utterly neglecting to thrust, and persuading themselves that the thrust serves to amaze one man only, but those edge blows are of force to encounter many."[112]

---

[108] See e.g. *Hofkleiderbuch (Abbildung und Beschreibung der Hof-Livreen) des Herzogs Wilhelm IV. und Albrecht V.* 1508–1551

[109] Some of these were also likely older than they claimed to be, desiring to retain their status as valuable men, despite their old age. See TLUSTY 2011: 24.

[110] The terms *Schlachtschwerter* and *Bidenhänder* are neologisms, i.e. non-contemporary words, only invented in the mid-19th century.

[111] See, e.g.: Klein, Dionysius. *Kriegs-Institution.* Stuttgart: Marx Fürstern, 1598, which makes plentiful remarks on the *Schlachschwert* and notes how soldiers should train in the art of fencing.

[112] *Il spadone al modo ch'oggi s'usa con quattro palmi di mani co et più et con quella croce grande non è stato ritrouato affine di adoprarlo da solo a solo a ugual partito come l'altre arme delle quali habbiamo trattato, ma per poter con esso solo a guisa d'un galeone fra molte galere resistere a molte spade o altre arme percio nelle guerre s'usa di porlo alla difesa delle insegne per che possa contrastado con molti difender l'insegne, et per le città si suol portar la notte et il giorno quado auiene che pochi debbano resistere a molti et perche il suo peso et la sua gradezza richiede molta forza pero a quest'arma so dedicati*

Twenty-five years later, Englishman George Silver (1599), described the advantages and dis-advantages of the two-hand sword compared to various types of swords and polearms. Like-wise, Portuguese general Diogo Gomes de Figueyredo taught a number of drills and tactics for such contexts in his *Memorial Da Prattica do Montante* (F) of 1651. Achille Marozzo had already taught the handling of the *spada da due mani* in 1536, while *Francesco Fernando Alfieri* did the same 117 years later in 1653.

Late 16th century German military strategy also prescribed the inclusion of the Schlacht-schwert in their Fähnlein (infantry company). For example, in 1570, an imperial Fähnlein of 400 men should have 100 pikemen armed with pistols, 50 experienced men with two-handed swords or halberds "to guard the standards", 50 unarmoured pikemen, and 200 "shot"[113] (with good rapiers and helmets).[114]

Balancing the composition somewhat differently for a Fähnlein of 300 men, Wintzenberger[115] prescribed 82 Doppelsöldner with harness and pikes, 14 Doppelsöldner with harness and Schlachtschwerter, 18 halberdiers, 42 musketeers, and 144 arquebusiers, while Olnitz and Lützen-kirchen[116], for a Fähnlein of 500 men, prescribe 200 arquebusiers, 200 pikemen, 45 halberdiers, 45 men with boarspears, and 10 men with Schlachtschwerter.

As can be seen from these ordinances, the halberd and especially the pike were the primary bladed weapons on the battlefields of the time, be it in a field or in a city street. But according to several of these regulations, the Schlachtschwerter were also an acceptable alternative to the halberd. The Schlachtschwert is also commonly depicted used in the front line by Doppelsöldner, and the Verloren Hoop—the Lost Bunch, who received double pay for the greater risks taken, often also receiving positions as officers, if they survived. Dagger fighting and grappling were also necessary skills, as was fighting in half or full armour, and for some burghers and nobles, combat on horseback.

Original treatises on group combat in this time period are scarce, with most, like Machiavelli in his *L'arte della Guerra* of 1521, referring back to Roman warfare and Vegetius' *De Re Mili-tari*, aimed at military commanders rather than troops. Individual fencing was primarily learned and trained in the fencing guilds, at university, from hired fencing masters (in private or in public), or from trained relatives.

***

Having familiarized ourselves with the weapons of war of Meyer's time, let's now dive into the fencing culture and how the humanist influence can be traced in it.

---

coloro che sono grandi di uista, et di membri rebusti è forti di gran cuore, i quali douendo soli resistere a molti per esser piu sicuri di ferire et per spauentare con la furia del spadone, tutti usano di adoprarlo a gran mandritti et riuersi di tutto tondo, fermandosi hora s'un piede hora su laltro. Lasciandosi quasi in tutto il ferir di punta come quello che puo ferire et spauentare un solo, et essi uogliono opporsi a molti. Grassi (1570): 93–94; trans. by I. G.

[113] Armed with arquebus or musket.

[114] See, e.g., *Besondere und geheime Kriegs-Nachrichten des Fürsten Raymundi Montecuculi*. Leipzig: Verlegt in dem Weidmannischen Buchladen, 1736. pp. 9–10.

[115] Wintzenberger, Daniel. *Beschreibung einer Kriegs-Ordnung, zu Roß und Fueß, sampt der Artalerey und zugehörigen Munition*. Dresden: Berg, 1588.

[116] Olnitz, Adam. *Junghans von der, Lützenkirchen, Wilhelm. Kriegs ordnung zu Wasser und Landt*. Zu Cöllen: Bey Wilhelm Lützenkirchen, 1598.

THE HUMANIST LOVE OF ANTIQUITY AND ITS INFLUENCE ON FENCING CULTURE

Meyer's world was highly affected by the city of Constantinople falling into the hands of Ottoman Sultan Mehmet II more than a century earlier in 1453. The related escape of Roman scholars into the West—scholars who brought with them rare ancient Greek and Roman texts—boosted the ongoing Italian, humanist Renaissance,[117] building up a profound fascination with pagan societies and a deep admiration for their cultural produce. It also caused a sense of concern for some, with the Fall of Rome serving as a cautionary tale for their own time to take heed and take warning from. A delightful, contemporary example of this humanist admiration of Antiquity is given to us in 1532, when Niccolò Machiavelli shares his habits and feelings towards it, in a passage from *The Prince*:

"When evening comes, I go back home, and go to my study. On the threshold, I take off my work clothes, covered in mud and filth, and I put on the clothes an ambassador would wear. Decently dressed, I enter the ancient courts of rulers who have long since died. There, I am warmly welcomed, and I feed on the only food I find nourishing and was born to savour. I am not ashamed to talk to them and ask them to explain their actions and they, out of kindness, answer me. Four hours go by without my feeling any anxiety. I forget every worry. I am no longer afraid of poverty or frightened of death. I live entirely through them."[118]

The same year, Francois Rabelais saw a great change with ancient knowledge being restored to its former glory but also men resisting, writing:

"What is the cause, most learned Tiraqueau, that in our time, so full of light, in which we see all branches of learning restored to their former estate, by some singular favour of the gods, we find men so constituted that they either will not, or cannot break out of the thick, almost Cimmerian fog of Gothic times, or raise their eyes towards the dazzling torch of the sun."[119]

In the study of Renaissance martial arts, the relation and the *perceived* (or at least *argued*) direct lineage to Antiquity and its martial arts and the connection between martial training, humanism, and the humanist desire to reinstate the ideal of civic virtue through education, are important. We see examples of this everywhere in Renaissance fencing culture. Viennese Frei-fechter and Trabant[120] Andre Paurenfeyndt goes so far as to have his whole 1516 fencing treatise[121] printed in the humanistic Roman typeface Antigua, rather than any of the contemporary

---

[117] The pre-Renaissance had started about 150 years earlier, with men like Petrarch, Giovanni Boccaccio, Coluccio Salutati, and Poggio Bracciolini amassing great collections of antique manuscripts, engaging in the study of pagan civilizations, and seeking to preserve Christianity via the teaching of classical virtues.

[118] WOOTTON 1995.

[119] FRAME 1999: 741.

[120] Paurñfeyndt served as a *Trabant* (bodyguard) to Matthäus Cardinal Lang von Wellenburg.

[121] Paurñfeyndt, Andre, Ergrundung Ritterlicher Kunst der Fechterey, Hieronymus Vietor, Vienna 1516.

Title page of Andre Paurenfeyndt 's 1516 treatise.

Gothic 𝔉𝔯𝔞𝔨𝔱𝔲𝔯 typefaces that were standard for works printed in German. Paurenfeyndt's treatise looks quite modern to us today, but it would have appeared antiquated and unusual in his time.[122]

In the 1540s, Meyer's fellow fencing enthusiast Paul Hektor Mayr had many texts of the Liechtenauer tradition translated into Latin and recorded in two of his three magnificent and lavishly illustrated fencing compilations.[123] In 1579, Heinrich von Gunterrodt also wrote his fencing treatise in Latin, and like Mayr and Meyer before him, made numerous references to Greco-Roman culture and history, with all three tracing the German martial arts back to Antiquity.

In a 1555 publication, Hans Sachs asked a hypothetical fencing master about the origin of the art of fencing, receiving the answer that it came from the Olympic Games, founded by Hercules. In 1589, Christoff Rösener, borrowing from Sachs, described the origin of the fencing

[122] The early Roman "Antique" and "Italic" typefaces were originally designed for printing in a form which emulated humanist miniscule handwriting and the text on ancient monuments and manuscripts, and were developed by Nicolas Jenson and Aldus Manutius just 20 years before the publication of Paurñfeyndt's treatise, in the 1470s and the 1490s, respectively.

[123] PMM and PMV, the latter of which also includes the original German text.

arts as starting with Hercules, linking it directly to the contemporary German fencing tournaments:

The Chivalric fencing art arose
And has its origins
Way back when Troy was burned
Some eleven hundred years ago
Before the birth of Christ
It was invented by Hercules
And with it came the Olympic Battle
In the land of Arcadia
At Olympus, that high mountain
In this Chivalric feat
Heroes fought unarmored on horseback
As Herodotus tells us
Those that fought in a Knightly way
And beat down others with their swords
They were awarded well
By the olive tree, with a beautiful wreath

—

Then the first Fencing School was held
As the ancients have testified
Diodorus and others besides
Held this as the greatest honor
When one fought for a wreath
Won fame and wealth, power and glory
And also from that came the combat games
In the mighty city of Rome[124]

Turning our attention to Meyer now, leafing through the first few pages of this fencing treatise, we are instantly met with Roman and Greek references: In the front-matter, we see Hercules fighting Cacus and Antaeus[125] above a fencing tournament scene, and on the sides of the coat-of-arms of the Cutlers' Guild, with its crown and three crossed swords, we also see

---

[124] Rösener (1589): A iv[V] – B ii; trans. by JORDAN THOMPSON.
[125] Paul Hektor Mayr also references Hercules fight with Antaeus in his treatise: "In the year of the world 1484 before the birth of Christ our saviour, Hercules did hold his duel with Antheus, he did lift him up from the ground by his strength and by embracing his abdomen did strongly crush him and asphyxiate him, as he might not otherwise have won." PMD: 6[V]; trans. by DIETER BACHMANN.
Likewise, Heinrich von Gunterrodt recounts the same myth in 1579: "Nevertheless it seems that this art... was invented much earlier: With this art Antaeus was defeated by Hercules... 'The son of Neptune and the proud offspring of Jupiter maintained their courage in the contest of a strong wrestling match. Their prizes were no wash dishes made from shining bronze, it was said: the life or death of the other! Antaeus fell down, the victory of the Greeks was beautiful, because Greece was the inventor of wrestling, not Lybia.'" Trans. from GEVAERT 2014: 18.

Gründtliche Beschrei-
bung / der freyen Ritterlichen vnnd
Adelichen kunst des Fechtens/ in allerley gebreuch-
lichen Wehren/ mit vil schönen vnd nützlichen Figu-
ren gezieret vnd fürgestellet.
Durch Joachim Meyer / Freyfechter zů Straßburg/
ANNO M. D. LXX.
Mit Römischer Keys. May. Freyheit/
Auff zehen Jar in keinerley gestalt nicht
nach zů Trucken.

Title page, frontispiece, and Rapier Figure A from Joachim Meyer's 1570 treatise.

two putti[126] holding laurel wreaths—the prize for the winners of the tournament, and also of the Pythian Games. Following this is the heraldry of Johann Kasimir and his motto "God knows the time",[127] and in each of the corners is a representation of one of Plato's Socratic cardinal virtues: Temperance, Prudence, Justice, and Fortitude.

Later in the rapier section, we also see Romanesque statues in Figure A teaching cutting lines and targeting areas on the human body. As can be clearly seen, only parts of the architecture illustrated are contemporary and German, and then commonly only in the backgrounds, with most of it idealized Roman architecture in a Mannerist style.

Looking at the text in the foreword, Meyer referred to the Roman general and statesman Scipio Africanus and briefly described some thoughts on the Fall of Rome and the birth of its extension, the Holy Roman Empire of the German Nation:

"...persons who embody intelligence and understanding among all peoples, and at every time in history, have devoted themselves so that their young people may also, in addition to other free and good virtues associated with honesty and virility, be instructed in this knightly art—how one should skillfully use all types of equipment and weapons, both on horseback and afoot according to need—as the old, credible history books of all peoples clearly and lucidly report and reveal, especially the Roman histories. From this, it should naturally

---

[126] *Putti* (singular *putto*) are chubby, sometimes winged, and commonly naked child figures in art, depicted both in the Antiquity and during the Renaissance, used both for symbolical and for decorative purposes.

[127] *Gott weiß die Zeit*.

follow that many brave heroes of knightly mien and true defenders of the Fatherland arose and were trained by those peoples, and thus the usefulness of applied efforts—while they are still young and before they have completely reached their age of adulthood—prevails. Such is especially apparent in the case of Scipio Africanus...

However, no proof is needed that it was customary for our ancestors and the old Germans to raise their children in knightly practice, in addition to other good arts, because this is intrinsically obvious as proven by their work. This is because, after the Romans opined that they had conquered the entire world and, as a people living in safety, that pleasure was to be considered more important than good arts, policies, and knightly practices, it was this view that led to the entire empire being attacked from everywhere and destroyed by enemies. It was then that the knightly Germans were called and promoted before all other peoples, to rescue [the empire], embrace it, and re-establish it.

This would have not occurred at all were the admirable Germans not practiced and experienced in all types of knightly play and the business of war..."[128]

This infusion of Greco-Roman culture and history is also integral to the contemporary fencing culture, with the Fechtschulen directly referencing the Pythian Games at Delphi,[129] a topic which we shall return to shortly. But before doing so, let us take a brief look at the German cities and some aspects of their martial culture.

FENCING GUILDS AND PUBLIC TOURNAMENTS

Before a fencing "guild" was established, a Freifechter (which Meyer called himself in several places) appears to have originally been a form of fencing teacher with recognized skills who would travel to teach and arrange the extremely popular Fechtschule. This is similar to the freemasters who practiced their craft outside of the guild system, simply because no guild existed for their trade. Similarly, among the craft guilds, it was customary for young craftsmen to be declared a Freygeselle (or "free man") after completing their years of apprenticeship and being freed of their obligations. As such, they were allowed to travel as journeymen to work and learn under masters of other cities. These journeymen were commonly organized in their own brotherhoods called Gesällverbänden, and they would elect leaders (called kings), demand dues of newly arrived journeymen, and organize which workshops they would work in.[130] The later Federfechter fencing guild may well have started out in similar fashion. With small and

---

[128] Meyer (1570): a ii^v–a iii. This is a crucial passage for understanding the worldviews of Joachim Meyer.

[129] Worth noting too, is that Heinrich von Gunterrodt (1579) also briefly mentions the Pythian Games: "The Pythian games of Apollo were established in honor of Apollo... The victor of this contest obtained a wreath of laurels with apples. Therefore Ovidius seems to tribute to the victor of the Pythian games a crown of oak in his 9th book of Metamorphoses, when he says: 'To make sure the long time could not delete his efforts, he founded the sacred games, with celebrated contests, called Pythia, after the name of the serpent he killed. Here each of the young men who had won with hand or feet or wheel, took the honor of leaves of oak.' That he did not write this from the vision of his mind, but to make a more appropriate transition to the story of Daphne, is clear from what he added: It was not yet a laurel tree". Trans. from GEVAERT 2014: 15.

[130] See FARR 2000.

even declining prospects for mastership among free fencers and fencer 'apprentices', a desire for independence may have evolved over time and caused a split.

In some places, crafts were also divided into geſchworne Handewerke ("vital sworn crafts"), which swore oaths to the city, and freien Kunſte ("free arts"), which did not. This included painters, printmakers, and manuscript illuminators.[131] Meyer of course calls his art "free, knightly, and noble",[132] and states that, "the free knightly exercise and art of fencing has, up until now, not received any particular focus while, at this same time, all other liberal arts have been described and have received consideration to an extent that they have ascended to the very heights",[133] in this very treatise. However, this may also be rooted in a desire to associate the art of fencing with the prestigious *artes liberales*, rather than the lesser *artes mechanicae*. Still, for now, the terms Freifechter and freye Kunſt des fechtens remain a bit unclear.[134]

Regardless, after Frederick III, Holy Roman Emperor, gave the Brotherhood of St. Mark a monopoly on certifying Meiſter des langen Schwert in 1487, the title Freifechter appears to have been put under their dominion, and was used as a title for what essentially was a Freigeſell, a journeyman fencer who was well on his way to becoming a maſter.

For at leaſt a half-century, this imperial privilege remained solely in the hands of the Marx-brüder, but in the mid-1500s we ſtart seeing artwork depicting fencers accompanied by a griffin. This mythical beaſt which would later become the official symbol of the Freifechter von der Feder (zum Greifenfels) and be included in their Wappen ("coat-of-arms")—which would include a handshake with the hands jointly gripping a quill, a crowned a griffin holding a Fechtschwert over its shoulder, and two crossed swords with wings.[135]

The Freifechter increasingly came to be rivals to the older Marxbrüder maſters, and in 1570, the Freifechter von der Feder guild was founded in Prague, with the Duke of Mecklenburg supposedly giving them their firſt charters and a small kleinod ("privilege"). This is the same duke who hired Meyer as fencing maſter in 1570, and it is therefore speculated that Meyer may have been involved in the formation of the early Federfechter guild. The city council of Frankfurt am Main (headquarters of the Marxbrüder) officially recognized the Federfechter in the city in 1575, and that same year, Johann Fischart added a passage into the German version of Rabelais' *Gargantua* mentioning the Federfechter. Then in 1579, Heinrich von Gunterrodt described the two groups in his fencing treatise, although without mentioning either by name:

"...let us turn to Germany, where two groups have arisen. From one part, they consiſt habitually of tanners and other craftsmen who are allied with them. Certain of them want to be considered maſters in the art, mainly in the art of using the sword. For they believe, based on a special privilege given by the Roman Emperors and Kings, that they excel in this art, (dwelling) on the markets of Frankfurt. And some people in their group want to declare this under oath and they diſtinguish themselves based on the same title. Others are usually opposed againſt them, ſtudying the good sciences and other arts but less experienced in dirty professions. // Certainly they are superior in the art of fighting than the others, though very

---

[131] See SMITH 1983.

[132] Meyer (1570): a i.

[133] Meyer (1570): b i.

[134] See GASSMANN & GASSMANN 2019.

[135] *Anzeiger für Kunde der deutschen Vorzeit*, Vol. 10–12. 1863. pp 461–64.

Fencing practice including a griffin. By unknown artist, ca. 1541.[136]

Four scenes of fencing practice, one of which includes griffin. By Virgil Solis, ca. 1550.

[136] This etching is in the collection of the British Museum <https://www.britishmuseum.org/collection/object/P_1848-0212-59>

few people can be found amongst them, who have a certain fundamental [knowledge] and who are able to teach their students with clear insights.[137]

It is interesting to note here that Gunterrodt associated the Sederfechter with cleaner, more intellectual activities, and believed they were therefore superior to the Marxbrüder (and seemingly considered himself above them all).

Still, it would take nearly another forty years before the Sederfechter fencing guild became officially approved to license masters of the long sword 'von der Feder' ("of the feather"). Rudolf II, Holy Roman Emperor, finally granted this privilege on 7 March 1607.

While we currently have somewhat limited primary sources on the customs surrounding the testing procedures of fencers aspiring to rise through the ranks of the German fencing guilds,[138] the Marxbrüder and the Sederfechter,[139] we do have very deep insight into the English Company of Meisters of the Noble Science of Defence through a manuscript which records their various practices,[140] and several of their traditions and practices may be quite similar. For example, the English oath for becoming a master was taken on crossed swords or crossed weapons in some form. This was also the German ritual, as described in contemporary German texts such as Sachs (and was also depicted satirically by Elias Baeck in 1720). The oaths sworn by masters and provosts in the English Company are preserved, and although they obviously would not be identical, we can speculate that there would at least have been similarities between these and the German Meister and Freifechter oaths.[141]

Included in the English masters' oaths we find the following:

- First, you shall swear that you shall uphold, maintain, and keep to your power all such articles as shall be here declared unto you and received in the presence of me your Master and these the rest of the Masters my brethren present here with me at this time.
- You shall be true to the Catholic church to augment and further the true faith of Christ to your power.
- You shall be true subject to our sovereign lady Queen Elizabeth and to her successors Kings of this realm of England.

---

[137] GEVAERT 2014: 19–20.

[138] The major early sources are the first section of HM (1491–1599) and Sachs (1555); Rösener (1589) is dependent on Sachs. These sources list the strikes and techniques which are required as well as the weapons to be tested in, and the formal procedures to follow to become a master. The ordinances of the *Federfechter* are described in p 461–64 in *Anzeiger für Kunde der deutschen Vorzeit*, Vol. 10–12. 1863. pp 461–64. Other important sources are RUDOLPH 1752 and SCHAER 1908.

Two Captains of the *Marxbrüder*, Peter Falkner and Anton Rast, also authored fencing treatises (PF in 1495 and AR before 1549), but they do not include any material on the brotherhood's customs or procedures.

[139] For more reading on this topic, see also VODIČKA 2019.

[140] See BERRY 1991 and ANGLIN 1984: 393–410.

[141] This oath of loyalty would be in addition to the oath of loyalty to their city which all citizens would swear on Schwörtag, repeated every year, both by citizens and their burghermeister. Likewise, different oaths for apprentices and masters of the craftsmen guilds, swearing loyalty and obedience to the guilds would have been sworn.

- You shall be true Master from this day forward to the last day of your life, loving truth, hating falsehood and not grudging or disdaining any Master of this science. And you shall always be ruled by your brethren who are Masters of this science, and especially any Master who is your elder.
- You shall not teach any suspect person, such as murderers, thieves, common drunkards or such as you know to be common quarrellers, nor keep company with them, but avoid all such as much as you can.
- If you come to any manner of prize or game or any kind of play at weapons concerning our science, you shall without respect, favour or hatred of either party, give true judgment of that which you see there.
- You shall take no scholar with the intention of teaching him or causing him to be taught, without the oath appropriate for a scholar. You will charge the same for his learning as other Masters are accustomed to do, not taking less than other Masters do, to spite or hinder any other Masters.
- You will not challenge any English Master. And especially you shall not challenge your own Master. Upon taking the Master's oath you shall pay your Master of all such debts, fees and demands which are due to him and you will love and honour him as your Master and elder.
- You shall be merciful and where you happen to have the upper hand over your enemy, that is to say without weapon or under your feet or his back towards you. And also, if you hear of any disagreement between other members, you shall do the best that you can to make them friends and always to keep the peace if you can.
- You shall help all Masters and Provosts of this science, all widows and fatherless children, and if you know of any Master of science that is fallen into sickness being in poverty, you shall put the masters in remembrance at all prizes and games and other assemblies.
- You shall not set forth any prize (nor keep more than one school in London) within twelve months and a day after the playing of your master's prize. Nor will you teach or cause to be taught any other Master's scholar, without the consent of his first Master and unless the said scholar has paid his first master all fees for his learning.
- You shall at no time set forwards any manner of Provost or Free Scholars prize without first seeking the consent of all the Masters. Having their consent, you shall give lawful warning where it shall be played and what day is appointed.
- You shall not, for monetary gain, set forwards any prize of a Master, Provost or Free Scholar without a lawful cause and taken by you and at least two masters more besides yourself. And at any prizes set forwards by you, you must see that every Master and Provost have his fee according to our ancient orders and rules.
- You shall not promise any person learning, unless you teach him or cause him to be taught as a Master ought of right to do. That is to say, a scholar like a scholar, a Provost like a Provost and a Master like a Master.
- You shall not cause any oath to be given in your name by any person under the degree of a Provost except it be your usher who is your deputy for that time as long as his contract lasts with you.
- You shall not allow or enable any Provosts licence without the consent of at least two Masters besides yourself. And you shall not agree with any person to keep a school for you or in your name so that they shall have the profit thereof.

"The Tournament Place in the City of Zwickau" (detail). By Paulus Reinhardt, 1573.

The 𝔐𝔞𝔯𝔵𝔟𝔯ü𝔡𝔢𝔯 and the 𝔉𝔯𝔢𝔦𝔣𝔢𝔠𝔥𝔱𝔢𝔯 𝔳𝔬𝔫 𝔡𝔢𝔯 𝔉𝔢𝔡𝔢𝔯 would continue as fierce rivals well into the late 18th century, licensing fencing masters, performing public displays of fencing, and arranging 𝔉𝔢𝔠𝔥𝔱𝔰𝔠𝔥𝔲𝔩𝔢𝔫 with dusack, sword, rapier, staff, and halberd. An example of a late occurrence would be in 1726, when GOTTFRIED RUDOLF,[142] mentions the use of two-handed swords in contemporary fencing schools. Likewise, in1735, the two guilds held a 𝔉𝔢𝔠𝔥𝔱𝔰𝔠𝔥𝔲𝔩𝔢 in Breslau with dusack, rapier and dagger, 𝔖𝔠𝔥𝔩𝔞𝔠𝔥𝔱𝔰𝔠𝔥𝔴𝔢𝔯𝔱, staff, and halberd. In Antwerpen, a public tournament with the two-handed sword was arranged by the *Sint-Michielsgilde* as late as the mid-1740s. While celebrating the coronation of Francis I as Holy Roman Emperor, a 𝔉𝔢𝔠𝔥𝔱𝔰𝔠𝔥𝔲𝔩𝔢 was arranged by the 𝔐𝔞𝔯𝔵𝔟𝔯ü𝔡𝔢𝔯 in 1745.[143] Finally, JOHANN BÜSCHING[144] stated in 1816 that at least a handful of skilled people still knew how to use the 𝔖𝔠𝔥𝔩𝔞𝔠𝔥𝔱𝔰𝔠𝔥𝔴𝔢𝔯𝔱, but that there were not many fencers trained by the few remaining 𝔐𝔞𝔯𝔵𝔟𝔯ü𝔡𝔢𝔯 and 𝔉𝔢𝔡𝔢𝔯𝔣𝔢𝔠𝔥𝔱𝔢𝔯.

Not all public fencing took shape in the form of a 𝔉𝔢𝔠𝔥𝔱𝔰𝔠𝔥𝔲𝔩𝔢 though, and regular brawls were also frequent. One such brawl involves the famous Heinrich Agrippa and is an amusing, but also interesting for several reasons, example. Heinrich Cornelius Agrippa von Nettesheim (15 September 1486 – 18 February 1535) was a German knight, ambassador, magician, occult writer, theologian, astrologer, alchemist, soldier, and as it appears, also a 𝔉𝔯𝔢𝔦𝔣𝔢𝔠𝔥𝔱𝔢𝔯. Agrippa's

[142] RUDOLF 1726.
[143] JUNG 1746: 14.
[144] BÜSCHING 1816.

history is fascinating in many ways, full of drama, war, free-thinking, controversy, magic, desperate poverty, jail sentences, and the deaths of two dearly loved and deeply mourned wives and several children. He was born in the Free Imperial City of Köln on 15 September 1486, almost exactly a month after Frederick III had given the 𝔐𝔞𝔯𝔵𝔟𝔯ü𝔡𝔢𝔯 their first privileges. There are numerous stories about how he used his rumoured knowledge of the black arts by summoning demons, raising the dead, and other similar acts, and below is a very interesting episode. This short passage describes an incident in Ingolstadt involving Agrippa, likely taking place sometime in the first two decades of the 1500s:

> "Likewise, there was a student who unfortunately got into godless company by the name of Heinricus Cornelius Agrippa, a fine student and 𝔉𝔯𝔢𝔦𝔣𝔢𝔠𝔥𝔱𝔢𝔯, who came to Ingolstadt just as the 𝔎𝔞𝔱𝔷𝔦𝔞𝔫𝔢𝔯𝔰 had invited the students to a fight. But although they could have all accepted, some students told the others they should stay calm. They knew someone who could master them all, him all the others followed. Heinricus Cornelius offered them an earnest fight, and told them to defend themselves, as he alone would defeat them all. Now as they came to strikes, he cast his black magic that could devour all swords. The furriers worried that if that happened with their fencing weapons it could even reach past their Katzbalgers, and they took to the heels and left. This Henricus always had a black dog with him."[145]

It appears that Agrippa's reputation was so fearsome that little fencing was actually needed, and a word and gesture were enough to strike fear into the hearts of his opponents, causing them to flee.

Several things in this short passage are interesting. The "furriers" is most likely a reference to the 𝔐𝔞𝔯𝔵𝔟𝔯ü𝔡𝔢𝔯. 𝔎𝔞𝔱𝔷𝔦𝔞𝔫𝔢𝔯 and 𝔨𝔞𝔱𝔷𝔢𝔫𝔯𝔦𝔱𝔱𝔢𝔯 ("cat knight") are also terms which appear in other texts mentioning the 𝔐𝔞𝔯𝔵𝔟𝔯ü𝔡𝔢𝔯 and the 𝔉𝔢𝔡𝔢𝔯𝔣𝔢𝔠𝔥𝔱𝔢𝔯, but the meaning is still unclear. According to FRIEDRICH KLUGE,[146] it was an insulting name for the 𝔐𝔞𝔯𝔵𝔟𝔯ü𝔡𝔢𝔯 derived from 𝔎𝔞𝔱𝔷𝔢𝔫𝔰𝔠𝔥𝔦𝔫𝔡𝔢𝔯 ("cat flayer"), used by students in Leipzig in the 1700s and originating from Michael Lindener's satirical *Katzipori* of 1558. Of course, the anonymous writer of the *Faustus* chapbook may also have learned the term after the events in Ingolstadt.[147]

With this small image of the fencing guilds and their tournaments in the Holy Roman Empire, let us now continue onwards, examining astrological concepts and mysticism as related to fencers and soldiers.

---

[145] SPIES 1587.
[146] KLUGE 1895.
[147] Imperial Army Commander Johann Katzianer should perhaps also be mentioned here. He took part in the 1527–28 Hungarian campaign against the voivode of Transylvania, John Zápolya. He also fought in the Battle of Tarcal in 1527, the Battle of Szine in 1528, and the Siege of Vienna in 1529 against the Ottomans. In 1537, he was commander of an army of 20,000 men. The so-called Katzianer Campaign, an attempt to drive the Ottomans out of Slavonia, ended in failure with his army slaughtered at the Battle of Gorjani. Worse, he fled the field, leaving his troops to die, and was consequently tried and found guilty of cowardice. He subsequently escaped prison, killed a peasant, and was caught and executed in 1539. If this in any way is relevant to the supposed derogatory term "Katzianer" is for now unclear.

𝔓lanetenkinder, 𝔉echtschulen, AND THE PYTHIAN GAMES

Associating the stars and the planets with earthly events is something humanity has been doing for at least 4,000 years. In Greece, the Sun—the life-giver—was given a symbolic and naturally important position with both Helios and Apollo (slayer of the serpent Python[148]) regarded as solar deities. Furthermore, Apollo was also the god of music, dance, and poetry, health and healing, and grappling, boxing, and archery, and was regarded as an ideal for the athletic youth. The Pythian games in Delphi,[149] celebrated for nearly a thousand years, were dedicated to Apollo and included four track sports (*stade*, *diaulos*, *dolichos*, and *hoplitodromos* or racing with Hoplite armour), wrestling, boxing, pankration, and pentathlon, as well as competitions in music, poetry, prose, drama, and painting. No monetary prizes were awarded to the winner. Instead, and as a symbol of Apollo's lost love for Daphne (who was turned into a laurel bush by his father Zeus), a laurel wreath was given to all winners.

"Then looking at him darkly Zeus who gathers the clouds spoke to him:
Do not sit beside me and whine, you double-faced liar.
To me you are the most hateful of all gods who hold Olympus.
Forever quarrelling is dear to your heart, wars and battles." [150]
    (Zeus, speaking to the god of war, Ares)

The Greek god of war and courage, Ares, was a more difficult deity which the Greeks felt a clear ambivalence, but overall negativity, towards. While the physical valour associated with him was sometimes positive, the brutality and bloodlust connected to the war god—and his frequently shifting allegiances and untrustworthiness—were naturally perceived as negative and dangerous, with most stories involving Ares also being humiliating to the deity. As an interesting example, a famous story involves the war god having an affair with Aphrodite, whose husband Hephaestus (god of the craftsmen, artisans, and smiths) discovers it through the sun god Helios and traps them in a net, shaming them under the ridicule of the other gods.

Also worth mentioning here is the Roman god Mithras, associated with soldiers, and Mithra—the Iranian god of covenant, light, oath, justice, and the sun. Mithras too was connected to the sun, as well as a mysterious and still poorly-understood winged lion-headed figure, entwined by a serpent and associated with time and change. Mithras and Apollo were both frequently given the epithet *Sol Invictus* ("the unconquered sun"). Similarities between the myth of Mithras and the stories of Jesus Christ, as well as an influence on Christian medieval art, have also been suggested but are difficult to validate. The fact that the 𝔐arkbrüder in 1487 chose St. Mark and his winged lion as their patron saint, and their links to the mercenary 𝔏andsknechte, is intriguing and difficult to fully ignore, but is likely purely coincidental.

Jumping forward, sometime after 832 CE, Abu Ma'shar al-Balkhi authored *Kitāb mawālīd*, a treatise on astrology which associated planets with certain activities and professions.[151] This

---

[148] This is the reason the Pythian Games were celebrated. See WEIR 2004.
[149] The Pythian Games were celebrated in Delphi starting from 582 BCE until at least 424 CE.
[150] LATTIMORE 2011: book 5, lines 798–891, 895–898.
[151] *Kitāb mawālīd al-rijāl wa-ʾl-nisāʾ* ("Book of nativities of men and women")

was widely circulated within the Islamic world. Although there are no original manuscripts preserved, the content was copied into other manuscripts, for example by Abd al-Ḥasan Iṣfāhānī in the *Kitāb al-Bulhān* in 1399.[152] Both of these depict the planets and their associated professions and pastimes, with the Sun associated with the lion, musicians, and swordsmen.[153]

Some of the earliest examples in the West of this astrological association between the planets and characters of men are found in the *Bellifortis* of Konrad Keyser (1405) and Christine de Pisan's *Épître d'Othéa* (1407).[154] However, it is not until the second half of the 1400s that we start seeing a great many similar treatises depicting what comes to be called 𝔓𝔩𝔞𝔫𝔢𝔱𝔢𝔫𝔨𝔦𝔫𝔡𝔢𝔯 ("The Children of the Planets").[155] This astrological concept connects all the (then seven) known "planets": Mercury, Mars, Jupiter, Saturn, Venus, the Sun, and the Moon, to different characters, traits and common professions and pastimes. Interestingly, we find one such text in the back of *Hans Talhoffer-København* (1459; HTK), written in German.[156] The following quotations are useful for contrasting views on two planets: the Sun and Mars:

Here Mars speaks about his type, which he has close to him with his bellicose and hateful manners, and yet they do not know why, or even how it is like this.

Mars is the third planet, and is hot and dry, unlucky and angry, and yet generous and measured in its course. And [it is] a planet of angry people, and they like to fight and steal, and are bald and have curly hair (and little of that). And among the planets, [it] is good in fights against stellae. [They] rob and burn and injure people, and [it] is therefore named Mars by the wise masters as an allegory, because Mars was named a god of battles by the unbelievers, and whenever the Romans wanted to fight, then they called upon Mars and brought him sacrifices in his temple and also carried him with them in the field where they then wanted to fight. And when the masters speak about it, he is thus called Mars because he reigns thus under the seven planets. Thus, there must be much fighting

I, Sun, say to you briefly that my brilliance is above all planets. My rising gives light to the day and my setting draws the stars delicately, and makes the human beings beautiful and light-hearted, which no other planet can do.

The Sun is the fourth planet, and is hot and dry and is joyful, an inflowing light to all those who ever lived on earth. It is a planet beautiful and joyful, and his face lights up human beings, and also illuminates people with all honorable thoughts, and [it] is then well with honorable people. The Sun is a kingly star and a light and eye of this world, and shines in itself and illuminates the other stars in all ways, And [it] is, among the seven planets, the middle one, and divides the time, and completes its course in exactly one year. It also makes human beings better, strengthening them in their bodies, and he makes the face beautiful and beautifully created, with large eyes and with a large beard and long hair, and makes the human beings to resemble their souls, and makes them wise according to

---

[152] Oxford, Bodleian Library, ms. Or 133.

[153] See, e.g., Paris, Bibliothèque national de France, ms. Arabe 2583: 15^V.

[154] *Bellifortis* appears in many fencing treatises, including BKE, BKW, HTG, HTK, and W01. It's likely that the version in BKW was created directly by Kyeser himself.

[155] For a deeper study in the evolution of this astrological concept, see KLINGNER 2017.

[156] HTK: 147^R–145^V. Trans. from GARBER & BACHMANN 2020: 39–41.

and war in those same years, and when Mars now moves in the Sun's course, then one can seldom see him, yet whenever he reigns, then the masters say that one sees him above the Sun. This thus signifies [a] great defeat against the nobility, [and] thus, that princes and lords, and also knights and squires, should not go to war in that same year because they will be laid low. However, the peasants have good fighting because all things go closely according to their desires. And therefore, the children who are then conceived when Mars reigns, they become quite bellicose people, and do not have the nature with those who are called Sanguine, because they are pugnacious and persevere often and mostly to their battles. However, if one views him [Mars] below the Sun, then he has a nature somewhat like those who are called Melancholic: they are quiet, silent fighters and they succeed well in their wars. And in the year when he reigns, then there a star usually reigns which is called Comet, and in whatever land it is seen, there will be, without doubt, great drought and hunger in that land. Because one cannot see it in all countries, because it is low in the heavens and close to the moon, thus the moon's shadow surrounds it, therefore one can see it well [when] the Sun is in the sign which is called Cancer or Leo, and in whatever year he reigns, then the Sun and the moon of the year are sickly. Whoever is born under the planet, he becomes red with any darkness, like those [who] are burned by the Sun, and also becomes unvirtuously bellicose and likes to foment dissatisfaction among people and has, among the 12 signs, Aries and Scorpio and their temperament and nature. And Mars fulfils its course in five hundred and thirty days.

other things and so that one loves them a lot and makes them rich in arts and talents and clever in all things. And the Sanguine are formed like the planet, because the same people are very talented in all things and arts, and are, however, dubious about divine things and items, and are also blindly passionate and often become easily angered, though it also quickly passes away from them. The child that is then born in the year when the Sun reigns, that [child] will have lovely flesh and a white color and with a little red mixed with that, and not a lot of hair, according to the similitude of the Sun. And [they] shine outwardly quite well and are yet loyal people according to their head, yet one opines that they are very wise people who are born under the Sun, and joyful, and become the enemies of evil people. Among the seven planets and among the twelve signs, the Sun has Leo with its temperament and nature, and completes its course in one entire, circumventing year."

ARTE·E·SENGNO·MASCVLIII·POSTO·NEL·5TO·CIELO·MOLTO·CALDO·FOCOSO·ET·HA·9STE·PPRIETA·DAMARE·M
A·bAT·TAGLE·ET·VCCISIONI·MALIGNO·DISORDINATO·DEMETALLI·HA·FERRO·DEGLI·HVMORI·LA·COLLER
TEMPI·LASTATE·EL·DI·2VO·E·ILMARTEDI·CONLAPRIMA·HORA·8·15·ETZZ·ETLASVAN·NOTTE
ABATO·ELSVO·AMICO·E·ILSOLE·EL·NIMICO·GIOVE·HA·DVE·AbITATIONI·ELDI·LARIETE·ELANOT
SCORPIONE·LAVITA·OVERO·ESALTATIONE·2VA·E·CAPRICORNO·LASVA·MORTE·OVERO
VMILIATIONE·I·ILCANCRO·ET·VA·E·12·SEGNI·IN·18·MESI·COMINCAIIDO·NELLO·SCORPIONI
I·NV·MESE·EMESO·CIOE·45·DI·IMSEGNO·40·MINVTI·PERDI·ET·PERORA·VN·MINVT
T·4·SECONDI

Mercenaries, rapists, murderers, and thieves—the Children of the Mars. By Baccio Baldini, 1464.

Pious fencers and athletes—the Children of the Sun. By Baccio Baldini, 1464.

The two planets are regarded as having very different influences: one positive, strengthening for body and soul, creating wise, loyal, and trustworthy "children"; and the other harmful to virtue and society, spreading thievery, hate, anger, and harm.

Around the same time, we also have the English Trinity College Cambridge manuscript O.1.77, which describes the planets and their children. And in the same country, one of the rare English books on fencing, *Man Yt Wol* (ca. 1440; MYW) appears in the same manuscript as an incomplete copy of *Mirror of Lights* (an alchemical text) which also includes alchemical associations between the planets and metal—naturally, gold for the Sun and iron for Mars.

We find yet another example of this world of thought in the *Wolfegg Hausbuch* of 1480.[157]

Mars

I am the third planet, Mars,
Wrathful, raging through the stars.
I'm full of hate—and hot and dry,
With my might I magnify
All who please me, and I will
Glorify those who war and kill.
My houses the Scorpion and the Ram.
If there is conflict, there I am.
In the Goat I'm lifted high,
In the Crab lose strength and die.
The twelve signs I travel through,
Not in one year, but in two.

All my trueborn children fight,
Murder, strive, and slay with might.
Angry, haughty, warlike, proud,
Liars, thieves, their boasts are loud.
Burning, cheating, robbing, hot,
Their quarrels may be just—or not.
Small teeth, small beards, tall and thin,
Noses sharp and hard rough skin.
Butchers of men, killers of swine,
Smiths and marshals, children mine.
Captains, gunners, doctors good,
All those who deal in fire and blood.

Sol

Men call me Sol, I am the sun,
The middle planet, on I run.
Beneficent and warm and dry,
By nature, my rays fill the sky.
The Lion's my house, therein I dwell,
And brightly shining I do well.
There I stand, fair and bold,
Against old Saturn's bitter cold.
In the Ram, I rule and reign,
But in the Maid, I fail, I wane.
And through the stars my way to wend,
Three hundred and sixty-five days I spend.

Noble and fortunate I am,
As are all my children.
Good beards, large foreheads, bodies fair,
Ruddy lips, of brains their share.
Happy, kindly, well-born, strong,
Fond of harps, viols, and song.
All morning long to God they pray,
And after noon they laugh and play.
They wrestle and they fence with swords,
They throw great stones and serve great lords.
Manly exercises are their sports,
They have good luck in princely courts.

As can be seen quite clearly, we here have two very different Renaissance perceptions of the celestial influence on men and their use of violence: one as orderly, good, and noble and the other as mostly disorderly and evil—one for the nurturing of virtues, and the other for criminal, uncivilized, and brutish violence (though also related to certain "good" but bloody professions, like butchers and doctors). This is further expanded in 15th and 16th century artwork showing the Children of the Sun as athletic and pious people, interested in music, wrestling

---

[157] See WALDBURG WOLFEGG 1998, BLAZEKOVIC 2003, and HERRERA 2012.

and practicing the art of fencing—as fencers—while the Children of Mars would murder, rape, steal, and pillage—as soldiers. Essentially, they represent the opposites of chaos and order.

At some point in the mid-1400s, Apollo seemingly became associated with high virtue, with kings, lions, and gold, and as a pagan symbol. Apollo had for some begun merging with Jesus—the Lion of Judah, the rising sun, and God incarnated. With the contemporary religious strife and the Greco-Roman influences via the humanists, the association of Jesus with Apollo, athleticism, piety, and virtue likely appeared self-evident, and seemingly came to be expressed with the laurel wreaths given to the winners of the fencing tournaments.[158]

In relation to this, it is also interesting to note how humanist Lorenzo Buonincontri da San Miniato (1410–ca. 1491), the court astrologer of Popes Sixtus IV and Innocentus VIII, sought to unite astrology with the Christian faith, claiming that God had created the planets as his "executive bodies" (*ministras executricesque*):

"In the beginning, to rule the world by laws, the Almighty Father set the stars and the planetary spheres high in the sky: he gave them numbers and names and gave them such a nature, that everything in the world happens at certain times. By which he determined the manners of men, their bodies, their every destiny, the coincidental occurrences, life as well as the last day, the consequence of destiny and the end of toil."

The popes openly confessed their belief in astrology as an important means of understanding God. KLINGNER also theorizes that papal recognition of astrology would have proven helpful to the Church, in that the stars and planets and their associated "ministers" or gods were closer and more "tangible" to people, and clarifying that this was not in contradiction to the Christian faith, but in fact, part of it—helping in the divination of God's will.

One example of this thinking is in the art of Baccio Baldini, which depicts the Children of the Sun and Mars in 1464. Numerous illustrations following the same recipe then follow over the 16th century, such as those by Virgil Solis and Georg Pencz on the following pages.

One final example should be mentioned: the alchemical treatise *Splendor Solis* from circa 1532–35.[159] One of its plates depicts the standard scene with Apollo/Jesus and the children of the Sun, although here with a crowned flask at the centre—a flask containing a three-headed dragon symbolizing three of the four stages of the process of the *Magnum Opus*: *nigredo*, *albedo*, *citrinitas*, and *rubedo*.

The three heads are facing the same direction and represent *harmonious development*, not just in matter, but also in body, mind, and soul, a central aspect of the "*great work*" in achieving elevation of all these, and also for the fencers, and the whole society, of the Renaissance. This is also not the only example of artwork on alchemy depicting fencers.[160]

Now that we have a good portion of our puzzle laid down, with pieces presenting a concern for the Fall of Rome as a cautionary tale for the militaristic German cities and their fencing

---

[158] See Price, David Hotchkiss, Albrecht Durer's Renaissance: Humanism, Reformation, and the Art of Faith, Ann Arbor: University of Michigan Press, 2003

[159] See, e.g., Paris, Bibliothèque national de France, ms. Allemand 113: 28[R].

[160] See for example Michael Maier's *Tripus Aureus* of 1618, which includes several plates depicting fencing, one with the Sun and the Moon riding lions and with a soldier in armour standing in the fencing stance *Zornhut* in the background, and another showing Mercury with a pair of fencers and the sun and the moon at his sides.

Planetenkinder: Marte. By Georg Pencz, 1531. (Formerly attributed to Hans Sebald Beham.)

Planetenkinder: Sonne. By Georg Pencz, 1531. (Formerly attributed to Hans Sebald Beham.)

Mars. By Virgil Solis (after Georg Pencz), 1530–62.

guilds, pieces with the lamentable gun harming the Art of Fencing and contemporary morals, and with some views on fencers versus mercenary soldiers as tied to ancient astrological concepts, let us focus our gaze on the dichotomy between having skilled fencers in a city, and society's need to control their capacity for violence.

Sol. By Virgil Solis (after Georg Pencz), 1530–62.

THE IMPORTANCE OF CONTROLLING VIOLENCE

Having a militia of trained, local soldiers is powerful for the defence of a city or a nation. Even today, men and women fiercely—and out of necessity—fighting to protect their homes and families are considered to have motives of a far higher level than those of an invader, with loyalty to the cause being more important than military professionalism. However, in peace-time that training and skill, combined with constantly being armed, can also cause problems, especially in societies where consumption of large amounts of alcohol is common and frequent. Add to this that some of these men would also go off to war as mercenaries, as German 𝔏𝔞𝔫𝔡𝔰- 𝔨𝔫𝔢𝔠𝔥𝔱𝔢, Swiss 𝔑𝔢𝔦𝔰𝔩ä𝔲𝔣𝔢𝔯, or Italian *condottieri*, brutally slaying and looting other cities, and likely found it difficult to adapt to regular, peaceful, civilian life afterwards, you get a vicious cycle that desperately needed to be broken. Violence was thus, as described also in the social framework of the 𝔓𝔩𝔞𝔫𝔢𝔱𝔢𝔫𝔨𝔦𝔫𝔡𝔢𝔯, and even among military strategists and commanders like Fronsperger, perceived as having two faces: one just, honourable, and good, in defence, and the other unjust and dishonourable, used for the harming of innocents and personal profit. Again, Machiavelli shares his thoughts on this, although this time in his *Dell'Arte della Guerra*, written in 1513 but not published until 1520:

> "Such evils do not result from anything else other than the existence of men who employ the practice of soldiering as their own profession. Do you not have a proverb which strengthens my argument, which says: 'War makes robbers, and peace hangs them'? For those who do not know how to live by another practice, and not finding any one who will support them in that, and not having so much virtue that they know how to come and live together honourably, are forced by necessity to roam the streets, and justice is forced to extinguish them... a good man was not able to undertake this practice because of his profession: the other, that a well-established Republic or Kingdom would never permit its subjects or citizens to employ it for their profession...
>
> "A well-ordered City, therefore, ought to desire that this training for war ought to be employed in times of peace as an exercise, and in times of war as a necessity and for glory, and allow the public only to use it as a profession, as Rome did. And any citizen who has other aims in (using) such exercises is not good, and any City which governs itself otherwise, is not well ordered." [161]

Rooted in his understanding of the ideal of the Roman State, Machiavelli regarded a popular militia [162] as far superior to professional, but unreliable, mercenaries—if not in combat skill, then at least from an ethical and social standpoint. These views formed the basis for the reforms he had sought to implement in Florence a decade and a half earlier.

Italian fencing master Ridolfo Capo Ferro da Cagli, too, touches upon this issue in 1610, stating that:

> "... the Roman soldiers, immediately upon arriving home, put down their arms together with their short uniforms, and soldiery, and assumed again their long civil robes, and attended to the studies and the arts of peace, because no Roman exercised the body (as says Salustius) without the mind, each one attending, beyond the studies of war, to every office of peace, and by such longing they endured the burdens of war, and therefore immediately upon the end of war, no more was heard of captain, nor of soldier, nor of military wages.

---

[161] NEVILLE 1695.

[162] A militia consisting of people living in the region, and not just citizens of a city.

In these times soldiers are a greater burden to Princes and to Lords, and more so to the populace in times of peace than in war, and because they are not trained in other studies than those of war, they hate peace, and much of the time they are the authors of turbulence and wretched counsel."[163]

This very much echoes the words of Imperial commander Lazarus Schwendi who in in his " Zustand des heil. Reiches",[164] sent to Maximilian II in 1570[165] (the same year Meyer published his book), saw a dangerous decline in German masculinity, military power, and virtue. For this reason, he proposed disallowing enlistment as and use of mercenaries, while also proposing a standing, centralized Imperial army, in particular to defend against the French use of mercenaries and the Ottoman threat. This served the purpose of preventing treasonous fellow countrymen from serving as mercenaries for foreign powers against their own, while also removing the corrupting influence that leading such a life had. He also proposed re-enlisting the knightly orders of the aristocrats in the wars against the Ottomans, to "re-establish honour, manliness, self-restraint, and military discipline among the nobility again".[166]

Similarly, Leonhard Fronsperger argued that military life under virtuous leadership bred friendship and unity, arguing like Machiavelli that fellow countrymen were far better soldiers than the untrustworthy—and often underpaid—mercenaries who only fought for money and booty, basing this on Roman and Swiss models.

In the same line of thinking we also find Paul Hektor Mayr, who by the 1540s had already written the following in his *Opus Amplissimum de Arte Athletica*:

"The ancient and modern Greek, Latin and German historians have put much zeal and effort into the question, in numerous points and articles, from which basis and causes, in which time, and in what land and situation, and by whose instigation, the knightly sport of fencing first took its origin and source, but in the cause and the tale of years, or in place and situation, they all are in agreement, that this knightly art of fencing was in the beginning established with the purpose of serving to the honour, virtue and stimulation to the youth of both high and low birth, and also to the protection and preservation of the fatherland...

"Finally, the knightly art has come to this, that it developed the Brotherhood known as Saint Mark's, which the most eminent, most mighty Roman emperors of most praiseworthy memory, Frederick, the third of this name, Maximilian, and now the invincible prince, Charles himself all three nobly born of the old and praiseworthy house of Austria, so that this knightly exercise may not decay completely, and at this time may once again be aided with privileges and liberties, presented furnished with their best and most gracious sympathy. Namely such that on every autumn's fair in Frankfurt, those who would be or become master of the sword, must first be examined in their masterly test in the iron-run[?] and golden art by ordained and sworn masters of Saint Mark's Brotherhood, and all that is pertinent to the knightly fencing with all virtue they must request and confirm their completion under oath. These [masters of the sword] may then hold schools as far as does extend the

---

[163] Capo Ferro (1610): 3. Trans. by WILLIAM WILSON and W. JHEREK SWANGER.
[164] "State of the Holy Empire".
[165] Supposedly, Schwendi also proposed this to Emperor Maximilian II at the Diet of Speyer in 1570.
[166] See BRUGH 2019.

Roman Empire of the German Nation and may teach to other men on their request the proper manner of the sword.

But for two reasons this exercise is held in low regard, the first, that gamblers, drunkards, usurers and lovers of beautiful women have been given much room in the highest places, such as princely courts and in the chief cities of the empire, and knightly art and exercise may not thrive due to them, but must always stay hidden behind the door, and the second, that numerous masters of the sword and otherwise masters, also Freifechter and numerous other fencers, to their own damage and disadvantage and that of the praiseworthy ancient art, act ineptly and ignominiously, foster envy and hatred in their fencing in schools, such that the young would hold the old in contempt at the fencing-school, and force on them their wantonness, and in addition abuse themselves of too much wine, from which results much frivolous discord, much to the chagrin of the authorities, and causes a considerable decline in the noble art. All this then serves more the extirpation than to the taking root of the noble and ancient knightly art, as pains me at the heart, and I would rather (in the interest of the noble art) hide it by my silence than to report it, especially as these vices are also hardly conductive to civil order. So I would gladly instruct all pious honest fencers to the profit and benefit of the fatherland, in the hope that they should be rewarded for it by everlasting praise."[167]

The true and proper art of fencing is thus not merely the practice of martial art, but also a means to protect and preserve society through the building of character, discipline, and honour in young men, while preparing them for the defending of society with skilled and trained use of just violence in inevitable military conflict.

With this, we leave behind the larger picture of the world Meyer lived in and ask ourselves where the art of fencing would have been put to use in Meyer's time.

WHERE WAS THE ART PUT TO USE?

What, then, were the various contexts which Meyer's martial art was designed to be used in? War was a very common context, and in his writings, Meyer was quite clear that he taught fencing for self-defence, war, and tournament, teaching both soldiers and civilians.[168] He mentioned war specifically in JMM: "...this is a good old thrust which also serves you well in battle..."[169] and in JML when describing the sixth Treiben with the dusack: "...particularly for someone who is strong with a Battlesword" (Schlachtschwert).[170] The pike and the halberd were incredibly important weapons in Meyer's time and elaborate artwork by Frans Hogenberg and the treatises on war by Johann Jacobi von Wallhausen and Niccolò Machiavelli give us an idea of how intricately choreographed such warfare could be—at least in the planning stage before the opposing sides clashed together. Complex formations were designed to be adaptable to different circumstances, with three front layers of pikes always being able to reach the enemy.

---

[167] PMD: 2^R, 13^R–14^R; trans. by DIETER BACHMANN.
[168] Similarly, Fabian von Auerswald (1539) explicitly stated that he teaches men of "high and low standing".
[169] JMM: 19^V; trans. from MAURER 2022.
[170] JML: 64^R; trans. from FORGENG 2016.

The placement of troops enabled both easy retreat for exhausted fighters, with others advancing to face new threats, as well as for forming a defensive "hedgehog" square for when outnumbered and surrounded.

The fields of battle were not the only places for war though. Just as important was urban warfare, with town militia fighting to protect their city, their neighbourhood, and their homes from mercenary soldiers. Townspeople defended their homes both within and outside of the city walls, as well as in the streets—sometimes with pike combat in a narrow street. The enemy would be caught under a hail of roof tiles, stones, and hot water thrown from windows and balconies by townsfolk and house-bound women, with shots from firearms and crossbow arrows raining down from above. Patricians and high-ranking guildsmen here often served as officers in this town militia, with the latter as quarter or lane captains who normally lived in the same neighbourhood they were tasked with defending.

Naturally, self-defence was another important context for fencing, not least when travelling. Likewise, duelling was a risk most men would have to consider and accept under certain circumstances, some more than others.

All these contexts were regulated by different laws and customs, restricting what one was allowed, expected, and required to do, shaping actions that sometimes would appear surprising to us—and here I am directing myself to historical fencers—as they might seem dishonourable in our "modern" minds, while in other cases they might simply appear stupid, as we fail to understand the historical mindset and considerations. For this reason, to truly understand these martial arts it is not enough to just study the weapons or the techniques, but also the customs and how people reasoned and behaved in these contexts, and why. Otherwise, as historical fencers, we risk ending up with empty mimicking of movements with no understanding of why an action is performed in a certain manner, thus ending up performing it incorrectly.

LAWS, CUSTOMS, MARTIAL TRAINING, AND USE OF VIOLENCE

With this in mind, a brief note also on laws and customs. Like today, laws regulated people's lives and conduct, but laws commonly varied between cities and often also between different neighbourhoods in the same city. Laws sometimes even varied depending on which religion one professed to be or one's social position, and it was not uncommon for custom to supersede law. This was especially true for violent altercations in defence of honour, body, family, or home. It was of great importance to be able to prove that one sought to not escalate the conflict oneself. One's behaviour and proper and honourable conduct was especially important in conflicts involving violence. Even in the frequent Fechtschulen, where—though it was unusual—fencers sometimes received fatal injuries, such incidents were closely examined for proper conduct similarly to any lethal incident in society, although also with a certain generosity in interpretation given the context.

This is also how we can understand some of Meyer's advice better, in the expectations for men involved in violent conflict to try to avoid escalating the level of violence. Learning to first strike the opponent with the flat of the blade was of vital importance,[171] as was to avoid

---

[171] See, e.g., the duel between Stefan Beurle and Georg Mayr in 1657, described in TLUSTY 2011: 113. Here the authorities, among other aspects of the conflict, investigated whether Mayr had struck with the flat or the sharp edge of his blade when he allegedly attacked Beurle.

thrusting against one's fellow countrymen, which in Meyer's time was held in deep contempt, both in civilian and military contexts and had serious legal implications.

However, the intentional use of flat, edge, and point goes even further back in time, as expressed, for example, in 1330 in *Le Pèlerinage de la Vie Humaine* by Guillaume de Deguileville:

"Now you have heard how you should /
according to various cases and obligations /
aim with the flat or the edge / and with the point,
while judging well. /
For one time you should condemn /
in another punish,
and in another persuade."[172]

That this custom extended outside of the fencing grounds is well described by ANN TLUSTY:

"The customary rules of military honour also dictated the standards of proper fighting. Few rules existed for fighting enemy troops in early modern Europe, but codes governing ethical fighting among soldiers on the same side, or between soldiers and civilians who were not officially enemies, could be detailed.

Military ordinances also threatened harsher punishment for unfair fighting among soldiers than civic ordinances did for civilian brawls. Hitting a man while down, for example, or stabbing at someone rather than striking with the flat of the sword in accordance with the customary rules of fencing could result in corporal punishment even if the fight did not lead to any injury."[173]

TLUSTY also describes the following incident between two civilians, one acting as town guard:

"Grocer Hans Müller from Mindelheim was teased by Christof Lang from Kirchdorf, for guarding the town gate without a gun, he responded by saying, 'here's weapon enough', and attacked Lang with the flat of his sword."[174]

Meyer in turn, mentions this custom briefly in 1570, and this particular passage needs to be considered in light of the customs described above.

"Due to the fact that although our ancestors permitted thrusting in serious matters against a common enemy, they did not permit the same in trivial exercises nor did they permit any kind of use of this by their mutually sworn men-at-arms or others who had come into

[172] *avez dont comment povez / selon divers cas et devez / viser du plat ou du taillant / et de la pointe en bien jugant / car une fois devez juger / l'autre pugnir l'autre prescher.* Paris, Bibliothèque national de France, ms. Français 829: 12ᵛ.
[173] TLUSTY 2011: 93.
[174] TLUSTY 2011: 53.

conflict outside of the mutual enemy, and the same should be maintained even today by honorable men-at-arms and other German citizens."[175]

This may partly explain why he teaches striking with the flat in the section on the 𝔓𝔯𝔢𝔩𝔩𝔥𝔞𝔲𝔴 ("Rebound Cut"): "as soon as it glances off, pull your sword around your head and hit at their ear from your left with the outward, inverted flat...",[176] and when describing striking to the four openings: "be diligent so that you can quickly conceive of and can fence at all four openings in a free-flying manner, because you only have three ways to cut and to strike (that is, with the long and short edges and with the flat), from which all fencing is composed together in harmony with the four parts of the opponent".[177]

What motivated Meyer to teach and write, and what did he wish to achieve?

Naturally, establishing himself as a skilled and knowledgeable fencing master among the rich families in the hopes of receiving their patronage would have been a strong driving factor for Meyer—a man so passionate about the art of fencing that he practiced daily, and while working as a cutler also managed to not only teach and arrange tournaments, but also to write several treatises on the topic of fencing. As we shall see later, these ambitions were also successful. Beyond this, we can also see deeper motivations rooted in a perception of issues contemporary to Meyer, centered on grave religious conflicts, a nobility perceived as having grown too comfortable and which relied on untrustworthy mercenaries to fight their wars, and with said mercenaries being profoundly immoral, fighting for anyone who would pay them, killing their fellow countrymen, and leaving behind a trail of rape, pillage, and murder in their wake. Meyer touched on all these things in his foreword, using the Fall of Rome as a cautionary tale, while also expressing a certain hopeful patriotic pride:

"After the Romans opined that they had conquered the entire world and, as a people living in safety, that pleasure was to be considered more important than good arts, policies, and knightly practices, it was this view that led to the entire empire being attacked from everywhere and destroyed by enemies. It was then that the knightly Germans were called and promoted before all other peoples, to rescue [the empire], embrace it, and re-establish it. This would have not occurred at all were the admirable Germans not practiced and experienced in all types of knightly play and the business of war..."[178]

As already expressed in various earlier examples, Meyer is very much a man of his time in seeing disciplined training for combat and war as the solution for the perceived moral corruption of his time. 54 years earlier, Paurenfeyndt had already touched upon such concerns in his printed treatise:

---

[175] Meyer (1570): II.50.
[176] Meyer (1570): I.13.
[177] Meyer (1570): I.60. Compare JML 31^V.
[178] Meyer (1570): a iii.

"...After a great deterioration and lack of attention became apparent to me, for this reason, in the name of impressionable youth amid daily practice, I have decided to concisely record the Knightly Art of Fighting and thoroughly explain the 3eðel in order to avoid gambling, debauchery, bad company, etc."[179]

Meyer writes somewhat more extensively, expressing a similar sentiment:

"Yet I have no doubt, if anyone had revealed this art and described it in a comprehensive and good organization before our times, then not only would the noble art not have fallen off in the case of so many, but also many abuses would have been completely avoided ([abuses] that must now be forcefully eradicated).

Therefore, I remain in the hope that regardless of whether my writings may be considered by some few as being of little worth, many honest journeymen and young fencers will step forward and not merely keep themselves with diligence and guard themselves from the disordered life—gluttony, drinking to excess, blasphemy against God, cursing, whoring, dicing, and all of that—through which this noble art has been fouled (since this knightly art has been used by many as a cloak to hide all types of vice and sloth, which many honor-loving people, and all honest fencers, additionally lament to the highest heavens). Instead, [that they] give thought to foundationally understanding this art and learning to use [it] in correct, honest seriousness, in order to free themselves from the bonds of useless, boorish tumult, and thus to endeavor in all virile, proper behavior and respectability so that, when they themselves have learned this art correctly and well and they lead an honorable life, they can subsequently stand before others—and particularly the youth—and may thus be considered to be hard working in their service."[180]

Meyer's solution to this difficult issue is thus simple: to have the population grow to become honourable warriors as Fechter, as fencers, learning discipline of body and mind. Furthermore, another underlying motivation which he expresses is that the printing of his treatise will contribute to the preservation of the art of fencing which, as already made clear, he believed to have a beneficial effect on society, while also protecting German lives through the teaching of ethical customs like not thrusting against one's fellow countrymen (as mercenaries might). Naturally, lives would also be preserved through the preparation for war:

"Therefore, it is the case that, in addition to the gun, other equipment, weapons, and arms useful in war are necessary, in current times as well as for our ancestors. And also, as is known to many, this includes not only good equipment, weapons, and arms (like harness, breastplates, swords, halberds, pikes and the like) but even more so daily practice, which can be correctly and skillfully used to the benefit of oneself and to the harm and disadvantage of the enemy. Therefore, it will be completely and absolutely necessary to learn this, as daily

---

[179] *Nach dem mir gsehen virdt grosser abgang und mangel unaufmerckung halben von wegen der zarten iugent und teglicher ubung, hab ich mir furgnumen kurczlich zu beschreyben dye Ritterlich kunst der Fechterey und gruntlichs auslegung der zetel fierer ursach halben vermeidung, spilss, prasserei poser geselschaft, und noch ains c&[!].* Trans. by CHRISTIAN TROSCLAIR.
[180] Meyer (1570): b iᵛ.

experience [shows] that for many, even when they are provided with the best, their equipment, weapons, and arms become more of a hindrance than a protection of their bodies and lives if they do not know how to use them nor how to thoughtfully defend themselves."[181]

Having thus concluded the puzzle which started with the question of what motivated Meyer to write this book, let us turn to his legacy.

MEYER'S LEGACY

Son of a paper maker, husband of Appolonia, citizen of Straßburg, master cutler and treasurer, Freifechter and fencing master—and a man deeply passionate about the art of fencing, to the degree that he practiced it daily, teaching nobility and commoners, boys and men, soldiers and civilians alike, striving to spread the knowledge of an art he saw as in dangerous decline. In doing so, Meyer hoped he would contribute not only to saving the art, but also to a better society where people behaved with more consideration for each other, with honour, dignity, and respect. His premature death naturally hinders this, but despite this, his name and his works in their own regard came to be known for over a century after his time, although he eventually would fall into obscurity. Still, before his slipping into the shadows, several fencers remarked on or used Meyer's material long after his passing. A new edition of his treatise was made in Augsburg in 1600, and Jakob Sutor von Baden copied large parts of Meyer's treatise of 1570 into his own printed treatise of 1612, with Theodor Verolini doing the same in 1679. Around 1618, Spanish fencing master Luis Pacheco de Narváez wrote in a letter:

> "And with this certainty I say, lord: That Carranza was not the first, nor the only one, that in Europe wrote concerning the *destreza* [fencing], that prior to him were Iayme Pons de Perpiñan, with a seniority of 135 years; Achile Marozo, and Camilo Agripa with 83 years; and afterward in their imitation Angelo Vizani, Giacomo de Grasi, Nicoleto Giganti, Salvador de Fabres, Federico Grisillero, Ioachim Meyer, Maestre Bico, Maestre Claso, Babote, and Marco Diocilini, all foreign masters."[182]

In 1670, Giuseppe Morsicato Pallavicini likewise listed Joachim Meyer alongside other famous masters like Monte, Marozzo, Docciolini, Agrippa, Dall'Agoccie, Carranza, Fabris, and Capo Ferro, while Francesco Antonio Marcelli does likewise in 1686.[183] Quite clearly, Meyer was known among fencing masters around central Europe well over a century after his death.

After this, the ghost of Meyer led a humble existence until the middle of the 1800s when the European pioneers of historical fencing, often military men and fencers, brought his printed work back to light, trying to understand it and even train the art of fencing described therein. They hoped to restore this fencing by bringing back a modern variant of the Olympic Games, at the time planned to include epee and dagger as well as single-stick fencing. These pioneers made numerous public performances using either antique swords or replicas thereof, including large two-hand war swords. The beautiful Veldenz treatise (PMM) was even displayed publicly

---

[181] Meyer (1570): a ii^{v}.

[182] Trans. from CURTIS 2004.

[183] See CHIDESTER 2023 for these and other 17th century references to Meyer.

in München in 1868, exhibited as part of the time period of Albrecht V, Duke of Bayern (1528–1579), in the "Fugger Room"[184] of the National Museum in the same city. It is described as follows:

> "On the smaller table opposite lies a very valuable manuscript with a handsome imprinted polychrome leather binding. This is the fencing book of Joachim Meyer from 1561 (published in print in Straßburg in 1570), dedicated to the 'luminous, high-born Lord Hern Görg Hansen, Count Palatine near the Rhine, Herțog in Bavaria, Counts of Veldenz and Hern zu Lutțelștein, my gracious prince and lord', with illuminated original hand-drawings for sword fencing, dagger fencing and grappling examples, ștaff-fencing, fencing with long pike, fencing and fighting in armour, etc."[185]

It remained in the permanent exhibition until at least 1887, but possibly longer.[186] After this, the treatise was "lost" for 140 years.[187]

Hiștorical fencing in this period remained ștrictly a niche, if international and sometimes public interest, and it eventually died out again largely due to the terrors of World War I. Once again, Meyer was lost in the shadows.

With the growth of modern hiștorical fencing and Historical European Martial Arts (HEMA for short) in the beginning of the 21 șt century—and with the great aid of the internet and online communities—Meyer's art has finally seen a quickly growing interest and love in the fast-expanding hiștorical fencing community, as well as in other groups with an interest in the topic. His name and works are now spread worldwide, over all continents (except perhaps Antarctica[188]), with tens of thousands, if not more, quite familiar with his name, the associated artwork, and his teachings. A whole international fencing guild with many hundreds of members globally, the *Meyer Freifechter Guild*, even takes its name from him, dedicating all their attention to reconștructing his teachings.

For a man born in a city of some 10,000 people, and who for most of his adult life lived, worked, and fenced in a city of only twice that population, this would have been beyond his wildeșt dreams and expectations. And yet here we are, with a legacy that keeps growing, with

---

[184] Albrecht was close friends with Augsburg patrician Hans Jakob Fugger, who to this day remains one of the richest men to ever live.

[185] *Auf dem gegenüberstehenden kleineren Tische liegt ein sehr werthvolles Manuscript mit einem hübschen gepressten polychromischen Originalledereinband. Es ist dieses das Fechtkunstbuch von Joachim Meyer 1561 (Im Druck erschienen zu Strassburg 1570), gewidmet dem "durchleuchtigen hochgebornen Hern Hern Görg Hansen Pfalzgraf bei Rhein, Hertzog in Bayern, Graffen zu Veldenz und Hern zu Lutzelstein, meinem gnedigen Fursten und Hern", mit illuminirten Originalhandzeichnungen für das Schwertfechten, Dolchfechten und Ringstück, Stangenfechten, Fechten im langen Spiess, Fechten und Kämpfen im Harnisch u. s. w. Das bayerische Nationalmuseum Mit Abbildungen und Plänen.* München: Kgl. Hofbuchdruckerei von Dr. C. Wolf & Sohn, 1868. p213

[186] MAYER, JOSEPH ALOYS. *Katalog der Büchersammlung.* München: Rieger'sche Universitätbuchhandlung, 1887. p35; *Führer durch das Königlich bayerische national-museum in München.* München: Bayeische Nationalsmuseum, 1901. p116.

[187] It remained lost until June 2021, when OLIVIER DUPUIS relocated it in the collection of the Nationalmuseum of München in 2021. This discovery was then presented in DUPUIS 2021.

[188] As far as I know.

new discoveries, transcriptions, and translations, and with martial artists the world over studying and trying to understand Meyer's words and ideas and apply them in a context which
mimics the Renaissance Fechtschulen (to the degree that is reasonable in modern society).

Already, modern historical fencing has achieved quite a lot, and for this reason, a list of the
current material relating to Meyer will here be presented. Several translations, transcriptions,
and semi-facsimiles have been made of Meyer's treatises over the past two decades.

- 2003 – Partial English translation of the 1570 (sword) by Mike Rasmusson, posted on
  Schielhau.org.[189]
- 2006 – Full translation of the 1570 by Jeffrey Forgeng, published as *The Art of Combat.*[190]
- 2012 – Partial modern German translation of the 1570 (sword and dusack) by Alexander Kiermayer, published as *Joachim Meyers Kunst Des Fechtens.*[191]
- 2012 – Full translation of JML by Kevin Maurer, circulated online.[192]
- 2014 – Partial English translation of the 1570 (staff) by Jon Pellett, posted on his
  homepage.[193]
- 2015 – Low-quality semi-facsimile of the 1570 published by Fines Mundi Verlag.
- 2016 – Full translation of JML and partial translation of JMR by Jeffrey Forgeng,
  published as *The Art of Sword Combat.*[194]
- 2021 – Full transcription of the 1570 by Michael Chidester, posted on Wiktenauer (later
  revised and published as *Foundational Description of the Art of Fencing* in 2023).
- 2022 – Full transcription of JMM by Olivier Dupuis, posted on the ELSAMHE site.
- 2022 – Partial translation of JMM (sword, dusack, rapier, dagger, and polearms) by Kevin Maurer, circulated online.
- 2022 – High-quality semi-facsimile of the 1570 treatise, published by HEMA Bookshelf.
- 2023 – Translation of the 1570 by Rebecca L. R. Garber, published as *Foundational
  Description of the Art of Fencing.*[195]

III. A brief linguistic and pedagogical examination

Oral and written traditions: Rhetorical and mnemonic de-vices

The first proper transition to a written culture in Europe began with the Greek philosophers
of Antiquity, changing from the oral tradition that underlaid formulaic and *cliché* epic poems
like the *Odyssey* and the *Iliad.* The wider and deeper transition, and the resulting sea change
on language, would take much longer. With the arrival of the printing press with moveable
types in the mid-15th century, we begin seeing an accelerated change in how books were

[189] Schielhau.org went offline in 2008, but the translation is preserved on Wiktenauer.
[190] Forgeng 2006.
[191] Kiermayer 2012.
[192] The text is also available on Wiktenauer.
[193] This site went offline in 2022, but the translation is preserved on Wiktenauer.
[194] Forgeng 2016.
[195] The book you're currently reading, as well as the Reference and Prestige Editions.

written, and in the language used in them, switching progressively from a form of writing rooted in the oral culture and tradition to one where writing was increasingly, though not fully, independent of verbal transfer of knowledge. Still, for a long time there was a great overlap between these two traditions, lasting for centuries, and like with medieval manuscripts, early printed books too have distinctively oral characteristics to them. Over time, the language evolved to become considerably more complex, and already in the 16th century, we see a great difference to 15th century sources. This is where a great shift happened in the fencing treatises.

Walter J. Ong states that:

"Oral habits of thought and expression, including massive use of formulaic elements, sustained in use largely by the teaching of the old classical rhetoric, still marked prose style of almost every sort in Tudor England some two thousand years after Plato's campaign against oral poets. They were effectively obliterated in English, for the most part, only with the Romantic Movement two centuries later."[196]

Now, texts tightly rooted in the oral tradition commonly have distinguishing features, often using old rhetorical and mnemonic devices designed to secure both the preservation, but also the transfer of information. Ong again states that:

"Early written poetry everywhere, it seems, is at first necessarily a mimicking in script of oral performance. The mind has initially no properly chirographic resources. You scratch out on a surface words you imagine yourself saying aloud in some realizable oral setting. Only very gradually does writing become composition in writing, a kind of discourse—poetic or otherwise—that is put together without a feeling that the one writing is actually speaking aloud (as early writers may well have done in composing)."[197]

While written text depends on certain physical conditions in the reading environment—light, stability, and distance—the oral tradition developed means of securing the transfer of information based on its own particular conditions. A listener may not actually hear or understand a word, so for that reason redundancy and repetition are used, either repeating whole concepts, or simply just adding one or two words of similar meaning. From this, various rhetorical devices developed, for example hendiadys, hendiatris, accumulatio, and tricolon, stacking descriptive words together in pairs or groups of three. For example, we see this quite clearly in the *Nicolas Pol Hausbuch* (N):

First and foremost, you should notice and remember that there is only one art of the sword, and it was discovered and developed hundreds of years ago, and it is the foundation and core of all martial arts. Master Liechtenauer understood and practiced this art completely and correctly; he did not discover or invent it himself, as has been written previously, but rather travelled through many lands and searched for the true and correct art for the sake of experiencing and knowing it. And this art is serious, correct, and complete, and

[196] ONG 1982. For more reading on oral cultures and traditions, see ONG 1982 and also MACKAY 1999.
[197] ONG 1982.

everything that proceeds from it goes toward whatever is nearest by the shortest way, **simply** *and* **directly**...[198]

Looking instead at the same text in the original German, we find the following 12 pairs and two triples.

dingen und sachen	durchfaren und gesucht
merken und wissen	rechtvertigen und worhaftigen
funden und irdocht	irvaren und wissen
grunt und kern	ernst, gancz und rechtvertik
vertik und gerecht	neheste und körtzste
gehabt und gekunst	slecht und gerade
funden und irdocht	nehesten, kortzsten und endlichsten

Rhyming was also used as a mnemonic device, making it easier to remember by having a sentence sound similar to the one that came before. This we see of course in Liechtenauer's Zedel,[199] which even Meyer used. Such texts were also commonly written as "flow of thoughts" and structurally, such treatises therefore commonly used simple additive conjunctions, commonly simply "und" (Ger.) "and" (Eng.) or "e" (Ita.), and high repetition of prepositions.

In Liechtenauer's Zedel, as presented in N (a text likely written sometime in the early 1400s), we see the very core of this art in a passage frequently repeated in many other later fencing treatises:

Vor, noch, dy czwey dink
syn allen kunsten eyn orsprink.
Swach unde sterke,
indes, das wort mete merke.
So machstu leren
mit kunst und erbeit dich weren.
Irschrikstu gerne,
keyn fechten nymmer lerne.[200]

---

[198] *Und vor allen dingen und sachen saltu merken und wissen, das nür eyne kunst ist des swertes, und dy mag vor manchen hundert jaren seyn funden und irdocht, und dy ist eyn grunt und kern aller künsten des fechtens. Und dy hat meister Lichtnawer gancz vertik und gerecht gehabt und gekunst. Nicht das her sy selber habe funden und irdocht, als vor ist geschreben, sonder her hat manche lant durchfaren und gesucht durch der selben rechtvertigen und worhaftigen kunst wille, das her dy io irvaren und wissen wolde. Und dy selbe kunst ist ernst, gancz und rechtvertik und get of das aller neheste und körtzste, slecht und gerade czu...* Trans. from CHIDESTER & HAGEDORN 2021.

[199] A didactic poem consisting of rhyming verses on fencing. The most common one was devised by patriarchal fencing master Johannes Liechtenauer around the turn of the 15th century. This evolved into a tradition of later fencing masters repeating these together with commentaries on their meaning, with examples of application.

[200] N: 18V. See also, e.g., JMM: 16V: *Vor vnnd noch zwej ding / Sinnd aller kunst ein vrsprunng / Schwech vnnd sterck / Inndes das wortt domit merck / So magstu Lernen kunst / Domit dich Ewren kannst Erschrückestu gernn / keinn Fechtten nit lehr.*

Occasionally we also see a form of *alliteration*, with words beginning with the same phoneme or prefix:

mosse, **vor**borgenheit,
vornunft, **vor**betrachtunge

Fencing master Fiore Furlano de'i Liberi too, in his *Fior di Battaglia* from 1404 (FLG), writes:

Io son posta longa e achosi te **aspetto**.
E in la presa che tu mi voray **fare**.
Lo mio brazo dritto che sta in **erto**.
Sotto lo tuo stancho lo mettero per **certo**.
E intrero in lo primo zogho de **Abrazare**.
E cum tal presa in terra ti faro **andare**.
E si aquella presa mi venisse a **manchare**.
In le altre prese che seguen vigniro **intrare**...

In Porta di ferro io ti aspetto senza **mossa**.
Per guadagnar le prese a tutta mia **possa**.
Lo zogho de Abrazare aquella e mia **arte**.
E di lanza, Azza, Spada, e daga o grande **parte**.
Porta di ferro son di malicie **piena**.
Chi contra mi fa, sempre gli do briga e **pena**.
E a ti che contra mi voii le prese **guadagnare**.
Cum le forte prese io ti faro in terra **andare**.[201]

While oral tradition seeks to preserve and conserve knowledge through the use of formalized language, with good memory crucial, it is also prone to mutation and a high degree of change in the very transfer of knowledge to other people, especially over generations.[202]

This process of change in oral information transfer is also apparent in the numerous fencing treatises that include Liechtenauer's Zedel,[203] the mnemonic verses that were used to preserve the European art of fencing, with distinct personal deviations already apparent in the very first few verses. From these deviations, we can see how some of those sources are likely to have been written by writers who were taught the verses in person, orally, while others simply copied a written text from another source, with the latter category showing a stricter consistency and fidelity to other texts, and the former showing great variation, with words and phrases omitted, changed, or added, with the verses actively recreated, rather than repeated from memory—a memory which by nature is not perfectly reliable.

In part, this process of change can also be explained by how associations and connotations to words and terminology shift over time, with meanings sometimes even getting lost. This is

---

[201] FLG: 6[R].

[202] See, e.g., the children's game 'telephone', which exemplifies the inherent flaws in a transmission chain, with messages mutating when travelling from one person to another.

[203] See, e.g., G, HTG, HTK, JWA, JWM, JWW N, PD, PKB, SR, and W01.

the case in 16th century fencing culture where the true meaning of "𝔄lber/𝔒lber" causes some confusion, which Meyer reasoned briefly about as follows:

"In my opinion, the Fool [𝔒lber] is named according to the word 𝔄lber (which means something like 'simple-minded') because no completely-finished strike can be achieved from this guard unless a fencer recovers with a new [strike] (after receiving the opponent's strike by setting it off). This is truly the measure of a fool and a simple-minded person: to allow themselves to be attacked without a prepared counter strike."[204]

𝔒lber or 𝔒lberbaum is a noun meaning "poplar tree", whereas 𝔄lber is an adjective meaning "foolish", but Meyer clearly believed the 'wrong' word was commonly in use. It should here be noted that the association with "fool" may also stem from older Middle High German alwære or Old German alawari, meaning "friendly" and "trusting", perhaps over time evolving into the meanings of simple shepherds, sheep, and fools; ein alber shäfer/shaf/narr.

25–30 years earlier and with a different perspective, Paul Hektor Mayr translated 𝔄lber using the Latin term *populus arbor* ("the silver/white Poplar tree");[205] curiously, in another place he also translated it as *pastor*, having or desiring to express connotations of shepherds and their "leaning on their staff".[206]

***

Unlike the subordinating style of the written narration, organizing sentences with subclauses, oral traditions have additive tools for connecting thoughts and ideas. One of the more renowned examples of how this is done, comes from the Bible, where in the 1610 Douay version (which is still close to the Hebrew original and the Latin), the text reads:

"In the beginning God created heaven and earth. And the earth was void and empty, and darkness was upon the face of the deep; and the spirit of God moved over the waters. And God said: Be light made. And light was made. And God saw the light that it was good; and he divided the light from the darkness. And he called the light Day, and the darkness Night; and there was evening and morning one day."[207]

Returning to N, focusing specifically on the structure and use of conjunctions, we see some very close similarities:

𝔙nd vor allen dingen und sachen saltu merken und wissen, das nûr eyne kunst ist des swertes, und dy mag vor manchen hundert jaren seyn funden und irdocht, und dy ist eyn grunt und kern aller kûnsten des fechtens. 𝔙nd dy hat meister Lichtnawer gancz vertik und gerecht gehabt und gekunst. Nicht das her sy selber habe funden und irdocht, als vor ist geschreben, sonder her hat manche lant durchfaren und gesucht durch der selben rechtvertigen und worhaftigen kunst wille, das her dy so irvaren und wissen wolde. 𝔙nd dy selbe kunst ist ernst, gancz und rechtvertik

---

[204] Meyer (1570): I.7^v.

[205] PMM: 90^R.

[206] *Habitus a similitudine pastorum qui dum armenta pascunt, fustibus innituntur.* PMM: 21^v.

[207] Douay 1610: Gen. 1:1–5.

und get of das aller neheste und kőrtzste, slecht und gerade czu, recht zam wen eyner eynen hauen aber stechen wolde, und das man im denne eynen vaden aber snure an seynen ort aber sneyde des swertes bünde, und leytet aber czőge den selben ort aber sneide of ienes blőssen, den her hauen aber stechen solde noch dem aller nehesten, kortzsten und endlichsten, als man das nűr dar brengen mochte[208]

Looking at the repetitions, the und ("and") alone makes up 13% of the text, with aber ("or"), adding another 3%. The conjunctions "and" and "or" thus make up 19% of all the words in this short piece of text. We may here also make a very brief comparison to the language of *A Propre new booke of Cokery* from 1545, on the topic of how to dress a crab:

Fyrſte take awaye all the legges **and** the
heades, **and** then take all the fysh out of
the shelle, **and** make the shell as cleane as
ye canne, **and** putte the meate into a dysche,
and butter it uppon a chafyng dysche of coles
and putte therto synamon **and** suger **and** a
lytle vyneger, **and** when ye haue chafed it
and seasoned it, then putte the meate in the
shelle agayne **and** bruse the heades, **and** set
them upon the dysche syde **and** serue it. [209]

We will now instead look at an example from *Johan Liechtnawers Fechtbuch* (SR) from ca. 1504–19, noting in particular the same, but also other, repeated words. As was the standard of this period, the text is centered around the 3edel rhymes of Liechtenauer, with associated comments and examples to help explain them:

Willtu kunst schowen
Sich linck gen vñ recht mitt hawen
Vñ linck mitt rechtem /
ist dz du starck gerst fechten

Glosa Merck dz ist die erst lere des langes schwercz dz du die hew von bayden sytten recht solt lernen hawen Ist dz du annders starck vñ gerecht fechten wilt Dz ver nym allso Wenn du wilt howen von der rechten sytten So sich dz dein lincker fu°ß vor stee Vnd wenn du wilt howen von der lincken sytten So sich dz dein rechter fu°ß vor stee Haw Hästu dann den ober haw° von der rechten sytten So folg dem haw nach mitt dem rechten fuß tu°st du dz nicht / So ist der how falsch vnd vngerecht wann dein rechte syten pleibpt dahinden Darum ist der haw zu° kurcz

[208] Trans. from Chidester & Hagedorn 2021.

[209] See a full transcript at the web site of *Justus-Liebig-Universität Giessen* <https://www.uni-giessen.de/fbz/fb05/germanistik/absprache/sprachverwendung/gloning/tx/bookecok.htm>. Retrieved 19 Aug 2022.

vnd mag sein rechten gang vndersich zu° der rechten syten anderen sytten vor dem lincken fu°ß nicht gehaben[210]

We can again see how certain functional words are repeated in a still very dense text, with those words making up about 24% of the text.[211] We also see two examples of stacking of adjectives in (starck vñ gerecht and falsch vnd vngerecht), and we see how phrases are repeated, mirrored for left and right, in a formal pattern:

Wenn du wilt howen von der rechten sytten So sich dz dein k lincker fu°ß vor stee...
...wenn du wilt howen vö der lincken sytten so sich dz dein rechter fu°ß vor stee...

Looking instead at the first *printed* fencing treatise in German, that of Andre Paurenfeyndt (1516), we still see many of the characteristics presented above. Here, the same repeated words make up twenty percent of the text:

Leg dich gegen ym wie for, greif mit deiner lincken handt in dein schwerdtz klingen in der mit und stich ym gegen seinem gsicht, So muesz er sich verseczen, und den stich austragen, So folg ym nach mit dem trit, und lasz dein lincke handt vom schwerdt, greif mit deinemm knopf uber sein ped hendt, und leg ym dein schneid an halsz, und leg yn in die schwech So wirfstu in...

Wan du mit ainem fichst, und nahendt czu ym kumst, So kum in den phflug, und treib den behendlich mit wenden von ainer seiten czu der andren, und daß dein ort alweg vor dir pleib, auß dem magstu treiben daß verseczen, daß ist die nech, und yn dem magstu stercken mit der langen schneidt, und dar auß treiben alle vor geende stuch, auch magstu hew und stich abseczen und die flechlichen prechen, und mit dem ort die pleß suchen...

Pistn dan unden auf deiner lincken seitten, und er sticht oben gegen dir und hast dein schwert in der rechten handt, undt dein ort in der anderen handt So versecz mit deinem halben schwert dasz dein ort ubersich kum an dem verseczen, So windt ym ein mit deinem knopf in sein lincken armen, und mit dem ort aussen an sein rechten armm, let ers So greif mit deinem knopf zwischen seine pain, und truck oben von dir, dasz magstu thun alsz oft du ein windest, sunder trit alweg in deinem einwinden hindersich, begreifstu dan dein schwerdt in die lincken handt und dein ort in die rechten, und er sticht von oben gegen dir, So versecz mit deinen halben schwert dasz dein ort undersich kum an dem verseczen, So windt ym mit deinem ort in sein lincken arm und mit deinem knopf aussen an sein rechten armm und zuck yn fursich, in dem vordren stuck, ker dein spicz ubersich, So hastu die einwinden und durcschiessen unden und oben, und ob ainer auff dich schlecht oder sticht, So wart desz abreissen oder desz einwinden[212]

Moving forward another hundred years in time, looking at the 1619 German translation of Salvator Fabris' fencing treatise entitled *Des Kunstreichen Italiänische Fechtkunst*, we find a quite radical change, with the same words making up only eight percent of the text:

---

[210] Trans. by Dierk Hagedorn.
[211] Here focusing on words like *mit, und, daß, wilt, vor, wenn, so, du*, and *daß*, while ignoring possessive pronouns, prepositions, verbs, and nouns.
[212] Trans. by Michael Chidester.

Die folgende figur welches eine quarta ist / Ist gar verscherden von den zwoen vorhergehenden / dan wie man an der figur sihet / So entblosset sie dem seind die brust / und stehet mit schlimmer und uberqueren schritt / im willens nach der einen oder nach der ander seitten zu tretten der gelegenheit nach / man kan auch die brust nicht verletzen noch das haupt / dieweil seine bein nicht seyn / das aeine auff der einen seitten / und das ander auff der ander seitten deß Rapiers desselbigen der da verletzen wil / also wan er eins auff heber / ist sein leib allezeit auß der gegenwart / kan man dann verletzen / mit gesagter quarta, tertien und secunden / nach dem es das tempo die notturfft und die gelegenheit erfordert / diese guardia ist entblosset von ausserthalb / und locket den seind / damit er verletze / weil dasselbige theil der starckeste / und machet einen solchen natürlichen winckel / auff die weisse / daß wann gesagter seind ihn verletzen wolte / er als dann mit dem lincken fuß in eine rechte lineam trette / samt außgestrecktem Arm / doch daß die hand am selbigen ort bleibe / so wird er von unden verleze der seind in die rechte seitten oder aber oben / doch daß er den winckel grosser mache / und die hand hoch halte / wie dann auch dieselbige schulder: Also wird er solche grosse stöcke haben / das wie der seind mehr wird suchen zu pariren / desto mehr wird er verlezet werden / und so der feind sich zuviel nahet ohne resolution / sol dieser die hand auß der quarta in die secundam wenden / und sich selben coop- eriren / das is / das haupt bedecken mit vorsetzung deß linken fusses / und also mit dem leib und dem Rapier auff ihn hineyn passiren / wird also auff die brust verletzen / mit gesagter secunda, aber welcher diese guardia gebrauchet / der muß achtung geben.[213]

Fabris' text here can be seen as an example of much more evolved *written* tradition, a tradition characterized by its persistence and easy transmission to masses of people, which helped structure the very reasoning by making ideas visible to the eye, moving away from formalized mnemonic devices to "prose." As a result, it uses complex structures, with distinctly longer sentences consisting of many subclauses and a varied choice of conjuncts, subjuncts, disjuncts, and conjunctions to tie words, clauses, sentences, and paragraphs together. The coordinating conjunctions expand to become more varied, and correlative and subordinating conjunctions become very common and important, marking the education, intelligence, and wit of the author.

How, then, does Meyer's treatise compare to the previously presented sources? As expected, we find him *in between*, in a transitional period.

Vnd erstlich wann du fur deinen Mann kommest, und also durch auffstreichen oder sonst mit auffziehen (zu einem Oberhauw) mit deinem Schwerdt in die höhe kommen werest, und er Hauwet dir in dessen gegen deiner Lincken zum Kopff, so spring wol auß seinem Hauw gegen seiner Lincken, etwas zu ihm umb, und schlag mit außwendiger flech gegen seinem herfliegenden streich, das du sein Schwerdt in die sterck antreffest, unnd das also starck, auff das sich dein vordertheil deiner klingen in solchem schlag, über seinem Schwerdt zu seinem Kopff einschwinge, welche dann gewis triffst, wann du mit ihm zugleich schlechst, unnd doch mit deinem Schwerdt oberhalb des seinen kommest, auff solchen Hauw er hab getroffen oder nit, so zuck dein Schwerdt wider übersich ab, und Hauwe übereck dargegen über, von Unden zu seinem Rechten Arm, in solchem Hauw trit mit deinem Lincken fuß wol aus gegen seiner Rechten, und bucke dich mit deinem Kopff wol hinder dein Schwerdts klingen, von dannen zucke behend wider übersich, und wincke

---

ihm mit kurtzer schneide zu seinem Lincken ohr, ersihestu das er ihm nach wischet, so lasse nicht antreffen sonder fehl ablauffen, und verschrencke bald dein hend in der lufft (die Recht über die Lincke) und schlag ihm mit kurtzer schneide dieff zu seinem Rechten ohr, als bald 3witrch umb und ziech ab, unnd merck hie, wann er dir auff deinen obgelehrten Underhauw, so behend nach folgen, unnd so hart auff dem tach sein würde, also das du zu dem ablauffen nicht kommen kanst, so hab acht in dem er von deinem Schwerdt abzuckt, so folge ihm mit dem Schnit nach auff die arm, &c.[214]

Clearly, there is a significant difference compared to Paurenfeyndt, but the language is still not as rich as that of Fabris. Still, it points in a direction which had started with certain didactic choices Meyer made for his printed treatise, in deciding to speak more plainly, for the ease of comprehension for the reader.[215] While Meyer clearly was a man of his time, his writing was distinctly shaped by a new language designed for the written text. This specific topic we will return to soon.

SAFEGUARDING KNOWLEDGE BY HIDING THINGS IN PLAIN SIGHT

In the case of the old German fencing treatises, sprung out of an oral tradition, the Zedel verses and the knowledge therein were also specifically designed to be kept *secret* and *hidden* from the uninitiated, and certain rhetorical tools may have been used for this particular reason. Synonyms, similes, and metaphors may, in direct teaching under a master, have been given specific meanings that would be obscured by other, common, associations to anyone not initiated in the art of fencing and specifically the Liechtenauer tradition. Likewise, the use of synecdoche,[216] *pars pro toto*, polysemy,[217] and enthymeme may have been used in similar manner. These too serve as mnemonic devices and aid the recall of large texts or orally transmitted teachings. Commonly it also used known imagery from the daily life of the reader. Looking here at the Fechtbücher, we see, for example, the stances vom Tag ("from the Roof/Day/Sky"), Ochs ("ox"), Stier ("bull"), Pflug ("plow"), Bastey ("bulwark"), Eber ("boar"), and Steürhut ("paddle guard"), and the techniques Pfobenczagel ("peacock's tail"), Wecker(hau) ("awakener"), Krawthacke ("cabbage hoe"), Vidilpoge ("fiddle bow"), Schielhauw ("squinter cut"), Schaitler ("scalper"), Rädlin ("wheel"), Schlaudern ("catapult"), and the Rose ("rose") (to mention just a few). These would be used to make associations with certain postures, movements, directions, trajectories, and even targeting.

Meyer to referred to this, for example, when saying "...the five Master Cuts... were actually invented as another division of the art, and named with their different names, by those learned in this art for more diligent and useful examination so that the art, which is hidden and

---

[214] Trans. by MICHAEL CHIDESTER.

[215] A deeper linguistical analysis is of course needed here, with a larger and more representative material for study. For now, this will have to be left for another day.

[216] Synechdoche is a rhetorical figure where a part of something is used to represent the whole, for example boot representing a new cop or soldier, suits representing businessmen, or number 10 representing The Office of the British Prime Minister.

[217] With polysemy, a symbol, morpheme, word, or phrase is chosen for having multiple meanings

wrapped in another, can be more readily and more easily understood, grasped, and considered in different ways".[218]

MacKay describes similar ideas in Greek oral tradition:

"...with Homer's own concept of signs, or *semata* as he calls them. Those objects or events that he labels 'signs' have in common the property of serving as keys to an otherwise hidden reality. Whether a *sema* describes one of Zeus' omens, Teiresias' prophecy of the oar or winnowing shovel, Odysseus' telltale scar, or the olive-tree bed, it dependably designates an emergent reality, a prolepsis, a secret known only to a chosen few. But it does so obliquely, by indexical reference, rather than by simply naming its subject. To those not 'in the know', the bed is just a bed, nothing more and nothing less; bird-omens require interpretation by a qualified seer and can otherwise be impenetrable; the scar does not explain itself unless its viewer knows the background of Odysseus' youthful engagement with the boar... Homer's signs are always replete with significance, but become transparent only when one fluently understands their language...

...formulaic phrases like "swift-footed Achilles" or "ox-eyed Hera" become keys that unlock traditional realities, automatically alluding to complex ideas... by citation of the agreed-upon code. In this fashion the sign refers not to the everyday, objective meaning of its components, but to a traditional referentiality...

...*semata* or signs of other types and sizes can also encode idiomatic meanings that only the properly prepared audience is equipped to understand."[219]

Another way of hiding the meaning of a word is to add physical actions which have to take place in person, and which would have to be learned by the speaker, and the listener/reader both. The width of this possibility is briefly described in *The Cambridge Companion to Textual Scholarship:*

"Whereas a text is fixed on the page and, in principle is infinitely reproducible, the defining trait of oral art forms is that no song is ever sung (nor is any story ever told) in the same way twice. Moreover, the words of a work of verbal art may be only one element in an embodied event holding such factors as intonation, facial or bodily gesture, melody, instrumental accompaniment, dance, ritual actions, audience participation, and so forth."[220]

It is entirely possible that the original verses of Liechtenauer's 3𝔢𝔡𝔢𝔩 were supposed to be accompanied by such actions—actions which would give deeper, or even *different*, meaning to the words of the 3𝔢𝔡𝔢𝔩. In fact, it is also possible that the listeners too were supposed to perform certain actions or movements.

---

[218] Meyer (1570): I.15. Hans Medel's gloss emphasizes this even further, saying that each of the five strikes has a secret name to hides its common name, e.g., *Zornhaw* is the secret name for the *Oberhaw* to conceal its technique from the uninitiated (HM: 23ᵛ).

[219] MacKay 1999.

[220] Fraistat & Flanders 2013

All these described methods for obscuring the meaning of spoken or written words make interpretation for the uninitiated extremely difficult, since the text has the appearance of plain words, with the true meaning is hidden in plain sight.

When it comes to the commentaries written by later fencing maſters in more than a few fencing manuscripts preserved from the 15ᵗʰ century, they may have been written in less secretive words, with the writers perhaps even speaking openly. However, this is difficult to know with any degree of certainty. They too would have had motives for keeping the written words obscure and connected to direct, oral transfer to their ſtudents and patrons.

FREEING THE SECRET AND HIDDEN ART OF FENCING

German Sechtbücher, before Paurenfeyndt in 1516 and Meyer in 1570, were written by hand in single copies. Such books, even though they commonly included extensive commentaries, were never meant for public consumption and inſtead aimed at specific initiated people: nobility, patricians, aspiring fencing maſters, and others willing to inveſt money and time into their learning from a fencing maſter, thus making the readers a very small group of people indeed. Paurenfeyndt and Meyer both chose a diſtinctly different and even radical direction here by getting their words into print.

Meyer appears to have wanted to break free from the old Liechtenauer tradition of keeping words secret and hidden, which is reminiscent of the change that came with Proteſtant ambitions to ſtrip away hierarchies that reſtricted men's access to the words of God, for inſtance by the printing of the Bible in people's own language inſtead of Latin.

It also reminds us of the ongoing societal changes with spreading literacy via the Neben⸗ or Winkelſchulen,[221] which were independent and commonly Lutheran schools for laborers' children, sanctioned by neither the Church nor city authorities. Literacy too would bring people closer to God, enabling them to read the words of God themselves without a prieſthood reading them to their congregations in Latin (a language many would not underſtand at all). In the same vein, there was a growing importance of self-learning via books and inſtruction manuals in particular, with topics ranging from cooking, manners, rhetoric, letter-writing, and dancing, to medical treatment, warfare and of course, the art of fencing.

Additionally, we may here consider the shift away from medieval scholaſticism and the old use of rhetorical and mnemonic tricks, and how Petrus Ramus argued that everything Ariſtotle had said was inconsiſtent simply because the ideas were not properly syſtemized and could only be recalled via arbitrary mnemonic decides.[222] Inſtead, for better pedagogical teaching, he recommended the use of dialectics, outlines, summaries, headings, citations, and examples. His thoughts, although very controversial and blasphemous to many, would become extremely influential in late 16ᵗʰ and early 17ᵗʰ Europe, not leaſt in the German-speaking countries and among Proteſtants of different kinds.

Along the same line of thinking, inſtead of dense and cryptic texts filled with obscure terminology, Meyer often intentionally leaves out terminology, describing actions mechanically, such as telling us to put the pommel under the elbow inſtead of telling us to do a Krumphauw. For simplicity and brevity, he ſtill uses the old nomenclature, but as a result the text is

---

221 See Maurer & Vanslambrouck 2013.
222 See Ong 1958: 46–47.

considerably easier to underſtand for a new fencer, even today. We already see this pedagogical ambition expressed in JMM, where Meyer says:

> "I have been caused to assemble the entire fencing art... and have produced it solely by the limited ſtucken of the same... giving their proper titles and names... in good part so that the teaching can be clearly underſtood, and brought to this point, that some ſtucken are so completely incomprehensible for and to the hand, that I myself may scarcely underſtand again their same proper titles and reverence, not to mention where the honorific words should remain, so that it might be of use to someone, that thus not intentionally, but rather without obscuring the art, the pieces have been written with general words."[223]

Still, keeping in line with the Fechtbuch tradition, he also chooses to include parts Liechtenauer's Zedel and adds his own commentaries and examples. This can be regarded as in conflict with his apparent greater ambitions, but may have come from the book being a bit rushed, forcing him to reuse older material prepared for JML. A number of flaws in the text and art, as well as pedagogical ideas introduced but later largely ignored, may also point to this.

PEDAGOGICAL AND DIDACTIC PARTICULARS OF THE PRINTED 1570 TREATISE

In contraſt to how teaching and military combat and gymnaſtics drills would come to be performed in the 19th and large parts of the 20th century, Meyer treats all ſtudents as individuals who muſt learn to fence in ways that fit them. This skill comes out of reflection and ſtudy of given examples that teach many things,[224] and from experience gained through daily training and practice. For this learning, actual underſtanding is a prioritized goal, rather than mere simple mimicry of movements or drilled responses without underſtanding. These views shine through when Meyer says:

> "Since everyone thinks differently than others, and likewise they will act differently in fencing... fencing is actually a type of exercise through which the body is directed to guide the weapons in all kinds of quick actions; and because, when one exerts oneself in this, one muſt then adapt one's self in the work and guide the weapon as the circumſtances demand; thus, the better one is practiced in this, the better one is able to encounter each opportunity as it occurs."[225]

---

[223] JMM: iiiR; trans. from MAURER 2022.
[224] For example, "Dear Reader to whom I have granted my didactic fencing poem, which, based on the Authorities, I have composed, improved, and correctly organized: I also wanted to explain it to some extent, using many beautiful and clever sequences and examples (so that many can gain more use from it) and to provide a small introduction in order to understand it. Because it is so rich in techniques and all types of cleverness, the longer you ponder this introduction, the more techniques you can gain from it. The fact that rhymes without an explanation are not very useful is clear from other preceding fencing treatises." (Meyer (1570): I.45V). Also, "you can learn many counters from the preceding sequences through deliberation..." (Meyer (1570): I.62).
[225] Meyer (1570): b iv–b ivV.

Furthermore, also when he states that,

"At this point, the cuts align themselves according to the type, nature, strength, and ability of the fencer, because the weak must seek a different advantage in the cuts than the strong, and this leads to strength",[226] and "...students are masterfully provoked to deep consideration about how to correctly use all types of advantages..."[227]

Daily training appears to have been both part of Meyer's own studies and something he considered vital for any serious student of the art of fencing: "...daily practice, which can be correctly and skillfully used to the benefit of oneself...",[228] again expressed in "those youth... practice each day at the time they have scheduled",[229] and again, "because experience (which can only be learned through daily practice) must inform the majority of such techniques".[230] The reason behind this is simple. As Meyer himself says, "the practiced always overcome the untrained".[231] If you want to be good at something, you have to practice, a lot. Consequently, truly learning the art of fencing could not be a simple pastime, but was something which required real dedication and effort, every day.[232]

Something important to understand when considering his text is that he directs himself towards several different groups of readers, ranging from inexperienced youths, for example when teaching the fundamentals of terminology: "so that the youth who want to dedicate themselves to this art would not be confused by those words that are unknown to them",[233] and in particular the whole dusack teachings "I have dealt with this weapon extensively because the youth are generally instructed too quickly in it",[234] but also more experienced and advanced students: "in the second part, a more extensive and sufficient report will be provided about the Master Sequences and that which is helpful for increased quickness related to this weapon so that this book can be useful to both the beginning student as well as to those more experienced in this art".[235] Other parts are intended for outright masters: "I have undertaken to write more, both for great masters and even more for students".[236] He even mentions students who already have had teachers, "...anyone who has read, diligently attended to, contemplated, and seriously practiced the same can understand and learn it readily (assuming that they have previously had a master)".[237] Finally, in the foreword to the reader, he also in numerous places refers to experienced soldiers (𝕶𝖗𝖎𝖊𝖌𝖘𝖑𝖊𝖚𝖙𝖊, literally "people of war"), stating that they can judge the value of his arguments.

---

[226] Meyer (1570): b iii^V.
[227] Meyer (1570): b iii.
[228] Meyer (1570): a ii^V.
[229] Meyer (1570): b i.
[230] Meyer (1570): II.67^V.
[231] Meyer (157): a iii^V.
[232] Worth noting here is that in the English Company of Meisters of the Science of Defence, when it started, it took about 15–17 years of daily study under a master to become a master oneself.
[233] Meyer (1570): I.1^V.
[234] Meyer (1570): II.49^V.
[235] Meyer (1570): I.2^V.
[236] Meyer (1570): I.55.
[237] Meyer (1570): II.72.

Comparing to the teaching methodology and pedagogics of fencing manuals that preceded his printed treatise, this work of his ſtands out. While he includes old mnemonic verses from Liechtenauer, he rewrites and rearranges them. Still, for the moſt part, this old, traditional way of teaching is set aside, in lieu of more modern methods for teaching, in line with the recommendations of Ramus. It is here very important to underſtand that Meyer does not juſt teach how to *fence*, but also how to *learn* how to fence. In contraſt to moſt earlier treatises, he also gives plentiful of personal advice, teaching the ſtudent both how to think, as well as fundamental fencing psychology, about human charaĉter types and their effeĉts on people's fencing.[238]

Method-wise, Meyer, like Paurenfeyndt before him, uses one weapon to teach another: the sword is the foundation for everything, the dusack for the rapier, and the ſtaff for the halberd and pike. Through these we learn body and weapon mechanics which can be applied to the other weapons. Meyer expresses this outright: "fencing with the sword is both an originating point and source for all other fencing",[239] and, "after the foundation was laid previously in the [seĉtion on] fencing with the sword, the dusack now follows, which takes its origin from the sword as the correĉt source of all fencing (both that direĉted with one hand and that with two). Therefore, I want to place this here—not only because it is the moſt commonly-used [weapon]

---

[238] Occasionally, we also get rare and even unique insights into not just techniques, but also how frequently they were in use, for example, when Meyer comments on the *Zwerchhauw*, saying, "You should know that if the Crosswise Cut did not exist (as it is currently used at the present time), only half of fencing would be possible. [This would be] especially true when you are inside and under the opponent's sword, as you cannot fence through the Cross with long cuts." Meyer (1570): I.55.

[239] Meyer (1570): I.1.

for us Germans (following the sword) but also because it is a beginning and foundation for all weapons that are used in one hand".[240] Then, in the dagger section, "the fourth part of this book deals with fencing with daggers so that you can learn how a fencer should use all types of similar short weapons",[241] and finally in the polearms sections: "...I will thus initially take the short staff in hand as the foundation for all long weapons...".[242]

To reach his goals, Meyer used a range of literary and pedagogical devices and tools. Yet it is evident that Meyer wrestled with how to be able to properly teach, feeling somewhat restricted by the limitations of a manual intended to be read on its own. As he puts it in his foreword to the reader:

"In the first place, it is due to the fact that this knightly art must be seized with both fists and learned through the cooperation of the entire body—that is, it must be taught more through physical experience than from books... The second reason is this, namely that this knightly art of fencing can only be written about in books or recorded in written format with difficulty, as it can only be directed in action through the practice of the entire body. Because I have experienced this reason myself, I consider it to be the most important and significant (as others of great comprehension can judge), and as much as the first [reason] demands, I, like others, must acknowledge that (as mentioned above) every art can be understood with less effort when written out in a well-organized manner, and can be grasped by those learning through practice of the body or the hand."[243]

At one point he even outright states that a particular example simply can't be properly put into writing, saying that, "it is lovely and fast work which is not easy to describe, but [can be] shown with a living body".[244]

Yet, he continues, explaining the added value of written manuals, "...it is also certain and true that this art, just as many others, is better imagined in the thoughts of learners if they have it written out and placed before their eyes (in addition to good instruction which is set together in the correct order). Consequently, it can also be taught and understood even better through exercise of the body than if it were explained in poor, oral narratives and received piecemeal."[245]

There is indeed a rich palette of tools used, stemming from Meyer's inventiveness in trying to convey his understanding of the art of fencing. He uses summaries, both of what the student will learn and when repeating what has been learned. He lists learnings, explaining how he has "written about this as a reminder to you so that you should diligently consider it, and so that when one of these is mentioned subsequently in the sequences, you can understand and take note of it more quickly and can therefore also grasp the sequence itself faster".[246]

---

[240] Meyer (1570): II.1.

[241] Meyer (1570): III.1.

[242] Meyer (1570): III.16.

[243] Meyer (1570): b i–b iᵛ.

[244] Meyer (1570): I.54.

[245] Meyer (1570): b i.

[246] Meyer (1570): II.72.

Furthermore, he also explores using diagrams teaching cutting patterns and sequences,[247] something he had already explored years before in JMM and JML. Likewise Romanesque sculptures are displayed in the rapier section showing both cutting lines and where to target on the body: the neck, below the ribs, and the knees. The floor patterns and square floor tiles are added as very important aids for the student, for the understanding of both the angles of the feet, hips, and torsos, and also the relative positions and angles between the fencers. Without them, the perspectives might deceive the student into assuming wider angles between the feet than intended. Similarly, shadows underneath the feet, or lack thereof, help understanding timing and stepping, and whether the feet are planted firmly on the ground or not. He also briefly experiments with showing footsteps in both the rapier and staff sections, but quickly abandons it, perhaps for lack of time.

Looking at the underlying pedagogical order for the sword, Meyer presents it in three main parts: "Firstly, to discuss the associated terminology and types—as were invented by masters of this art with particular diligence—so that a fencer can understand and learn the secret and skill behind the same more swiftly and easily. After, to explain and interpret this terminology so that everyone can understand what should be understood by this type of speech. Then, thirdly, to present the exercise of the art in itself: how it should be directed in the work from the explained cuts and stances."[248] His pedagogical reasoning is to start with the core elements, aiming to build an understanding as fluid as putting thoughts into ink on paper.

"If you want to write a word correctly so that the letters that are useful and proper for this flow in an orderly way, one after the other from your pen, then you need to hold all of the letters in your thoughts and memory and fundamentally know the type and property of each one. Likewise, you should grasp the previously-described techniques well and hold them in your mind to such an extent that, whenever you approach another to fence, these appear then in your mind when they are needed. However, because not all letters can be used for each and every word, it is also impossible to want to make use of all of the techniques in each sequence as they have been explained. For that reason, you should pay attention to how your counterpart positions themselves against you (as events require). Also observe how the person is—whether they are fast or slow, large or small—and know how to use your work accordingly and to oppose them."[249]

Then, with the foundation laid, he sets out to teach more experienced students: "In the second part, a more extensive and sufficient report will be provided about the Master Sequences and that which is helpful for increased quickness related to this weapon so that this book can be useful to both the beginning student as well as to those more experienced in this art."[250] By page I.25^V, he declares having completed the basics, saying, "these techniques discussed and

---

[247] Some of these diagrams also, somewhat more subtly, teach us fundamentals about openings to target in our opponent's defence, and about openings we reveal ourselves as we move, as well as how to cover them better as we fence. This is reminiscent of the teachings of famous swordsmen like Miyamoto Musashi and George Silver both.

[248] Meyer (1570): I.1.

[249] Meyer (1570): I.25^V.

[250] Meyer (1570): I.2^V.

explained previously are actually nothing more than a beginning and an elemental foundation from which all fencing sequences of the sword can be gathered...".[251]

Finally, despite its many literary and pedagogical devices and the magnificent and quite cleverly designed illustrations, at times the treatise gives off an impression of being somewhat rushed and unfinished. Sometimes an illustration designed for another purpose is used to describe something else, with Meyer saying that one should act similarly, but not quite as de-

picted. This is true for example when he describes Hangetort: "The portrait on the right side in the abovementioned Figure [F] teaches how you should correctly initiate the Hanging Point (except that the arms are not shown as sufficiently straight here)."[252] Arguably, the same could be said for Sword Figure D, used to teach both Krump‹ hauw and Verkehren. Elsewhere, like in the dagger and the pole-arms, whole pairs of fencers are left unexplained and without written example for learning. There is also a distinct inconsistency, even confusion, regarding the terminology, in particular in the polearms, where it appears as if he had not completed his pedagogical thoughts on the redefinition and use of terminology for these weapons, and thus confuses his own meaning of the terms. Likewise, the use of footprints on the ground, only occurring twice, once in the rapier and once in the staff, seems haphazard and leaves us wanting more.

IV. A LOOK AT MEYER'S ART OF FENCING

MODERN CONCEPTS OF HONOUR, FENCING IN EARNEST, AND TOURNAMENT

Through the past couple of decades, a lot of discussion regarding the intended contexts for Meyer's art of fencing has been had, a lot of it rather uninformed, and while one camp has argued that his teaching was intended exclusively for "sport", there is little base to support such an argument. In direct contrast, Meyer himself speaks about war and repeatedly about fencing in earnest.[253] Even the tournament fencing of the time served to prepare for war and was both bloody and could cause both maiming and deaths. The confusion likely stems firstly

---

[251] Meyer (1570): I.25^V.

[252] Meyer (1570): I.9.

[253] Again, see the foreword, where Meyer says, "And now, for the sake of brevity, I will skirt the issue that the art of fencing is numbered as a noble part of war practices. Due to this practice, students are masterfully provoked to deep consideration about how to correctly use all types of advantages, including other, quite useful applications of these practices." (Meyer (1570): b ii^V).

from a superficial reading of his teachings, secondly, from the fact that he also teaches how to learn how to fence and how to practice fencing, and thirdly, from a misunderstanding regarding the two-hand sword, where it has been regarded as an outdated weapon no longer in use outside of "sport", which is provably wrong by about a century. It is also worth noting here in particular how Meyer in numerous places teaches techniques that would have been allowed neither in training, nor in the tournaments, and which we today might even find dishonourable and brutal. Examples of this are attacks to the groin, throat, and neck, breaking of fingers, arms, and elbows, and foot stomping, cutting off fingers, as well as eye gouging—all banned in the contemporary rules for Fechtschule tournaments. JMM also describes targeting openings in armour with dagger: the neck/head, the throat, armpit, hip joint, hand thrust, knee joint, and the groin, as well as,

"Wrestling and the Forbidden Attacks that shall not be allowed in the Open Schools, namely the murder thrust to the temple*, to the groin, arm breaking, leg breaking, knee thrusts finger severing* and breaking, neck thrusts, throat thrusts and eye gouging."[254]

In the same treatise, under the rapier section, Meyer describes the main thrusts as:

"...the face thrust, the second is the heart thrust, third is the throat thrust, the fourth is the groin thrust, the fifth is the joint thrust, the sixth is a double thrust and the seventh is armpit thrust."[255]

And in the dagger section:

"Item: if he is strong and you cannot take his weapon, then grab onto the ring, use that to your advantage so you overtake him, grab him by the throat or jab a thumb in his eye, or drive up under his nose, or punch him with your thumb under his chin." [256]

Finally, in the halberd section, again, we read:

"I must take the time here to set these six cuts for the halberd at the beginning. They are useful both for practice—in that they can aid your body in learning greater quickness—and (indeed even more so) because each of them is necessary for those who want to become skilled in this weapon."[257]

The 1570 treatise, however, has these aspects toned down, but there are still plenty of more examples that clearly indicate that they are ultimately meant for fencing in earnest, not tournament. It also makes complete sense, since the target groups for the book, according to Meyer himself, are both experienced soldiers and young boys, to be used for learning on their own without a teacher, thus requiring more safety in practice.

---

[254] JMM: 73ᵛ; trans. from MAURER 2022.
[255] JMM: 48ᴿ; trans. from MAURER 2022.
[256] JMM: 65ᴿ; trans. from MAURER 2022.
[257] Meyer (1570): III.32

CHARACTERISTICS OF MEYER'S ART OF FENCING

I would now like to transition to a brief presentation of themes in Meyer's teachings. Interpretation of fencing treatises is a very subjective process, but this is the paradigm that I find useful for understanding and applying Meyer's teachings.

VERSATILITY

Most fundamentally, the fencing art that Meyer taught was meant to create versatile fencers. It was one system, working with particular principles, but applied to many different weapons, with techniques and principles that often, but not always, extended over all of them, together teaching you to fence with all contemporary weapon types, including large knives, swords, spears, and more. Essentially, you were taught to fence with any weapon at hand, or just your own body, including the use of what today would be considered "dirty" tricks, like attacking the testicles and eyes or breaking fingers. This fencing was intended to be performed on foot or on horse, in and out of armour.

WEAPON GROUPS—ONE TEACHES ANOTHER

The structure in this, his most complete and complex treatise, appears to be the same as what the fencing guilds followed in their teaching progression, and is roughly the same as in Paurenfeyndt's treatise. In order, it teaches sword, dusack, rapier, rapier and dagger, dagger, grappling, staff, halberd, and pike. The sword lays the foundation for everything, and especially the dusack. The dusack, in turn, lays the foundation for the rapier, and the rapier then ties it all together. The staff likewise lays the foundation for the halberd and the pike.

All of them are also weapons in their own right, used both in civilian and military context, still a hundred years after Meyer's time. In the case of the sword, which became more and more specialized in the larger shape of the Schlachtschwert (the giant two-hand sword), it was used in outnumbered scenarios to protect dignitaries, banners, and cannons, and used as late as in the late 17th century by town guards, as well as carried by ship captains fighting as the last man standing on deck. Still, in Meyer's time it is not uncommon to see middle-aged men depicted with two-handed swords[258], and even forty to fifty years after Meyer's death we see bodyguards carrying complex-hilted two-handed swords protecting dignitaries and men fighting in the fields or cities swinging swords.

---

[258] The examples are numerous. See for example *Hofkleiderbuch (Abbildung und Beschreibung der Hof-Livreen) des Herzogs Wilhelm IV. und Albrecht V. 1508–1551.* See also the depiction of the Battle of Polotsk, by G. Mack, Nuremberg 1579, which appears to depict German mercenaries with two-handed swords on their hips. See also the complex-hilt two-handed swords carried by bodyguards in Mattheus Merian d. Ä.'s 1610 depiction of the funeral of Charles III, Duke of Lorraine in 1608. Likewise, see his Beisetzung Karls-III in der- Franziskanerkirche zu Nancy, 1611AD, his Zug Herzog Heinrichs II von Lothringen zur Kirche and his Vier Gruppen zu Fuß, Les Eglises Parochiales.

DIFFERENCES BETWEEN THE WEAPONS

While the individual weapons are used according to shared principles, they are also all different, requiring different handling, and the principles are embodied differently in different techniques. Underneath of it all, at least at higher levels, lies a core understanding of what connects it all, and what parameters are at play when working with the principles. Here, fundamental terms like Vor ("Before"), Nach ("After"), Fühlen ("feeling"), Sterck ("Strong") and Swech ("Weak"), Hert ("Hard"), Wech ("Soft"), Indes ("Inside"), and Nachreissen ("Pursuit") are central to the understanding of leverage, "feeling" with the weapon, and timing, terms which are also elusive and hard to grasp for the beginner, but understanding and application comes with (daily) practice and study.

BODY AND WEAPON MECHANICS

While the manipulation of bladed weapons can be fairly easily forced using muscle power, this is not true for the polearms, where instead one simply has to use proper body mechanics, using posture and "crossroads", which are essentially key stances which allow for easy changing of the motion of the weapon. This particular way of moving, in my opinion, runs underneath of all of the teachings in the treatise, not just the polearms, including sword, dusack, rapier, and dagger. It is a single, cohesive system and way of moving in fencing, regardless of weapon, meaning you move more or less the same way with the sword as you do with the halberd. It is quite possible that earlier masters also taught this similarly, but few earlier sources are as explicit or as pedagogical as Meyer is with his treatises, and therefore, we cannot tell as clearly with those. Looking at certain sources however, like the treatises of Hans Talhoffer and Albrecht Dürer, would seem to indicate strong similarities even in this particular area.

ECONOMY OF MOVEMENT

Quite characteristic of Meyer's system, and often misunderstood, is the economy of movement. While certain strikes can appear to be wide, they are not meant to be anything but economical and fast. Furthermore, all predefined and named stances are connected through actions: through cuts, strikes, thrusts, and through using different edges of the bladed weapons. All actions, in turn, are connected through the stances. This means that as you attack, defend, deflect, or have your own attack deflected, you should commonly strive to return to the closest stance, letting the weapon move by itself towards it, switching between cuts, thrusts, and deflections as suitable and needed. This preserves your stamina and strength, while also saving you time for a counter.

WEIGHT SHIFTING, FOOTWORK, AND EXTENSION

Absolutely central to learning how to move this way is to learn to use weight shifting. This is similar to how you move when you shovel snow, reaching back, and then scraping forward, ending by throwing the shovel and snow over your shoulder and side. It is also similar to how you move with a scythe, shifting the weight between the left and right foot. With Meyer, however, you do this in a low stance, and instead of from side to side, the shifting is front and back, with the feet commonly kept in an L-shape at 90 degrees.

This weight-shifting is also part of the footwork, where the rear leg is extended, but where the foot is kept on the ground, accelerating the whole body for as long as possible, before the foot actually leaves the ground, much like motocross riders sometimes break just before the jump, as once in the air, acceleration is no longer possible. It also provides very good stability, keeping your feet in contact with the ground throughout the movement, significantly lessening the risk of slipping or stumbling.

Pay special attention to the floor tiles and shadows in the artwork, as they teach you how to angle your feet, and when they touch or leave the ground. Keep in mind that the perspective twists the 90-degree angles in the artwork.

CORE ROTATION

To be able to move in a controlled fashion, you need to be able to move in balance, and this is where the concept of core rotation comes in. Simply put, for a large part of the movement, when you step, the torso rotates around your core, as if a pole ran through your skull and whole torso. Doing this correctly, allows you to fence in any direction of the compass, even spinning 90, 180 or 360 degrees.[259]

TWISTING—*FIGURA SERPENTINATA*

Another central aspect to the body mechanics is twisting, where the leading foot is twisted out on the heel, as the rear heel is raised, leaving the toes in the ground. This twisting enables you to extend the rear shoulder further forwards by several inches, compared to when leaving the foot straight forward, thereby giving you longer reach towards your opponent, while also liberating your whole torso for freer movement with the

weapon, as you pass from, through and to the various stances. This also prepares for the step, as the twisting of the leading foot provides better stability, setting it near the angle it should have at the end of the passing step done with the rear foot. Again, this manner of moving appears quite similar in the fencing treatises of both Hans Talhoffer (especially HTK and HTM) and Albrecht Dürer (AD).

LEANING AND TILTING

At the end of the various steps you make, you can complete the weight shifting by leaning your whole torso forwards to reach farther, or back or to the side to get a better angle or to

---

[259] 270 degrees is generally not needed, as instead you just twist 90 degrees in the other direction. Also, a step backwards or forwards might be necessary.

protect your upper body, but always only as far as you can with maintained control of your body. Never go lower than you can manage in real combat, but train to go as low as needed in real combat. This is specifically advised by Meyer. The head, in turn, can also be tilted away from the opponent's weapon, as it is a common target, and therefore of course should be kept as safe as possible.

GRIPS AND GRIP SHIFTING

Another fundamental skill taught in the system is varied gripping and grip shifts. This is too complex and varied a topic for the different weapons to go into here, and I will therefore instead refer to close study of the artwork in the various sections.

STRIKING AND STRIKING LINES

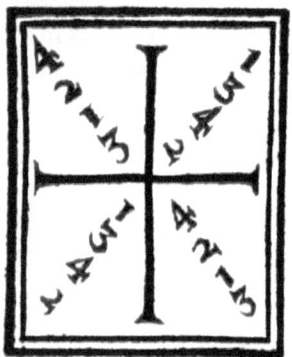 The most fundamental movement and exercise to practice is cross-cutting, cutting in the shape of an X, from above and below. This can be done with all weapons using both "edges", and even the flats of the weapon, and as half cuts to a stance extended straight forward, or fully, all the way up or down, into the stances Ochs and Pflug, or their equivalents, on both sides. To this, one can add different, specialized strikes. Meyer here also includes examples of striking sequences, both with illustrations and text, which more subtly also teaches us the fastest way of covering the openings we reveal as we perform a strike.

STANCES AND GUARDS

Regardless of weapon, there are a number of stances and guards that are used, and from which you learn to perform different techniques, so that you can respond to an attack wherever you find yourself in your movement between attack and defence. The polearms in particular are distinctly asymmetric in the use of guards, meaning the left and right stances can be distinctly different despite having the same name.

Important to note here is that we are not meant to remain in a particular stance for very long, only doing so for the "blink of an eye", as we consider what to do next. Ideally, the decision is made already before arriving in a new stance.

THE BEGINNING, THE WAR AND THE WITHDRAWAL

Simplified, the concepts of Zufechten, Krieg, and Abzug revolve around the entering of the distance where at least one of the fencers is able to attack the opponent. This is commonly done with long strikes or thrusts at the furthermost distance. At closer distance, the techniques that are suitable changes. For instance, Meyer here mentions how the Zwerchhauw makes up half the fencing. Finally, once you have successfully landed an attack, you also have to withdraw safely, expecting your opponent to try to attack you as you do so, preferably controlling the opponent's weapon as you do so. This would be true even for fencing in earnest, as even a lethal attack might not kill or stop the opponent instantaneously.

TARGETING

Also closely related to striking lines is targeting. First of all, the opponent is divided into four quarters: left and right, high and low, with the horizontal line placed roughly at heart height. These are called the Four Openings. Similarly, the head is also divided the same way, with the horizontal line placed below the eyes, right where the neck starts.

Much of the targeting taught with bladed weapons, aims at three less protected areas: the neck, the flank below the ribs, and above the knee. Naturally, the power of the polearms means that even with armour, the impact of a strike or thrust would be massive. Common for foot soldiers in his time period was half armour, where the lower legs were unprotected and therefore good targets. The face and the hands too were commonly unprotected for practical reasons, and consequently, the two were also important targets.

DECEIVING, PROVOKING, TAKING, HITTING

Quite central to Meyer's style of fencing are the words Reitzen, Nehmen, Treffen ("Provoking, Taking, Hitting"). Simply put, this means that you can provoke your opponent with an attack using your weapon against an opening, or using body language, to cause your opponent to move in a certain way and direction, letting you exploit the opening that is created as the opponent does so. Other strikes and thrusts displace the opponent's weapon, using power, structure, angle, or leverage to gain an advantage. Yet other strikes simply hit the opponent. A strike may also do several of these at the same time, displacing the opponent's weapon and hitting him or her in the same motion. A feint which the opponent does not aim to defend is completed as a hitter. Of course, to students of earlier masters, this emphasis on deception is largely new and unfamiliar, but vital to understanding Meyer.

TACTICS AND PSYCHOLOGY

Unlike Pietro Monte and his 1509 treatise *Collecteana*, where he divides fighters and their mental and spiritual natures by the four humours: blood and sanguine, active and enthusiastic; yellow bile and choleric, angry and uncontrolled; black bile and melancholic; and phlegm and apathetic, Meyer defines fencers by a seemingly more practical list of characteristics, roughly here translated as:

1. overly aggressive (and a bit stupid) fencers;
2. inexperienced but artful opportunity-seeking fencers;
3. safety-first fencers, who only take a target when they are sure to have it and know they can retreat safely;
4. passive and apathetic fencers, who just wait for the opponent to act.

Meyer favours the third style in his own fencing, however, stating that:

"The Third don't cut at an opening if they are not assured of it; instead, they pay more attention as to whether they can recover from the extension into the cut safely back into a counterposture or to Defensive Strikes. (I mostly hold myself with these [fencers], but it depends on my counterfencer.)"[260]

Now, Meyer teaches us what approaches and methods to use against all these for types of fencers, but not only that, he also tells us that we should learn to appear to belong to all four, in order to deceive our opponent, and depending on the opponent's particular nature.

"Now, as the First [are] impetuous and perhaps foolhardy and, as one might say, unreasonable; the Second cunning and sharp; the Third cautious and deceitful; the Fourth are similar to fools. Therefore, you must yourself mimic and act appropriately according to all four, so that you can deceive the opponent—perhaps with impetuousness, perhaps with cunning, perhaps with cautious observation, or also provoking with foolish body language. You thus mislead and deceive them with regards to their intended sequence, and you also make space and expose openings so that you can more safely touch and hit them."[261]

A general guiding principle for all fencing is to go with your weapon from the opponent's body to the weapon, controlling it so you are kept safe from harm. Likewise, you move from the weapon to the body, applying pressure to your opponent so you can exploit a given or created opening. Finally, you should make sure to retreat safely, commonly with a threat, like a strike.

Another guiding principle is to go from a strike to a thrust and a thrust to a strike, again sticking to economy of movement, going a short and fast route to your target.

Finally, Meyer adds a whole layer of tactics and strategies, working with psychology and mind games, something which is often missing from earlier treatises.

To sum up his tactical and strategical advice, we see the following:

- **Attack first.** Take the initiative and don't wait for the opponent to attack you. It is better to force him to respond to what you do than the opposite.
- **There are three different types of attacks: Provoking, Taking, and Hitting.** Most exchanges include at least two of these, in any order. Some strikes and thrusts take (displace) and hit at the same time. Others just perform one of these functions at a time but can be used in all three ways. Provoking is a way of controlling your opponent by giving him an idea about what is "right" to do. That way you are ahead of your opponent already and can treat him in whatever way you like.
- **Transform a cut into a thrust and a thrust into a cut.** It is quicker and better to move the shortest or quickest way. From a cut, it is better to keep the point in line and thrust from the inside or the outside. However, from a failed thrust it is often better to use the momentum of the opponent's parry to strike around, especially for one-handed weapons.
- **If he tends to cut wide: void and counterattack with quick cuts or thrusts.** For opponents who strike overly hard and wide; just step out of range and as his sword is overextended, then attack quickly to whatever body part you can reach.

[260] Meyer (1570): II.99.
[261] Meyer (1570): II.99–99ᵛ.

- **If he tends to void and counterattack: provoke and feint.** Likewise, if you face an opponent who likes to fence like the above, then instead work with provocations; threaten with thrusts and strikes in, or slightly out of range to force the opponent to move in a certain fashion, thus revealing a new opening that he will find it hard to defend quickly, which you then attack.

- **Always cut from the sword to the body and the body to the sword.** You always have to be ready to control the opponent's weapon and body after you land an attack, since you can't trust your attack to incapacitate him immediately. Likewise, as you strike or parry, you will be revealing an opening that is difficult to protect quickly. Consequently, this is the opening that is the most attractive for your opponent and you have to be ready to protect it.

- **Withdraw with a threat.** Following from the above, you also have to make sure that you not only bind your opponent's weapon after a successful attack, but also that you withdraw in a manner that discourages the opponent from attacking you, either with extended sword, a thrust, or a strike.

- **Don't do things you can't do in a real, dangerous situation.** Meyer states that we all think differently and thus fences differently. There simply isn't a single best fencing style and there are as many styles as there are fencers. Not only that, but he also discourages us from using techniques that are difficult for us in a situation where we need to rely on them, since not all techniques are for everyone. Our bodies and lack of training will hinder us in executing some of them properly, which means it would be suicidal to use them in a real fight.

***

With this, I end my first attempt at sharing in print my thoughts and ideas from a decade and a half of intensive studying of Meyer's art of fencing and the times and places he lived in. As with everything, this is a snapshot of the current state of things, and it is likely to change with time, as new discoveries are made and new epiphanies are had.

Again, I feel incredibly grateful to a whole lot of researchers and fencers whom I have met and crossed blades, staves, and arms with over the past fifteen years. The memories are plentiful, and fantastic, and will remain with me for the rest of my life. I hope what I have put down on paper here will help and provide a useful contribution, despite whatever flaws I have been unable to clean out.

Thank you.
ROGER

ABOUT

Until 2020, ROGER NORLING was an instructor on Joachim Meÿer's 𝔥alben 𝔖tangen (Quarter-staff) with the Gothenburg Free Fencer's Guild (GFFG). He started with the Gothenburg Historical Fencing School (GHFS) in 2008, and he was a member of the GFFG from 2015. His main focus in his research is the 𝔎unst ðes 𝔉echtens and primarily the sword, dussack, and polearms.

He has been focusing on the works of Joachim Meÿer since 2009. In this he has enjoyed collaborating with the Meyer Frei Fechter Guild (MFFG). In May 2013, he became a 𝔉echter of the MFFG; in 2016, he received the rank of Research Scholar of the same; and he was finally appointed as 𝔘nterhauptmann in 2019.

Currently, he is writing several books which will explore the teachings of Joachim Meyer as well as pedagogics for teaching martial arts. He is the creator behind the three sister sites *HROARR.com; Water on a Rock*, an online journal on philosophical ponderings; and *Northernbush.com*, and shares his experiences and knowledge in articles on both sites.

Johan Kasimir, Count Palatine of Simmern, leading a Calvinist army past Straßburg to support the persecuted Huguenots of France in 1576. By Frans Hogenberg, 1576–78.

Translator's Introduction

I sat my first Middle High German seminar in 1984. It was an elective, alongside lectures and seminars in other periods of German literature. But while the others were interesting, I fell in love with the language, history, and literature of the medieval period. I loved reading texts that were hundreds of years old; hearing that poetry and learning about the customs and history. I found joy in crusader lyrics that spoke of duty and loss, canon epics and romances that depicted single combat and warfare, courtly poetry that claimed to present both sides of a love affair. I followed that love to graduate school and a doctorate.

The internet of the 1990s was in its infancy; the dictionary of Early New High German, started by the brothers GRIMM in the 18th century, had been finished on typewriters and then computers by scholars at the University of Dresden; it was not hosted online until the 21st century. Medieval manuscripts remained in their libraries, where only vetted scholars were allowed to read them—photography was strictly limited and facsimiles were prohibitively expensive. One of my most memorable fangirl moments was sitting in Wolfenbüttel at the desk next to the scholars examining the recently returned *Evangeliar Heinrich des Löwen* ("Gospels of Henry the Lion"), at the time the most expensive book in the world. This was my world and I reveled in it: researching and writing about medieval literature, and especially the niche I had found in reading texts by women writers, most of whom were religious women of various denominations and degree.

Physical sword fighting came later (the *Nibelungenlied* and its many battles came first), and not through HEMA or the SCA but through stage combat. I confess that I like the story of fighting more than actually getting hit, and stage combat offered a means to practice with swords and other weapons. It was a hobby, play-fighting where all participants knew that it wasn't real, and it didn't intersect my professional work until 2006, when I was invited to join one of the HEMA translation groups.

The Cambridge Historical European Martial Arts Study group (CHEMAS, now SCHE-MATA) met at MIT, and I joined by Skype. I contributed expertise in medieval German language, linguistics, dialects, and paleography, and learned about historical combat in exchange. Through them, I met my partner in Latin crime, KENDRA BROWN, when we worked on a first draft translation of the *Florius de Arte Luctandi* (FLP), which was published in 2015 and which we are currently revising, and subsequently sections of the writings of Paul Hektor Mayr (primarily PMM).

Through all these textual journeys, my focus remained on the language and a translation that faithfully reflected what the medieval authors had written; not what I wanted or expected, but what the words meant; not as read through a lens of modern fencing, but through the context of the described actions. If I have done my job correctly, the reader will not be aware that these texts share the same translator at all.

This has only intensified during my work on this translation, which is the result of numerous dives through dictionaries: tracing the linguistic developments of spellings, contexts, and meanings; juxtaposing verb and noun forms; comparing the interpretation of verb stems with the meanings of their various separable and inseparable prefix forms; using other contemporary translations to gain greater understanding of the actions described; following the evolution of technical terminology across geography and time. Once I felt I understood what the Middle High German (MHG) or Early New High German (ENHG) terms meant, I sought

English vocabulary to convey the same content. These were often not cognates, as German and English have many 'false friends' that don't share meanings and were also often different from modern equivalents since the languages have evolved differently. In addition, words of equivalent meaning might convey different levels of urgency or timing, or carry different types of baggage, or have links to modern jargon that obscures older meanings. Consistency of translation can also be problematic: for example, the German term krumm/ krump/krumpf can mean 'crooked', 'bent', 'curved', 'arched', 'arced', etc. in English, depending on context. The edge of a blade described as krumm and the Master Cut krumphauw can't be translated with the same English term, since a 'curved cut' is a meaningless description while a 'crooked edge' would describe, not the curve of the dussack, but an edge out of true.

Another level of difficulty in this translation were the two, somewhat contradictory stylistic requests for the English: that it remain as close as possible in tone, rhetoric, and vocabulary to the original, and the sequences should be easy to follow with clear instructions and inclusive pronouns. Thus, the German references to a singular man or Mann are consistently translated as 'opponent', and Meyer's various means of referring to the person on the other side of the weapon are translated to match his terminology.

The result is a text that varies in rhetorical style following the original, in which, for example, the introduction to the text is far more florid than the explanatory sections. Since German does not have a grammatical concept of run-on sentences—most of the sequences are a single sentence in the original—I have employed English grammatical rules for fluency. However, I have maintained sentence fragments in some cases in order to specifically and rhetorically mark that a statement is not part of the instruction.

These decisions highly alter the original text as-printed, which appears as justified blocks of text with paragraphs indicated by indentation, by an empty line, by tabs within the text block, or merely by Item.

The sequences are presented as bulleted texts, with the initial starting point set in-line with the left textual margin and the first movement from that stance as the first bullet point. Movements by the opponent are in *italics* to clarify who is doing what to whom, or who is reacting to what action. The pronouns are the inclusive 'they/them', both to recognize instruction in groups, and to include everyone who wishes to participate. Exceptions are explained in the footnotes.

It is my greatest hope that you, the reader, find this text useful. It is, after all, a Sachbuch: a technical text meant to be used. I don't expect everyone to agree with my translation choices. There were several readers and editors involved in creating this text, yet in the end, any errors in the translation are mine. I can only hope that they are few, and do not distract from Meyer's incredible contribution to fencing knowledge.

Dr. REBECCA L. R. GARBER, Ph. D.

A FEW BRIEF NOTES ON SOME SPECIFIC TERMINOLOGY

The only fencing term that is not translated is **Indes**. This is because Meyer has a three-paragraph explanation about what the term means to him, which appears on page I.25 in the Sword Section. **Indes** is a small word fraught with meaning. In that moment in which you strike, you should simultaneously observe the other openings that you could also hit, and also all of the sequences your opponent can use to strike or counteract your action. It is therefore a point of hyperfocus on what you are doing and what you and your opponent could be doing at the same time and afterward. Meyer states that he will use **Indes** as a shortened form to call all of this to mind.

Within his text, Meyer tends to use **Indes** to indicate a disruptive action—that is, because he is describing actions sequentially that actually occur simultaneously or disruptively. Thus, *while* you or your opponent are performing the first action, execute the second action *at the same time* and *in the same space*.

Meyer distinguishes **Indes** from **indem**: the latter indicates either that moment in which the opponent executes X or that moment in which you observe or perceive Y. **Indem** is thus a point of timing your immediate action, instead of a disruptive act that is always already supposed to have taken place while the preceding action was happening.

Versatzung, the noun (as distinguished from the verb **versetzen**), is also a difficult term.

At this point in history, the word **versetzen** amalgamates **vorsetzen**, **fürsetzen**, and **versetzen**, with the different prefixes indicating the meanings of "place in front", "replace/substitute", "move in error", and "move from one place to another", all of which makes translating **versetzen** akin to interpretive dance. In fencing treatises, **Versetzung** is commonly used as a response to an action, and while this response usually places the sword somewhere in front, the response itself may be defensive or offensive. "Counteract" is an English term broad enough to cover all of the possible movements that works in context and includes the concepts of (offensive) defense.

Meyer uses the term **Versatzung** to refer to a previously-mentioned stance or as the noun equivalent of **versetzen**. In the Royal Armouries MS 1.33, **Versatzung** (*obsessio* = barricade/obstruction) is the response to the **Hut** (*custo* = guard). In Paul Hektor Mayr's Latin translation of Lew's gloss, **Versatzung** is translated with '*permunitio* = completely-walled, defensive position' or '*castrorum* = defended military encampment'. Therefore, **Versatzung** by itself, as a reference to a specified or unspecified stance, is understood here as a defended position in response to a particular action by the opponent, thus a counterposture that one adopts in the middle of a sequence.

With regard to a guard fully introduced in the Dusack Section: the **gerade Versatzung** is a starting point and thus not a response. As it is linked in form and vertical movement to the **Bogen** (or "Arch"), I also looked at architecture to see if there was a similar impetus or imagery. The terminology I chose, "Simple Brace", comes from the diagonal braces used in half-timbered building construction, which are called **Versatz/Versatzung** in German and Brace in English. The entire architectural structure that is a **Versatzung** includes a vertical post, a horizontal beam, and a diagonal connection as a support or brace. A **gerader Versatz[ung]** has no additional offsetting wedge to affect its angulation, thus it is 'simple' as opposed to complex. Both the Arch and the Brace reflect common architectural terms that would have been known by Meyer's contemporaries.

While Meyer refers to fencing itself as an art (𝔎unſt), he refers to the action of doing so as "work" (𝔄rbeit), not "practice" (Übung). 𝔄rbeit specifically referred to manual labor, as distinguished from mental or spiritual labor. He thus labels the central part of a fencing encounter (between the onset and the withdrawal) as 𝔥andarbeit—that is, labor with one's hands or handwork. I have maintained Meyer's term throughout my translation.

Finally, I have rendered 𝔷edel as 'didactic poem'. Historically, a 𝔷edel was a written instrument; in MHG it was specifically a legal instrument—that is, a record or complaint/pleading. ENHG both reduces and expands this original meaning. In the judicial sense, it is reduced to information presented as a list that was less informative than a legal brief. However, the concept is also expanded to include an abbreviated or abridged way of communicating that is contextually comprehensible (the implication being that it is incomprehensible without this context) and written using terms that are limited to specific locations or to specific technical language.

The fencing 𝔷edel (recitations, didactic poems) meet this definition in that they often contain lists, they are highly abbreviated or abridged, they are contextually comprehensible, and they use highly specified terminology.

|a i| **Foundational Description of the Free, Knightly, and Noble Art of Fencing, with all commonly used weapons,** decorated with many beautiful and useful figures, and presented by Joachim Meyer, Free Fencer in Strasbourg,

in the year 1570.

By privilege of his Roman Imperial Majesty: not to be reprinted in any fashion for ten years.

|a iᵛ| [In a banner, initialized:] *God knows the time*

|a ii| **To the illustrious, high-born prince and lord, Lord Johan Casimir, Count Palatine of the Rhine, Duke of Bavaria, my gracious Prince and Lord.**

Illustrious, high-born Prince, my completely voluntary service, to my utmost ability, is offered as a vassal to Your Princely Grace. Gracious Prince and Lord, whereas the knightly and noble art of fencing has devolved in current times due to many and various reasons, yet there is no doubt that the greatest and most eminent cause, among others, is namely the recent emergence of the harmful gun, which has taken the upper hand to such an extent that the most virile and courageous heroes are robbed and deprived of their lives by the same, sometimes even by the least and most timid, and at times both friends and enemies are unintentionally injured and aggrieved.

Therefore, it is no surprise that this practice of free knights[1] has not merely declined, but has even become somewhat contemptible and not lacking in prejudice toward the age-old, praiseworthy custom. Elsewhere, this aforementioned cause is claimed to be sufficient and occurs among knowledgeable fighting men, principally because, in the case of guns, nothing functions correctly without other equipment, weapons, and arms,[2] as occasionally the entire battle must be stopped due to such fragile weapons and handguns (because the guns cannot be used due to some reason), as knowledgeable fighting men attest.

|a ii^v| Therefore, it is the case that, in addition to the gun, other equipment, weapons, and arms useful in war are necessary, in current times as well as for our ancestors. And also, as is known to many, this includes not only good equipment, weapons, and arms (like harness, breastplates, swords, halberds, pikes and the like) but even more so daily practice, which can be correctly and skillfully used to the benefit of oneself and to the harm and disadvantage of the enemy. Therefore, it will be completely and absolutely necessary to learn this, as daily experience [shows] that for many, even when they are provided with the best, their equipment, weapons, and arms become more of a hindrance than a protection of their bodies and lives if they do not know how to use them nor how to thoughtfully defend themselves.

And because this knightly art (and indeed all arts) is difficult to grasp and learn correctly and productively as an adult, therefore persons who embody intelligence and understanding among all peoples, and at every time in history, have devoted themselves so that their young people may also, in addition to other free and good virtues associated with honesty and virility, be instructed in this knightly art—how one should skillfully use all types of equipment and weapons, both on horseback and afoot according to need—as the old, credible history books of all peoples clearly and lucidly report and reveal, especially the Roman histories. From this, it should naturally follow that many brave heroes of knightly mien and true defenders of the Fatherland arose and were trained by those peoples, and thus the usefulness of applied efforts— while they are still young and before they have completely reached their age of adulthood— prevails. Such is especially apparent in the case of Scipio Africanus, |a iii| namely that when he was still young and 18 years old, he rescued his father—the governor and supreme field marshal—from enemies and kept him alive with particular skill that he gained from this knightly practice, in a battle that took place against Hannibal at the Ticino River.

However, no proof is needed that it was customary for our ancestors and the old Germans to raise their children in knightly practice, in addition to other good arts, because this is intrinsically obvious as proven by their work. This is because, after the Romans opined that they had conquered the entire world and, as a people living in safety, that pleasure was to be considered more important than good arts, policies, and knightly practices, it was this view that led to the entire empire being attacked from everywhere and destroyed by enemies. It was then that the knightly Germans were called and promoted before all other peoples, to rescue [the empire], embrace it, and re-establish it.

---

[1] *Freier Ritter* is a class indication for a low class of nobility who could fight with swords and originated from the *ministeriales*, who were initially unfree, trained servants and bureaucrats.
[2] Meyer appears to be using *gewehr* in its collective meaning of weapons or arms that are not projectile weapons. He is further distinguishing between 'traditional weapons = *waffen* = swords' (and polearms, etc.) and '*geschütze* = guns'. *Rüstung* can mean armor, but can also mean all of the equipment required for a fighting knight or man at arms.

This would have not occurred at all were the admirable Germans not practiced and experienced in all types of knightly play and the business of war, in addition to good policies, as is clear in the mighty deeds of many unconquerable German heroes such as Pepin, Charlemagne, Louis the Pious, and Henry I. These heroes were always adorned with well-practiced knighthood to such an extent that the previously-mentioned and praiseworthy Emperor Henry I, after he triumphed and emerged victorious from a dangerous battle due to his knightly Germans, did not merely marvel at their well organized and practiced knightly ways but also considered, from his true imperial nature, how to essentially maintain the same, not only at this level (as [the Germans], in their diligence, came before their Emperor to promote the same more laudably), but also how it would be extended and anchored in their descendants.

|a iiiᵛ| Therefore, he established the true school of knighthood, namely the praiseworthy German Tournament³ at Magdeburg, and thus left [this inheritance] to their descendants to maintain the same. Although the abovementioned tournament field has been removed, perhaps due to serious reasons, to the courts of many of our laudable German princes, tournaments are still held today that are not without some renown. This is all for the goal that young adolescents, high and praiseworthy princes, counts, lords, and knightly nobles would be practiced in all knightly endeavors, on horse and afoot, and (as the saying goes) "spurred and whetted to a point", so that in times of need they would be able to support the common Fatherland even more conducively—and also save their own bodies and lives so much better and, in contrast, impressively destroy the enemy (as the practiced always overcome the untrained).

Gracious Prince and Lord, because my mind and thoughts are so disposed that I might demonstrate my obligatory service to the common Fatherland, even with the minimal talent which the Almighty graciously granted to me, since I (without any fame to speak of) not only learned the noble knightly art of fencing from skillful and famous masters, but also have now practiced [this art] for many years and have instructed some young princes, counts, lords, and those of noble descent in it. And it has been requested multiple times, graciously and sympathetically, by you [and other] princes, counts, gracious lords, and noble [men], that I would draft the abovementioned praiseworthy art of fencing in a certain order and then publish it for public consumption as a printed text, so that the same [art] is revealed for the use of many people of our nation. Therefore, as I ought not set myself to resist such gracious and sympathetic entreaties very long, and have thus, in the name and through the fatherly grace of the Almighty, compiled all that which I have learned and experienced (with work and effort) in this aforementioned praiseworthy art over many years and included it in this treatise in the most comprehensible organization that is possible.

|a iv| It is my comforting hope that it should thus be good and quite serviceable to many persons of high and low standing, who have love, joy, and desire for the art of fencing, while taking under consideration the fact that, to my knowledge (without intending to diminish anyone), the same has never been published in the German language.

---

³ *Teutschen Turnier.*

Gracious Prince and Lord, I have undertaken to dedicate this work to Your Princely Grace as your servant, for which I have many types of significant reasons. Primarily, however, is this one:

Firstly, it can be seen in the same way as with Your Princely Grace, that our German nation should commendably have a reliable, cautious respect for the same (as to an especially courageous prince), as Your Princely Grace has shown and proven yourself to be—virile, princely and generous—even during your younger years in the arduous French wars of that time, about which I would make extensive reference to reports from reputable people.

Secondly, that I place no doubt upon the fact that, in addition to the appropriate subjects and other good [liberal] arts to which Your Princely Grace was raised in a princely manner (and with the greatest of diligence by your beloved Lord Father, the Serene, Nobly-Borne Prince and Lord, Lord Frederick, Count Palatinate on the Rhein of the Holy Roman Empire, Arch Seneschal and Electoral Prince, Duke of Bavaria, etc.), you, my Gracious Lord, were also instructed in this praiseworthy art of fencing and have not a little experience in it. And therefore, you will be able to graciously and best judge this, my negligible but nevertheless loyal and diligent work, from your elevated and intrinsic Princely understanding.

Due to this, and other more significant reasons, I neither should nor would want to rightly seek another patron for my work than Your Princely Grace. Therefore, Gracious Prince and Lord, I present to Your Princely Grace this, |a iv^v| my work and foundational description and explanation of the quite inspiring knightly art of fencing, compiled with all diligence and through long experience. I, your vassal, request with the greatest of diligence that Your Princely Grace would accept this and would take it up from me with all graces, and also be and remain the high Patron of the same (as Your Princely Grace remains my own), and my Gracious Prince and Lord. I humbly beseech the Almighty from my heart, that he will omnipotently grant Your Princely Grace, including the entire Electoral house of the laudable Palatinate, long-lasting peaceful government, and to all, temporal and eternal health.

By Your Princely Grace, graciously commanded to be your vassal. Dated Strasbourg, 24 February, Year 1570.

Your Princely Grace,
                Your vassal and obedient
                         Joachim Meyer,
                         Free Fencer[4] and citizen of Strasbourg.

---

[4] Grimm offers only *lanista privilegiatus* for 'Freifechter'. A *lanista* was a trainer of gladiators, not a gladiator himself; however, he was often a retired gladiator; *privilegiatus* is someone granted the privileges appertaining to a university degree, a clerical vocation, etc. Scherzii glosses *frie/frey* as *adel* (or noble), belonging to the nobility. Meyer uses this term as a title here and on the title page but does not use it elsewhere in his introduction to claim expertise and instead explains how he gained his knowledge and experience.

Foreword to the reader

Because the free knightly exercise and art of fencing has, up until now, not received any particular focus while, at this same time, all other liberal arts have been described and have received consideration to an extent that they have ascended to the very heights, there is no doubt that this can be ascribed to two reasons.

In the first place, it is due to the fact that this knightly art must be seized with both fists and learned through the cooperation of the entire body—that is, it must be taught more through physical experience than from books. Indeed, this reason held me up for a long time, and in weighing the great deal of effort and expense, I almost came to withdraw. However, it appears to me that many people who love honor are also engaged with this for these serious and moving reasons:

The first aspect is namely that this art (which was not prestigious yet still received great consideration) must be learned primarily through exercise of the body, but it is also certain and true that this art, just as many others, is better imagined in the thoughts of learners if they have it written out and placed before their eyes (in addition to good instruction which is set together in the correct order). Consequently, it can also be taught and understood even better through exercise of the body than if it were explained in poor, oral narratives and received piecemeal.

Secondly, the intellectual temperaments [of the students] are not so overburdened by retaining this through manifold considerations, and they can use the time (which they would otherwise waste with such efforts) to turn to their other subjects of study.

Thirdly, those youth, as they grow older and after they have learned from a true master—yet do not have the [master] by them at all times—can be reminded from this [book] and practice each day at the time they have scheduled, so that that which they have learned does not soon lapse from their attention (or as this then commonly occurs, that the larger part is forgotten).

Therefore, this [book] can provide no minor service and benefit for them—and especially for young lords and for others of noble birth, to whom this knightly art is suited more than for others and which they ought to learn.

The second reason is this, namely that this knightly art of fencing can only be written about in books or recorded in written format with difficulty, as it can only be directed in action through the practice of the entire body. Because I have experienced this reason myself, I consider it to be the most important and significant (as others of great comprehension can judge), |b iᵛ| and as much as the first [reason] demands, I, like others, must acknowledge that (as mentioned above) every art can be understood with less effort when written out in a well-organized manner, and can be grasped by those learning through practice of the body or the hand.

Yet I have no doubt, if anyone had revealed this art and described it in a comprehensive and good organization before our times, then not only would the noble art not have fallen off in the case of so many, but also many abuses would have been completely avoided ([abuses] that must now be forcefully eradicated).

Therefore, I remain in the hope that regardless of whether my writings may be considered by some few as being of little worth, many honest journeymen and young fencers will step forward and not merely keep themselves with diligence and guard themselves from the disordered life—gluttony, drinking to excess, blasphemy against God, cursing, whoring, dicing, and all of that—through which this noble art has been fouled (since this knightly art has been used by many as a cloak to hide all types of vice and sloth, which many honor-loving people, and all honest fencers, additionally lament to the highest heavens). Instead, [that they] give thought to foundationally understanding this art and learning to use [it] in correct, honest seriousness, in order to free themselves from the bonds of useless, boorish tumult, and thus to endeavor in all virile, proper behavior and respectability so that, when they themselves have learned this art correctly and well and they lead an honorable life, they can subsequently stand before others—and particularly the youth—and may thus be considered to be hard working in their service.

If I should then see and sense that my writing were to find a place with some, my efforts would be not merely eased by this fact, but indeed, should instead be provoked to bring more to the light of day (according to my little understanding).

And in order that these young fencers may thus better prepare themselves to knowledgeably use this book, I will additionally state that I compiled a short summary at the beginning in the first chapter of each of the first three weapons, describing the organization in which this same weapon can thus be properly taught and presented, and also placed before their eyes the entire art of fencing in a short summary, which then should serve to increase the understanding of this book.

And the entire art of fencing stands primarily on two principals. The first is contained in the cuts and thrusts, with which cuts and thrusts you intend to injure and lay out your enemies. |b ii| The second principal is the counteraction, which is a teaching about how you should rebuff and beat outward those aforementioned cuts when they are guided against you or have been directed toward you by your counterpart.

Now, regarding the first primary technique pertaining to the cuts: you should know that, regardless of how many cuts are presented and taught, there are actually intrinsically no more than four Primary and Principal cuts, from which the others all derive and have their origin, namely: the High Cut, Wrath Cut, Middle or Crossing Cut, and the Low Cut. And no cut, however odd, can be created that is not contained within these, because even though many names are used for the cuts (such as Change, Plummet, Squinter, Crooked and the like), they are not completed outside of the scope of the four. For the Plummet is, in itself, nothing other than a High Cut; however it is named thus because in the cut, it plummets over at the head. Likewise, the Change is also named thus because the sword changes from one side to the other during this cut; this can also occur in other ways, yet it is completed most properly through this diagonal cut. You will find the origin of all of the names and cuts, and how they are to be completed, explained in two weapons, namely the Sword Section and especially in the Dusack Section (at length in the fourth chapter).

The second primary technique is completed in two ways. Namely for the first: when your counterfencer crowds you with cuts and thrusts, you catch or deflect the same by offering your extended hilt or weapon, so that you are hastily and quickly equipped and prepared with counterstrikes before they have recovered back from the strike or thrust they have executed. Secondly, those of your counterpart's cuts and thrusts which they guide toward you can also be deflected and cut away from yourself using the first primary technique—that is, by using identical counterstrikes. Whenever your counterfencer cuts at you with a crossing strike, you can suppress the same (from above down to the ground) using a High Cut and the associated steps outward, and also use the same to cut at their head. In contrast, when they cut from above, you can take out their cut and deflect it away from you using a crossing cut.

The third [primary technique], which I have named the Middle or the Handwork, arises from these two primary techniques through practice. This third technique unites the first two primary techniques in use so |b ii^v| closely that the counteraction and injury can sometimes occur together in one strike.

Lastly, the practical aspect arises in and of itself, and teaches how one should use the mentioned techniques correctly and well against an opponent, how you should adapt each to their appropriate point, and how you should direct yourself in the work.

Therefore, if you now want to attack your counterpart with the techniques mentioned, then prior to that, you must divide them in an organized fashion into various parts so that you can use your sequences differently, moving from one point to another according to your advantage. You should also know how to apply the same according to your opponent: if they are tall, small, strong, weak, quick, or slow.

In order that you may always consider this better, stances are used, which are intrinsically nothing more than tarrying or holding the weapon at the outermost point to which you have arrived when pulling up to strike, where you still have space to consider in the middle of the same before you have completed the cut, whether to complete the same cut according to your first intention, or whether it would be more useful to turn to a different point.

So that you do not neglect a chance opportunity in the Before or After, but instead (while recalling the word Indes[5] to mind), you have quickly considered all advantages. And this is how, as mentioned, the stances or guards were created.

Not only are youthful adolescents instructed in these two abovementioned primary techniques of fencing (including their assorted circumstances) by their qualified and experienced fencing masters with the greatest diligence, but also all exemplary leaders among all peoples have seriously instructed the men at arms they have under [their authority] in these exercises. Whenever they have no enemies before them and are idle, they personally instruct their men at arms regarding how they should hold their equipment, arms, and weapons according to their advantage, and also to skillfully rebuff the enemy's strikes and thrusts and correctly set themselves against them, as such is then explicitly read about Hannibal. Because of this, in years just recently past, every proper fencing master was granted additional pay during military campaigns and otherwise.

And now, for the sake of brevity, I will skirt the issue that the art of fencing is numbered as a noble part of war practices. Due to this practice, students are masterfully provoked to deep

---

[5] On I.25 of the Sword Section, Meyer explains his interpretation of *Indes* to be a small word fraught with meaning: in that moment in which you strike, you should simultaneously observe the other openings that you could also hit, and *also* all of the sequences your opponent can use to strike or counteract your action. It is therefore a point of hyperfocus on what you are doing and what you and your opponent could be doing at the same time and subsequently. Meyer states that he will use *Indes* as a shortened form to call all of this to mind.

Within his text, Meyer tends to use *Indes* to indicate a disruptive action; that is, because he is describing actions sequentially that occur simultaneously or disruptively. Thus, *while* you are performing the above action, execute the following action *at the same time* and *in the same space*. Because *Indes* is an adverb of time in German, it will frequently appear adverbially in the translation, thus *Indesly*.

consideration about how to correctly use all types of advantages, including other, quite useful applications of these practices.

That is also related to the fact that a meeting of fencers |b ɪɪɪ| is a fine, short presentation of how one should conduct a military campaign against the enemy, as these can properly be compared with one another in this way.

Therefore, [in a manner that is] similar to how it is necessary for a wise, knowledgeable military leader to be best equipped and provided with good equipment for war, including all accessories, and he must have learned through good spy craft about the enemy's strength, equipment, and intentions, and also be informed about the place or location where the same [enemy] is staying. As a consequence of this, he has considered well beforehand how he would withstand all types of unexpected events and how he will encounter the enemy in the flesh and bring the war to an end. Therefore, it is intrinsically encouraged that [the reader be] practiced and learned in all types of practical aspects in enticing and provoking the enemy out of their advantage (in a cautious and virile way[6]) and then, as soon as he observes his advantage, to know to attack with quick earnestness (in a cautious and virile way) so that he does not unproductively miss out on a chance opportunity—that may perhaps never come to anything— through too long consideration. Consequently, [he must know] to also press seriously onward, and, if his enemy has been weakened, to not be too eager with chasing so that he does not arrive before [the enemy] due to his skill and virility, nor lose due to incautious following after, so he should well observe all opportunities so that he can finally withdraw with the victory.[7]

Likewise, any good fencer[8] (when they have been provided with their equipment) should also pay good attention to all circumstances surrounding the counterfencer (as well as their own), and not attack the [counterfencer] incautiously, but instead the fencer should practice with all types of techniques related to how they can entice and provoke the counterfencer from their advantage. As soon as the fencer sees their advantage, they should attack cautiously, quickly, and swiftly, follow after in a virile and prudent way, and forcefully crowd and frighten the counterfencer with all kinds of techniques to all parts of their body to such an extent that the counterfencer is unable to come to any fruitful work or counter-defense. Immediately, at the point when the fencer has brought their intention into effect, the fencer would withdraw carefully so that they would not receive any injury at the end due to carelessness. However, if it were to occur that the fencer could not gain their intention, the fencer should not tarry long in front of the counterfencer at this time such that the fencer should not work in vain, but

---

[6] *Fürsichtig und manlich* does occur twice in this sentence, in what is likely a typesetting error; however, the goal of the translation is to reproduce the meaning of the text, ergo, the potential errors as well.

[7] Note that this paragraph has focused on the practice of war while using similar concepts from fencing. Therefore, I have maintained both the use of 'enemy' instead of 'opponent', and masculine pronouns to denote the most common gender of men at arms and military leaders of Meyer's time. The following paragraph returns to a discussion of the practice of fencing, while using similar concepts from war. As such, the use of non-gendered language is reinstated to reflect the difference in focus.

[8] This paragraph describes the interactions between two masculine, singular persons. As the English would get bogged down between two unspecified persons, I am distinguishing them as the fencer and the counterfencer (as the term which Meyer uses), so that the interactions are easier to follow.

instead should observe how they could properly withdraw from the counterfencer so that the fencer could recover for a fresh attack, and thus skillfully equip themselves to deal with the fault that previously prevented the fencer from their intention.

In all of this, the fencer should diligently apply themselves |b iii^v| to gaining information about the counterfencer's type of fencing and not allow the fencer's own special techniques to be often viewed, but to keep them proprietary and secret. Thus, there is practically no technique proper to a man at arms that could not also be considered useful in fencing. Therefore, I will let experienced men at arms judge whether the examples of such aforementioned leaders of ancient peoples should be usefully followed today.

Accordingly, I have undertaken this work for the honor of the art, and to describe it inasmuch as my meager understanding allows. I have primarily insisted first on the cuts as the correct principal part of all fencing. Secondly to this, I have envisaged the opponent and their divisions, against which you should direct these cuts and thrusts. Following this as a third, I have desired to show how one should guide the cuts against the divided opponent (who will certainly not work in a clueless manner) through many types of examples. [This is done] not with the expectation that one must absolutely follow these examples, but instead as a view toward how one could be guided to learn and practice through these examples, so that, in a time of need, they would have learned to direct and guide the cuts according to their opportunities and the ability of their own body. Indeed, that the cuts—and how and when one should cut them—are not thus forced into a specific and certain shape, but instead 'only the market is permitted to be the teacher of the buyer'.[9]

Therefore, when you want to engage in an aspect of this art, it is my advice that you initially (as already often mentioned) first learn to powerfully cut the cuts or the thrusts, in and of themselves, correctly and with extended arms, and also with the addition of the strength of your whole body. Regarding this, you will find a useful pattern for beginning to learn these cuts in the third chapter of the Dusack Section, stipulated by four Rules. When you can now cut those [cuts] correctly and well (as stated), you should then learn the second—that is, to pull the [cuts] skillfully back or to let them fly in a complete motion or course so that you can turn any cut elsewhere in the same fluid motion (before the opponent is correctly aware of it and you would have been aware that it would be in vain at this location) so it should scarcely contact. When this has occurred, only now are you trained and capable of stepping onto the [fencing] space and to begin to learn to direct such cuts in practical exercises, and also in the work against your counterfencer. At this point, the cuts align themselves according to the type, nature, strength, and ability of the fencer, because the weak must seek a different advantage in the cuts than the strong, and this leads to strength.

|b iv| Now, when two come together with the abovementioned cuts, the two principal elements—that is, the cuts and counteractions—give rise to a marvelous battle because each one would rather cut than counteract, so that now this one strikes that one, then that one swiftly counteracts this one, both fencing with the same techniques around the Before and being above as the stronger.

The practical aspects of this are difficult to conceive of: namely, when and toward which target each cut may be applied and completed at the correct time. Therefore, I have paid assiduous attention through all techniques to the Before and After, such that I hope that the diligent

---

[9] This is somewhat akin to experience being the best teacher.

reader may receive a not insubstantial instruction in the practical applications, because there is no separation to be found between the three liminal points, namely between the Before, Simultaneous, and After, except that when two want to come together with weapons, one of the two will cut in the Before, that is first, from which it follows that the second cuts After in response, or both must cut Simultaneously.

Now the one who wants to cut the first strike may well consider whether they might not bring themselves into some risk with this and (if they do not know how to gain the advantage in this) be thus caught and rushed past in their own cut. Likewise, the second must consider, in the cut After in response, whether another advantage may occur by cutting Simultaneously, so that they do not both contact one another, as often happens. Therefore, I have (as stated) explained the differences with diligence and at length in all cuts and techniques, and I have especially taught how one should use some cuts to provoke, to thus bring the counterfencer out of their advantage; some to take, that is, when you have brought them to a cut by using the abovementioned provocation, that you turn that aside with a countercut or receive it with a counterposture; and then for the third, some to contact, as you will see this throughout the Dusack and Rapier [Sections].

And indeed, the true art is found herein, and the practical aspects, in which human rationality, acuity, swift deliberation, caution, skill, and virility become visible and can be distinguished, as this is the point where the skill depends on the person, such that a poor technique may be more productively carried out in the work by a cautious fencer than the best [techniques] may be fenced by a fool.

Since everyone thinks differently than others, and likewise they will act differently in fencing, I considered it to be quite good to treat the cuts in multiple ways (both how one should cut them, and turn aside the cuts which are cut against one) so that anyone, whether they are strong, weak, quick, or slow, |b iv^v| can take something from this that is useful for themselves. Because fencing is actually a type of exercise through which the body is directed to guide the weapons in all kinds of quick actions; and because, when one exerts oneself in this, one must then adapt one's self in the work and guide the weapon as the circumstances demand; thus, the better one is practiced in this, the better one is able to encounter each opportunity as it occurs. However, whatever usefulness regarding the balance and health of the body that is created from this exercise, I will cede to those with greater learning [in this aspect] to recognize.

Whatever is not explained elegantly according to the letter or shape, as it should well be, I would ask the reader to consider well my intention and give me credit, and thus to best receive and take note of my diligence and efforts.

Contents of the Entire Book

Namely, what and how many weapons are to be treated:

Firstly, the Sword as a foundation of all fencing
The Dusack
The Rapier
Dagger
Wrestling
Quarterstaff
Halberd
The Long Pike

Contents of the first book about fencing with a sword, and the order to be followed in the description of the same, and also the basis for this knightly art.

I have undertaken to most diligently and truly describe, to the best of my understanding and ability, the art of fencing with this knightly and manly weapon, which is common among us Germans at the current time. And, because it is obvious (and experience provides) that fencing with the sword is both an originating point and source for all other fencing, and also the most artful and manly before other weapons, I therefore considered it to be necessary and good to begin with this and to deal most briefly, yet also quite clearly, as occurs with other arts and practices.

Firstly, to discuss the associated terminology and types—as were invented by masters of this art with particular diligence—so that a fencer can understand and learn the secret and skill behind the same more swiftly and easily. After, to explain and interpret this terminology so that everyone can understand what should be understood by this type of speech.

Then, thirdly, to present the exercise of the art in itself: how it should be directed in the work from the explained cuts and stances so that the youth who want to dedicate themselves to this art would not be confused by those words that are unknown to them, nor |I.1ᵛ| would there be cause for [them to] scorn this art. Also, because it would be quite annoying to read if, in the middle of the art, it would be necessary to first explain the words mentioned. Also, so that those experienced can gather understanding that the practice of fencing has its pedigree in a correctly understandable foundation and does not lie in the frivolous burlesque of fools. After all, there is a vast difference between that type of buffoonery and fencing: the knightly art of fencing is held to be of great value by all highly experienced warriors (in particular the Romans), while in contrast, the fools are held to be the most unworthy and trifling in all the world.

Fencing with the sword is nothing more than an exercise in which two strive together with swords in the understood agreement[10] that fencers use cuts and other Handwork to prevail and defeat the other with care and quickness—using the sword artfully, properly, and in a virile way so that when it is necessary in serious cases, a fencer can be even more courageous and skilled and more careful in the protection of their body due to this practical exercise.

This may elegantly, properly, and well be divided into three parts: namely, into the Beginning, the Middle, and the End. These three parts should and must be kept attentively in mind in every technique that you undertake to fence so that you know with which cuts or from which stances you want to attack your counterpart. And then, when you are in the attack, how you should work in the Middle with the Handwork to fly freely at the openings in order to maintain your Before when you have overwhelmed them in the attack. Lastly, how you can withdraw from them properly and well—if with no damage to them, then at least without injury to yourself.

I call the Beginning the Onset:[11] when an opponent initially attacks against the opponent whom they face.

....

[10] A *Versatz* is a pledge or deposit to make good on a deal, the diagonal support in construction, a connection using rivets, the second tanning of hides, etc.

[11] The Onset, or *zufechten,* occurs at the Beginning with cuts from or out of the stances. GRIMM, vol. 32, col. 356. The Middle is the *handwerk*, and the Withdrawal is the *Abzug.*

The Middle is the secondary work or Handwork: when a fencer persists in the bind or longer in their work against the counterfencer and harries them with all speed.

The End |I.2| is the Withdrawal, how the fencer can cut away and apart from their counterpart without injury.

The Onset in the Beginning takes place using cuts from the stances, which are of two types: namely the Primary Stances and the Secondary Stances (which originate from the Primary Stances).

The Primary Stances are four [in number]: the Day or the High Guard, the Ox, the Fool,[12] and the Plow. The Secondary Stances are eight [in number]: Wrath Guard, Breaking Window, Longpoint, Crossed Guard, Unicorn, Key, Iron Gate, and Change [Guard].[13]

The cuts, as related to the sword, are of two types, which can generally be called the straight and the inverted cuts. The first are the Primary or Principal Cuts, from which all other cuts have their origin, and of which there are four: High, Low, Middle, and Wrath Cut. The others are called the Secondary or Derivative Cuts, of which there are twelve: namely, Squinter, Crooked, Short, Glancing, Rebound (Single and Double), Obscured, Twisting, Crown, Wrist, Plummet, Change Cut, etc.[14]

The true Master Cuts are taken from both of these [groups], and are so named because all masterful and skillful techniques with the sword are encompassed, carried out, and completed in the same. These are Wrath, Crooked, Crosswise, Squinter, and Hairline[15] Cut. I will clearly unveil how they should all be completed and carried out when I come to their description in the Onset and when I speak about the cuts.

The secondary or Handwork in the Middle comprises the greatest art and all of the speed which can take place in fencing, because it shows both how a fencer should let the sword bind on, twist, change, mislead, pursue, slice, double, and drop downward, and also in which configuration a fencer should strike around, catapult, shift forward, set off, pull and jerk, restrict,[16] wrestle, run in, throw, and crowd after.

|I.2ᵛ| It also contains the openings, which must be understood using the division of the opponent and the sword, and additionally correct positions and steps also belong to this, which shall be discussed at its place.

The Withdrawal at the End flows from the Middle and is of great use in practical applications. Therefore, at the end of every sequence, the Withdrawal associated with the same will be explained in an organized way. And this will all be explained completely in the first part related to sword fencing. In the second part, a more extensive and sufficient report will be

---

[12] Or "poplar tree".

[13] The lists of stances are, in order: *Tag* or *Oberhut, Ochs, Olber, Pflug*; then *Zornhut, Brechfenster, Langort, Sch[r]ankhut, Einhorn, Schlüssel, Eisenport, Wechsel*. The list on I.5ᵛ deletes the Breaking Window and Crossed Guard and adds Side Guard and Hanging Point. The stances and cuts are glossed in the text during Meyer's individual descriptions of the same and can also be found in the glossary to the text.

[14] The list of cuts includes: *Ober, Under, Mittel, Zornhauw*; and *Schiel, Krum, Kurtz, Glitz, Brell* (*Einfach* and *Doppel*), *Blendt, Wint, Kron, Knichel, Sturtz, Wechselhauw*.

[15] *Scheitel/Schedel* refers to both the skull and the point defined by the hairline and its central parting, as opposed to the crown of the head, which is, interestingly the back of the skull, and not the line defined by wearing a crown. Meyer uses *Scheitel* and *Kron* interchangeably at this point.

[16] This is *Verstillen* in the text, which is "restriction".

provided about the Master Sequences and that which is helpful for increased quickness related to this weapon so that this book can be useful to both the beginning student as well as to those more experienced in this art.

I considered it to be beneficial to provide this introduction so that this book would be understood even more easily and [readers] would know how to orient themselves herein if they understood from the beginning the order in which I would undertake [to explain] this knightly art. Therefore, in the first chapter, I will discuss the division of the opponent, and how they are divided into four quarters, as this is most useful to report on first.

|I.3ᵛ| **About the opponent**[17] **and the division of the same**
                                     Chapter 1

Although the division of the opponent (from which the openings and stances originate, toward which and from which one fences) actually belongs to the Middle or Handwork, and for this reason will also be discussed in the contents of that book, I want to describe it and set it out here at the beginning for a specific reason: that is, because it is initially necessary to know in fencing (and in all matters and arts) what it concerns and what the goal is; therefore, this must be discussed. Consequently, [I] wanted to raise this point of discussion at the beginning so that it would not be necessary to describe the openings when they were mentioned during the Middle or Handwork [Section], which would cause an interruption in the course of my writing.

The opponent is now divided into four quarters or parts: into the upper and lower, and each of these into the right and left. The what and the why of their intrinsic existence requires no more extensive description because a mere glance at a human body reveals what is the highest or lowest and the left or right part. However, the portrait on the right in the previous Figure [A] [is offered] to explain what I mean here for greater understanding.

And, while the aforementioned four parts of the opponent would be sufficient according to the customs of the Old German fencers (for whom thrusting was permitted as well as cutting), because we Germans of the current time often and most properly fence at the head (and especially in the Handwork with twisting) most, I will also divide it into the abovementioned four pieces (like the human being as a whole). Namely, into the upper, which relates to the hairline;

---

[17] The German word *Mann/man* will be translated as "opponent", and references to pronouns will be the gender-inclusive they/them, unless the pronoun refers to a specifically named Master. In addition, Meyer uses several other terms to refer to the opponent, which will be reflected in the translation.

|I.4| the lower, which is assigned to the chin and neck; and again the right and left part, which are generally assigned to the right or left ear because the ears are located on both sides.

Even though this division could be thought of as somewhat childish (as there are always more who criticize than improve), the well-intentioned [reader] muſt keep in mind that this is really only described because all other matters which are necessary and are part of fencing arise and flow from this division as from a spring or well.

Because again (as mentioned above), fencing is nothing other than when two persons fight with the same weapons againſt one another and how a fencer can injure the other with quickness or defend and proteƈt themselves carefully, my intention is also to present and explain them using certain sequences and brief summaries (as in the [seƈtions on] other weapons). Therefore, I neither can nor should betray [my intentions] to mention this division of the person myself, so that if a counterfencer fences from one or another part, an opponent can take note and know how to secure their defenses with the correƈt counteraƈtion, or, if a fencer were to expose their opening somewhat on one or the other side, then their opponent could place their ſtrike at [the opening] correƈtly and with advantage (because an opponent is either hit at one of four points due to exposed openings or muſt position themselves into the guard or counterpoſture to proteƈt the same again).

Therefore, that which is subsequently taught about the ſtances, cuts, and openings, can be gathered more easily because all of this flows out of this division of the opponent which is now sufficiently described.

|I.4ᵛ| **About the sword and its division**
Chapter 2

Since I have explained what fencing concerns—namely, how a fencer can injure the other in one part of their body (for example), or conversely, how to defend themselves in a knightly manner—therefore, the division of the human body was briefly explained. Thus, it is also necessary to show why and through what means the same are completed on another person. And even though the title or heading of this part clearly indicates this (namely, that nothing besides the sword will be mentioned here), yet it is proper to say something now about the division of the sword that is useful and appropriate for this art, because this does not occur in a single way but instead here with the short edge and there with the long edge, here with the Strong and there with the Weak. As regards the parts of the sword, relating to its shape and figure— its pommel, point, crossguard or hilt, grip or wrapping, and blade—it is unnecessary to use many words for what each of them is.

The blade has two different divisions, the firſt of them is the Strong and the Weak, the other is the short and long edges, which are the front and the rear.

The Strong of the sword is the part ſtarting from the crossguard or grip up to the middle of the blade. The Weak is ſtarting from the middle up to the point or end of the same. The distinƈtion of techniques into long and short arises from this.

The long edge is the complete edge facing outward from the fingers ſtraight toward your counterpart. The short or half edge is the [edge] toward the thumb or between the thumb and index or firſt finger, facing toward the fencer themselves (ſtated by way of comparison); |I.5| in other words, the back of the sword, as is clear from the previous Figure [A].

The correct, complete division of the sword (which is quite useful in fencing) originates from the abovementioned divisions: namely, that the sword is commonly also divided and distinguished into four parts (as is clear in the Figure [A] printed previously).[18]

Long Edge

Short Edge

The first is called the wrapping or grip, and includes the pommel and crossguard, and is useful for running in, wrestling, grappling, throwing, or other work.

The second, the Strong (as mentioned above) is useful to slice, twist, press and other [techniques] that are fenced from the Strong.

The third part is the Middle (taken equally from the Strong and Weak around the halfway point) and is suitable for changeable work and can always be used according to opportunity.

The fourth is the Weak, for the change through, flick, catapult, and whatever belongs to fencing from length [out of distance], as you will subsequently have many and sufficient examples and sequences.[19]

About the stances[20] or guards
Chapter 3

For the sake of increased usefulness and understanding on the part of those who intend to learn this knightly art, three primary points should rightly and properly be kept in mind in all fencing: firstly, what is the concern in fencing—namely, the opponent—and following that, with what should |I.5ᵛ| the fencing be carried out—which is with the sword.

Because enough has been discussed about these two matters in the two preceding chapters, the organization demands that an explanation should additionally be offered about the third part and matter, which is: in what way should the fencing be completed. This primarily occurs through the three sections—the Beginning, the Middle, and the End (as indicated above). Because the Beginning should and must be made using two different elements (as stated previously) originating in and from the stances—from which the cuts have their origin—I will explain how many of them there are and how they should be completed.

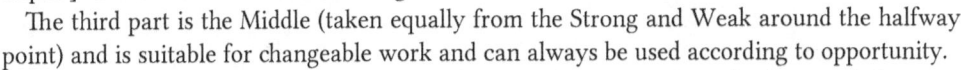

[18] This fencer has been reversed for clarity, since they are left-handed in the original.

[19] The techniques listed are, with the hilt: *Einlaufen, ringen, greifen, werfen*; the Strong: *schneiden, Winden, drucken*; the Middle: *Wechselbaren arbeit*; the Weak: *durchwechseln, schnellen, schlaudern*.

[20] *Leger* from '*sich legen* = to position oneself'. According to GRIMM, meaning 7) of the reflexive, *sich legen* can refer to armies that besiege or surround, or to '*schutz und schirm* = protection', which meaning is comparable to *lager, sich lagern*, which refer to military encampments. This is reflected in the Latin *castra* for a defensive position. Meaning 8) is linked to this, as it can mean 'to set oneself against another', 'to attack in a hostile manner', *gegen einen legen, feindlich angreifen*. The *leger* thus becomes the noun from the verb, 'the position into which one moves for protection'. In addition, *sich legen* point 2) also refers to movements of the torso that bring the upper body out of the normal, vertical position. GRIMM, vol. 12, column 536, meanings 7), 8), and 2).

The guards or ſtances are a splendid (yet also necessary) positioning and comportment of the entire body with the sword, into which the fencer[21] positions and guards themselves (if, as often occurs, the fencer arrives at the location [of combat] before their counterpart), so that the fencer is not unexpectedly overwhelmed and injured. Specifically, the fencer may pay attention and await their counterpart here, and as soon as the counterpart arrives, the fencer may attack and cut with advantage and a certain speed—and thus guard themselves againſt their counterpart so that the counterpart cannot cut at the fencer without injury. Inſtead, when the counterpart works at the fencer's openings then the counterpart muſt expose themselves, and the fencer can remove the opening which the counterpart expeĉted to have by advancing or ſtepping around. Or, at the very leaſt, if the fencer is provoked from their advantage by this, they can take out the counterpart's blade, hinder the counterpart, and spike the counterpart's work.

As mentioned above, the ſtances and guards have their origin in the division of the opponent. Because, juſt as the opponent is divided into four quarters (high, low, right, and left) so are there four openings where the counterfencer should properly be hit. And, like the four openings, there are also likewise four Primary Stances or guards from which the others arise from and originate: the Ox, Plow, Day, and Fool; the Secondary, or those that originate from these, are the Wrath Guard, Longpoint, Change [Guard], Side Guard, Iron Gate, Hanging Point, Key, Unicorn,[22] which will be dealt with briefly and in order.

|1.6ᵛ|

Ox [Ochs]

The upper part of the fencer is apportioned to the Ox, and because that has two quarters (the right and the left), a fencer can also partition the ſtance of the Ox into two parts like this (namely, the right and left).

The right Ox is now formed like this: ſtand with your left foot forward, hold the sword with the hilt next to your head high on the right side such that your front end[23] points toward your opponent's face.

Regarding the left Ox, position yourself contraſtingly to this: namely, ſtep forward with your right foot, hold your sword with the hilt next to your head on the left side as reported above.

Thus, you have both Ox guards or ſtances. [The right] ſtance is depiĉted on the left side in the Figure designated with the letter B.

---

[21] This entire section is a description of the actions of one against another, rendered in the German as two virtually indistinguishable masculine pronouns. I have labeled the first *er* as the fencer, and the second *er* as the counterpart, as using the singular "they" for both is as difficult to follow as the dueling "he"s of the original.

[22] The previous list included the Breaking Window and Cross Guard but omitted the Side Guard [*Nebenhut*] and Hanging Point [*Hängetort*], which are included in this list. Unicorn, Key, Iron Gate, and Change also appear here in a different order.

[23] The terms *vorderen ort* and *hinteren ort* come from the polearm section, where they refer to the two ends of the staff, but occasionally Meyer also uses them as synonyms for the tip and the pommel of other weapons.

Plow [Pflug]

The lower part of the fencer is apportioned to the Plow, and being of the same configuration as the preceding [guard], has two quarters or two sides (the right and the left); therefore, the Plow is also called the right and the left. Both are, in and of themselves, nothing more than a thrust upward from below.

Complete the right Plow as follows: stand with your right foot forward, hold your weapon with the hilt next to your forward knee, turn the tip or the point into your opponent's face (as if you wanted to thrust at them from below). Thus, you are in right Plow.

If you step forward with the left foot and execute the same at them, then you rest in the left [Plow], and the right Plow is thus also depicted on the right side in the aforementioned Figure [B].

Day [Tag]

The Guard of the Day, which is also otherwise named the High Guard, is completed in the following way: stand with your left foot forward, hold your sword high over your head so that the point extends straight upward (as the portrait teaches you on the left side in the Figure which is designated with the letter C).

|I.7$^\text{v}$| Whatever is then worked from above here is called fencing From the Day or from the High Guard; therefore, such stances are called the Day.

Fool [Olber]

In my opinion, the Fool is named according to the word Alber[24] (which means something like 'simple-minded') because no completely-finished strike can be achieved from this guard unless a fencer recovers with a new [strike] (after receiving their opponent's strike by setting it off). This is truly the measure of a fool and a simple-minded person: to allow themselves to be attacked without a prepared counter strike.

---

[24] The distinction in German is between *Olber* and *Alber*, with the former being a feminine noun meaning 'poplar tree' and the latter being an adjective/adverb meaning 'foolish'. The spelling distinction has collapsed by this point in time. Various authors make use of the pun, but Meyer does not.

It is formed like this: ſtand with the left foot forward, hold the sword with the point extended in front of you on the ground in front of your forward foot (so that the short edge faces upward and the long edge faces downward). Thus, you reſt correctly in this guard (as you can see on the right side in the aforementioned Figure [C]).

Wrath Guard [Zornḥut]

The Wrath Guard is so-named because such a ſtance indicates a wrathful body language[25] and is formed like this: ſtand with your left foot forward, hold your sword at the right shoulder such that the blade hangs down behind you (for the planned[26] ſtrike).

It should be noted here that all sequences that originate from the guard of the Ox can also be fenced from the Wrath Stance, except that the fencer should manifeſt dissimilar body language to mislead their opponent in this quarter, so either one can quickly be used (this now, the other then). Regarding this [guard], see the Figure designated with the letter E.

Longpoint [Langort]

Stand with your left foot forward, hold your weapon with arms extended long in front of your face so that your forward point extends toward your counterpart's face. Thus, you reſt in the guard of the Longpoint (as the portrait teaches in the Figure designated with A).

[25] *Geberde* includes all senses of one's attitude or demeanor and can be consciously or unconsciously expressed. You can consciously use it to deceive your opponent, and you can also take note of their unconscious *geberde* to ascertain their next likely moves. "Body language" expresses the interpretation of stance as well as facial expression, and also the potential for conscious and unconscious projection.

[26] *Gefasst* in the sense of being mentally equipped or prepared in order to do something. GRIMM, vol. 4, column 2134, d).

|I.8| Change [Guard] [Wechsel]

This guard is completed like this: stand with your right foot forward, hold your weapon with the point or Weak extended to the ground next to your side (so that the short edge faces your opponent), as you can see the same in the portrait [on the right] in the Figure marked with D.

Side Guard [Nebenhut]

Position yourself in this guard like this: stand with your left foot forward, hold your sword next to the right [side] with the point on the ground (so that the pommel extends upward and the short edge faces toward you).

Iron Gate [Eisenport]

You will find more extensive information as to what is the correct Iron Gate later in the Rapier [Section].[27] Because thrusting with the sword has been abolished for us Germans, this guard has also completely fallen into decline and has perished, though the Italians and other nations currently use it. These days it is fundamentally the Crossed Guard and is used by the inexperienced, as they lack information about the Iron Gate.

However, because there is a difference between them, I wanted to briefly explain both here. The Iron Gate is formed like this: stand with your right foot forward and hold your sword with the hilt in front of your knee with arms hanging straight down (so that your point extends upward out toward your opponent's face). You thus have your sword in front of you for defense like an Iron Gate. Then, when you stand with your feet wide so that the body sinks down, you can remove all cuts and thrusts away from you.

In contrast, the Crossed Guard is when you would hold your sword with crossed hands in front of yourself with the point on the ground (as is clear to see in the following Figure whose letter is F).

[27] See II.54 and the large scene in Rapier Figure C.

|I.9| **Hanging Point [Hängetort]**

The portrait on the right side in the above-mentioned Figure [F] teaches how you should correctly initiate the Hanging Point (except that the arms are not shown as sufficiently straight here).

Therefore, position yourself in the aforementioned guard like this: stand with your right foot forward, hold your weapon in front of yourself with arms extended (so that the blade hangs somewhat downward toward the earth). This stance is certainly almost identical to the Ox, except that you hold the arms straight up high in the Ox. Here, however, they should be extended [forward] exactly in front of your face, and you let the sword hang toward the ground, for which reason it is then also called Hanging Point.

Key [Schlüssel]

The Key is illustrated like this in the Figure which is designated with the letter D: if you stand with your left foot forward, and you hold your sword with the hilt and crossed hands in front of your chest (so that the short edge rests on top of your left arm and the point extends toward your opponent's face), then this stance or guard is formed correctly.

Unicorn [Einhorn]

Move into the Onset with your left foot forward, wing your elbows out on both sides as if you wanted [to position] yourself in the aforementioned Key stance, and lift your crossed hands upward to your right (so that the tip extends[28] outwardly upward). This is called the Unicorn, and you stand as you can see in the portrait on the right in the Figure marked with E.

---

[28] *Sehe* in the text has to be a typo for *stehe*.

|I.9ᵛ| This is a brief mention about the names and the number of the ſtances or guards, and how each one is initiated or completed. However, in all fencing, after you cut, work, counteraᶜt, or execute whatever work you want, you ought not to tarry in one ſtance. Inſtead, you muſt always move from one into another and transform one into another, because it is worth your while to pay diligent attention to how the ſtances liſted above follow one from the other, which I then want to explain in a few words with the cuts through the lines or paths.

Firſtly, if you execute a High Cut or Hairline Cut, then you find [yourself in] three ſtances: in the beginning, you reſt in the Day; in the middle, in Longpoint; and at the end, in Fool. Thus, you have three guards or ſtances in the ſtraight line from above downward from A to E. If you lift back up from below with crossed hands to the counteraᶜtion, then you are positioned in three ſtances once again: namely, in the beginning the Iron Gate; in the middle, the Hanging Point; at the end (extended fully upward), the Unicorn. If you pull your sword with the hilt in front of the cheſt, so that the half edge reſts on top of your left arm, then you ſtand in the Key. Thus, you move from one ſtance into another by lifting and falling in the line A to E.

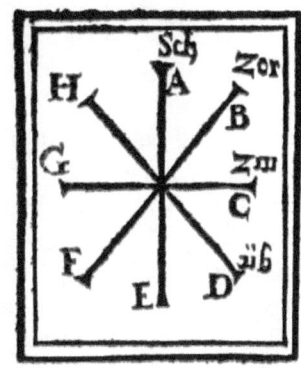

There are two of the other lines, which pass downward obliquely through the upright line. One passes from the upper right quarter, indicated by H and D, the other extends from the upper left part to the lower right part, marked B–F. You now cut through whichever one you want (those mentioned above and all those that I wanted to briefly repeat here), so that all cuts and ſtances can be completed on both sides, right and left, even though they are moſtly only described in one way for the sake of brevity. Thus, you move firſtly or initially into the Wrath Guard (from which the cut also receives the name, as it |I.10| is named the Wrath Cut due to the wrathful body language). The halfway point of the cut [is] into Longpoint, and the Change [Guard] is at the end. If you pull the ſtrike back up from there with the long edge, then you move back through three ſtances, as there is the Side Guard in the beginning, Longpoint again in the middle, and upward at the end is Unicorn. If you ſtrike through the abovementioned line, regardless of which side, you move out of the Change [Guard] through the Longpoint into the Wrath Guard. In ſtriking upward, you can also turn your sword into Hanging Point, from which you can lift farther upward into the guard of the Ox. Thus, you will always find at leaſt three ſtances as often as you follow one of the designated lines.

However, a good fencer should not become accuſtomed to waiting long in their ſtances, but inſtead, as soon as they can reach toward their counterpart, they should attack the same and fence through their intended sequence. [This is] because long waiting requires many counteraᶜtions, from which a fencer can only slowly move to ſtrikes (as is discussed further below regarding the counteraᶜtions).

The ſtances are also very useful with respeᶜt to the division of the sequences, because, if a fencer moves into a ſtance in the pre-ſtriking phase without danger, they can also quickly remember from there what kind of techniques can be fenced from this [ſtance]. They do not merely funᶜtion for elegant and proper changing from one ſtance into another, but also to mislead an opponent so that they become confused and cannot know what they should use to fence at you. And finally, this is also useful and good in that you can easily see and recognize the techniques of your counterpart, approximate how they can or will fence against you, and thus you can more properly counter them.

Much is said about the Beginning of this art—namely, about the Onset against an opponent, which takes place by cutting through the stances. Now, the second part of the entire art follows (which still pertains to the first), which is about the cuts.

|I.10ᵛ|

About the cuts
Chapter 4

One now advances to the art and boundless knightly exercise itself, namely the cuts, which are truly a principal part[29] of fencing (as such was mentioned at the beginning): how many of them there are, what each one is, and how it is formed and should be completed.

It is necessary to state something here at the outset, merely as a reminder to the friendly reader: as there is a large difference between fencing with the sword in our times and how it was used by our ancestors and the ancients, I will describe at this point only that which is currently customary, and as much as belongs to the sword and the cuts. However, I want to discuss much that is relevant to the older usage separately in its proper place, as they fenced with both sharp cuts and thrusts.

There are two different types of cuts which are now relevant to the sword, the straight and inverted cuts. I name "the straight cuts" those which are cut against an opponent with the long edge and extended arms, of which there are four: High, Wrath, Middle, and Low Cut.[30] All others originate from these, and there is nothing that can be conceived of or invented in the world that cannot properly be understood as belonging to one of them. They are also deservedly called the Main or Principal Cuts.

The inverted cuts are those in which the fencer inverts the hands with the sword in the cut, so that the fencer does not hit their opponent with the full or long edge but instead with the half edge, flat, or with a corner, as this occurs using the Glancing, Short, Crown, Squinter, Crooked, Crosswise, Rebound, Obscured, Twisting, Wrist, Plummet, and Change Cut.[31]

|I.11| Because these originate from the four cuts mentioned above, they are called Derivative Cuts.

Five cuts, which originate from and are selected from these two types, are called the Master Cuts, not because whoever can complete these correctly is quickly named a master of this art, but instead because from them originate all correct, skilled techniques which would well befit a master to know. And whoever can correctly fence and use them is considered a skilled fencer, because all Master Sequences are concealed within them and fencers cannot do without them. These are the Wrath, Crooked, Crosswise, Squinter, and Hairline Cut.[32]

---

[29] *Hauptstück.*

[30] The cuts in order are: *Ober, Zorn, Mittel,* and *Underhauw.*

[31] The cuts in order are as follows: *Glitz, Kurtz, Kron, Schiel, Krump, Zwerch, Brell, Blend, Wind, Knickel, Sturtz, Wechsel.* Terminology choices and discussions are found in association with the individual cuts.

[32] *Zorn, Krump, Zwerch, Schieler, Scheitel.*

I now want to show in an orderly, sequential way how they all should be formed, and I will discuss the straight cuts first, among which the first is the High Cut.

High Cut [Oberhauw]

The High Cut is a simple cut straight from above toward your counterpart's head at the part, for which reason it is also called the Hairline Cut.[33]

Wrath Cut [Zornhauw]

The Wrath Cut is an oblique cut from your right shoulder toward your counterpart's left ear, or through their face and chest (obliquely, as shown by the two lines which cross through the upright line and cross each other). This is the strongest of all, because within it lies all the force and virility of the fencer who is fighting and fencing against their enemy. For this reason, it was also called the Fighting Cut or the Father Strike by the ancients. (You will find more later about imaginary lines, etc.)

|I.11ᵛ| **Middle Cut [Mittel] or Crossing Cut [Uberzwerchhauw]**

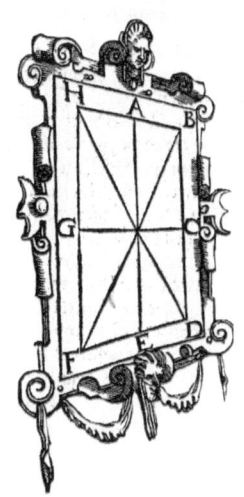

The Middle or Crossing Cut[34] can be formed in almost all ways like the Wrath Cut, except for this difference: that is, while the Wrath Cut is completed obliquely, this one is completed across, as can be seen in the crossing line (designated with the two letters G and C). You will subsequently find these lines in the Dusack [Section].[35]

Low Cut [Underhauw]

You form this one like this: use cuts so that you move into the right Ox (which was discussed in the immediately-preceding chapter), and as soon as you can reach the counterfencer, step and cut across (upward from below) at their left arm, so that you arrive with the cross-

---

[33] *Schedelhauw* is a problematic term in English. It refers to the top or crown of the head, but is lexically distinguished from the *Kronhauw* or "Crown Cut". It is sometimes spelled *Schädel*, which is the skull, and is linguistically linked to this term in ENHG (GRIMM, vol 14, col 2480, (1)a). It also indicates the "parting of the hair" or "hairline", which accurately reflects the target point. While *Scheitel* has been translated as "scalp" or "scalping", I find that term to be problematic as the scalp refers to all of the hair-covered area on the head and not specifically to the front or top of the head. The expression "scalping cut", which has appeared in other texts, could be considered offensive and will not be used in this translation.

[34] The term *überzwerch* means 'to cross over a center, vertical line, either horizontally or diagonally' (GRIMM, vol. 23, col. 697, A. (1)).

[35] The lines appear in Dusack Figure A and are explained in detail in Dusack Chapter 3.

guard high over your head. Thus, you have completed it. Regarding this, observe the small ten-ants[36] on the [right][37] side in the Figure with the B.

Squinter Cut [Schielhauw]

Squinter Cut is also a High Cut, but it is so named because it is cut together with a small skewed[38] bit, and is formed like this: position yourself into the Guard of the Day or Wrath (regarding this, see the third chapter) with your left foot forward. If an opponent cuts at you, then cut to counter; however, in the strike, turn your short edge against their strike and hit equally with them (with inverted hands). Step with your right foot well to their left side and move your head quickly with [the cut]. Thus, you have executed it correctly, and you stand like the larger scene on the left side in the next Figure indicated with G.

[36] *Bossen* are the two small figures (often animals) holding heraldic shields on either side; in English, they are known as "supporters" if they are animals or "tenants" if they are human. In this text, *Bossen* refer to the smaller fencers (usually pairs) to the sides or in the background of the main fencers in each Figure.

[37] The text says "left", but the action described fits the right tenants better.

[38] *Schielen*, according to GRIMM, col. 15, col. 14, means 'to squint', which appears to make sense as one squints along the blade to sight the target. It can also mean a 'skewed eye' or 'looking askew', which could describe the small movement at the end of the strike. The noun also refers to a splinter or other small bit that is torn off, or the lump that forms on a broken bone. It is also related to *Schele/schelen*, meaning 'peel', which could also describe the small movement at the end of the strike.

|I.12ᵛ| ## Crooked Cut [Ꝁrumpþauw]

This cut is completed like this: ſtand in the Wrath Guard with the left foot forward. If your opponent cuts at you, then ſtep with your right foot well toward their left side and out of their ſtrike. Cut with the long edge and crossed hands counter to their cut or between their head and blade, across at their hands, and let the blade shoot well paſt their arms (such as can be seen in the upper tenants on the right in the Figure with the D).

Crosswise [Ʒwerch] [Cut]

Comport yourself like this for the Crosswise Cut: in the Onset, position yourself in the Wrath Guard to the right (regarding this, see the previous chapter). Place your left foot forward and hold your sword at your right shoulder (as if you wanted to execute a Wrath Cut). If your opponent then cuts at you from the Day or from above, then simultaneously cut with the half edge across from below toward their cut. Maintain your crossguard high over your head, so that your head is blocked,[39] and ſtep well to their left side simultaneously with the cut. Thus you counteraðt [their cut] and you hit the weapons with one another (like the two tenants on the left side indicate in the Figure with the H).

---

[39] This is a reading of *versetzen* as '*vorsetzen* = to place something in front'. At this point in time, *versetzen* amalgamates *vorsetzen*, *fürsetzen*, and *versetzen*, with the different prefixes indicating the meanings of 'place in front', 'replace/substitute', 'move in error', and 'move from one place to another'. Which makes translating *versetzen* akin to interpretive dance.

How you would complete this Crosswise Cut on the left: you should direct yourself in the work toward their right side, with the exception that you should hit at their right with the long edge.

Short Cut [Kurtzhauw]

This is a secret move through and is executed like this: when your opponent cuts at you from above, position yourself as if you wanted to bind on their sword with a Crooked Cut (that is, with the short edge). Pass under it instead and quickly move through under their sword. Strike with the short edge and crossed arms |I.13| over their right arm and at their head. Thus, you have caught their sword with the long edge, and completed the Short Cut. At the end of the same, you stand like the small tenant on the right side of the upper left scene indicate (which Figure is designated with the letter B).

Glancing Cut [Glützhauw][40]

The Glancing Cut is completed in this way: if a fencer cuts toward you from above, then hit with reversed or inverted hands against their strike at the upper left opening. Allow your sword blade to glance off with the inverted flat down their blade so that the short edge hits the head in the sweeping motion over the hands.

Rebound Cut [Prellhauw]

This one has two forms: the first is called the single, the second the double. The single is executed like this: if your counterpart cuts at you from above, then oppose their strike with a Crosswise Cut. As soon as it glances off, pull your sword around your head and hit at their ear

---

[40] *Glitzen* can indicate both '*glänzen* = to spark or flash', and '*glitschen* = to slide or slip'. The Latin translation of Lew's gloss uses '*coruscare*, = to spark, flash, gleam'. In order for a cut to produce sparks, it strikes at an angle, slips along the edge, and moves off. While a "sparking cut" is not apparent to a modern reader, a "glancing blow" makes sense.

from your left with the outward, inverted flat[41]—but [do it] so that the sword rebounds back again (as the large scene shows on the right side in Figure K).

Pull the rebounded sweeping movement back around your head, cut with the Crosswise Cut at the left, then it is complete.

Execute the double like this: as soon as your counterpart brings their sword into the air to work in the Onset, position yourself in the right Ox (regarding this, see the immediately-preceding chapter), pull the sword around your head and cut strongly at their blade from your right with the inward flat (so that in the hit, your pommel touches [you] low on the forearm bone), as can be seen in the larger portrait on the left in the Figure designated as I.

In the strike, step with your right foot well around their left and as soon as it glances off or touches, jerk it upward and Indesly rip it outward and toward the left side. Quickly strike back |I.14| around at the same opening from the outside and with inverted hands (namely, with the reversed or inverted flat). If it strongly rebounds around in another rebounding hit, then you have executed it correctly.

[41] Note that the scene shows the Rebound Cut hitting the right ear with the inward flat.

Obscured Cut [Blendhauw][42]

Bind to your opponent on their sword from your right. In the bind, twist through below toward their left side using the hilt or grip. If your counterpart now wants to swipe to follow this twisting, then quickly flick the Weak (that is, the forward point) from your right toward their left at their head with crossed hands. Quickly twist back through, or rip out toward your left side with the half edge. Thus, you have completed the Obscured Cut. This Obscured Cut is executed in many ways; more about this in later sequences.

Twisting Cut [Windhauw][43]

The Twisting Cut is carried out in the following configuration: if your counterpart cuts at you from above, cut at their sword from below, from your left with crossed hands (so that your pommel peeks out[44] from under your right arm). As soon as it glances off, immediately step out with your left foot away from them (well on your left side). Pull your sword's pommel back up into a circle toward your left side, so that your long edge touches their head in a sweeping movement over their right arm (and behind their blade) or [you] hit over their right arm. (Regarding this, view the large portrait on the right in the following Figure designated with the H.)

And, following this, immediately let your sword fly out next to your side and quickly cut back counter to this through the cross. This is how it is executed.

---

[42] A *Blende* is a cover or means for obscuring eyesight. *Blenden* means 'to obscure one's eyes', or 'to obscure defensive aspects from a besieging army'. This is an attack at a target you can't necessarily see clearly.

[43] *Winden* and *wenden* are related words, and both can mean 'to twist or turn about an axis'; while a complete revolution is rare, a twist or turn of 90° to 180° is common. In addition, *winden* means 'to twist, wrap around, curve back and forth', and describes both the growth of vines and the serpentine movement of snakes. *Wenden* means 'to turn back in the direction from which one came, to reverse or change the course of something in motion'. They are consistently translated in this text as '*winden* = twist' and '*wenden* = turn'.

[44] '*Aussehen* = to appear, look like, here, to look out from'. The pommel can be seen sticking out from under the arm.

Crown Cut [Kronhauw]

This is conducted like this: when you stand in Plow (or otherwise fence up through a stance from below—this was discussed in the preceding chapter) and your counterpart cuts at you from above, then lift the crossguard horizontally, catch their strike in the air on your ricasso or crossguard, and, as soon as it glances off, quickly thrust the pommel upward and hit them behind their blade on the head using the half edge. Thus, you have completed the Crown Cut correctly.

|I.14ᵛ|
Wrist Cut [Kniechelhauw][45]

This one is named after the body part towards which it is directed. Complete it like this: after the first attack, when you have arrived under your counterfencer's sword with both hands high over your head and they hold their head between both arms, cut with the Crosswise Cut under their sword pommel, upward at their wrists or at the joint between their hand and arm. If they hold their hands too high, then cut from below upward at the knot [of bone] by the elbow (using the abovementioned Crosswise Cut). This is how it is carried out.

Plummet Cut [Sturzhauw]

Attention should be granted to the fact that there is a small difference between the [Plummet Cut] and the [Hairline Cut]. Although this cut is a High Cut, it is named the Plummet Cut because, in the cut through, it always Plummets headlong from above so that the point moves toward the counterpart's face in Ox. It is mostly used in the approach or in the Onset.

Change Cut [Wechselhauw]

The Change Cut is nothing more than changing, with cuts, from one side to the other, from above to below and back again, in front of your opponent in order to mislead them.

Flick [Schneller] or Tagging Contact [Zeckrühr]

A flick or a tagging contact is almost a concept, one in which the cuts are not cut but instead are flicked. They are completed in the middle of the work or when the work is complete, whenever you have a suitable opportunity. Namely, you allow the weapon to snap from above, from both sides, or from below at your counterpart with the flat or outermost part of the blade, or you flick over or under their blade in a fluid motion.

This is a brief summary of the actual description of the cuts, as they usually occur in sword fencing. Though |I.15| the same are described here with such strikes, steps, and cuts that they can only be easily used on one side and manner, yet they can be fenced on both sides, so I wanted to remind the gracious reader about this here. That is: with the same configuration,

[45] The *Knickel/Knichel* is the bone protruding on the side of your wrist that conveniently faces your opponent.

each of the preceding cuts, as is established and is executed from one side, can also be completed from the other neatly and suitably. I have then only skipped this in good faith because the extensive repetition and repeated description of the same is excessive.

Because there may be questions about many of the cuts and their usage—why they occur, when everything is sufficiently understood in the four Main Cuts as well as the Squinter (in which the other inverted cuts are contained)—I wanted to admonish the kind reader of this art that such cuts (as described previously) are all apprehended in the five Master Cuts since they originate from the same. However, they were actually invented as another division of the art, and named with their different names, by those learned in this art for more diligent and useful examination so that the art, which is hidden and wrapped in another, can be more readily and more easily understood, grasped, and considered in different ways.

A useful reminder about counteractions [Uersetzen] [46]
Chapter 5

As fencing is based on two noble concepts: namely, the cuts on the one hand—with which you desire to bring your enemy to a standstill—and the counteractions on the other hand—that is, how you remove the strikes which your enemy directs at you and how you should make [them] a powerless thing of no importance. The manner in which you complete the cuts and direct them in the work has been sufficiently explained earlier, and because each cut is used both for defense against your enemy's strike (in order to remove [it]) and also to injure their body, the cuts could not be taught without teaching about the counteractions. Therefore, as you previously learned to cut the |I.15^V| cuts, you will likewise have also been taught and instructed how you should move the cuts aside. This cannot be separated (whether or not it now really belongs to the cuts), yet it is also necessary to deal specifically with this using a different division.

Therefore, take note from the beginning that the counteraction is two-fold.

The first is that you only counteract in the common way, due to fear and without any specific advantage, in which you then do nothing except hold your weapon in opposition to your counterfencer, such that you catch the strikes as they occur from your opponent. Also, you do not desire to damage them; it is solely sufficient that you can withdraw from them without injury.

When you are compelled into this counteraction with force and haste, observe that you can still free yourself by stepping away and you can return with advantage in the Before. Liechtenauer also informs about this counteraction, when he states:

Guard yourself from counteraction / if this plight happens to you, it will burden you. [47]

---

[46] *Versetzen* at its most basic means 'to move or shift from one location to another'. Because of its links to *vorsetzen* and *fürsetzen*, there is a strong relationship with placing something in front of another item, yet the reflexive form merely moves the body to another location/position. It is commonly used in fencing treatises as a response to an action, and while this response usually places the sword somewhere in front, the response itself may be defensive or offensive. "Counteract" is an English term broad enough to cover all of the possible movements that works in context and includes the concepts of (offensive) defense.

[47] Verse 71 of Liechtenauer's didactic poem on the long sword.

He does not intend to absolutely forbid the counteraction with this, meaning that you should not learn anything except cutting. However, it is to your detriment (as mentioned above) if you were to accustom yourself to counteract to the point that it becomes an excess, such that you would be forced to only counteract. Thus, it is not useful that you scream loudly with strikes, nor that you would want to thoughtlessly cut in simultaneously with their strikes with your eyes closed. This is not fencing, but instead is consistent with a foolhardy peasant's brawl.

For the sake of greater usefulness, I want to separate out cuts and the counteractions that occur with a strike and to teach you solely how you should use such cuts for counteracting, which can also occur in two ways. Firstly, when you proceed to move your counterpart's strike aside or rebuff it with a cut, and after you take their frontal defense, you rush their body with a cut.

|I.16| The other type of counteraction is that you simultaneously counteract and injure your counterpart, which the Old Masters then praise as deservedly superb. The saying arose from this that, "a correct fencer doesn't [solely] counteract, but instead, if their opponent cuts, then they also cut; if their opponent steps, then they also step; if their opponent thrusts, then they also thrust."[48]

Regarding the first: you should now know that the High Cut suppresses all other cuts below itself from above (thus the Wrath, Middle or Crossing, and Low Cut). In this way, when you leap out of their cut and simultaneously cut at their cut in that moment in which they strongly let fly (such that you correctly hit the same), you weaken their weapon to an extent that you can well cut the second [strike] at their body before they correctly recover.

Now, as the High Cut (from above with a downward motion) suppresses all the other strikes, it would be driven off by the Wrath or Crossing Cut, and the Low Cut also takes out the High Cut by lifting the High Cut upward (if it is executed with strength and if it is supported by stepping out).

If two identical cuts with their assigned steps thus occur together—so that one fencer steps a little before or after the other, as that also occurs in the blink of an eye—the cuts also bring their counteraction into play. This counteraction is included in the first and straight cuts.

The second counteraction is that you simultaneously counteract and hit, which occurs with the inverted cuts like the Squinter, Glancing, Crown, and Crosswise Cuts. You heard above during the explanation of each cut how these are to be completed.

These inverted cuts were superbly invented for the purpose that using them simultaneously counteracts and hits. Therefore, so that I deceive in nothing and so that you can be well advised to greater understanding and instruction, I will provide an example of the counteraction with the Crosswise Cut.

|I.16ᵛ| Position yourself into the Wrath Guard. If you are cut at from above, step with your right foot toward your counterpart's side and cut a Crosswise Cut—that is, short edge across and inward simultaneously—such that you intercept their cut on the Strong of your blade close to their crossguard and you hit their left ear with the outer part of your sword. Thus, you have then simultaneously counteracted and hit, one with the other.

---

[48] This saying seems to be a paraphrase of Sigmund Ainringck's gloss of verses 70–71 of Liechtenauer's didactic poem, which mentions cuts and thrusts but not steps. Curiously, this part is not present in the fragmentary version of the gloss in JMR, leaving it unclear where Meyer heard or read it.

Regarding the other cuts which neither counteract nor hit,[49] such as the Short Cut and Misser, etc.: [they] are not actually considered to be part of fencing, but instead are admitted only due to circumstances or by chance so that fencers can use them to mislead, incite, anger their opponent, and drive their opponent from their advantage. This does not often occur without risk, so no counteraction can occur when using them.

Handwork [Handarbeit][50]

Following the previous, sufficient reports made in the first part of the fencing, both about the stances and cuts; these have brought you to this point where, by using the same, you will come under your opponent's sword. The correct seriousness and striving arise only at this point: namely, how you drive them, frighten them, and force them to such an extent with all the types of work in the Middle, like misleading, pursuit, changing, doubling, and pressing after, so that with swift work, you can achieve and obtain the prize (which is the goal that all fencers desire).[51]

It is by no means sufficient and correct to have begun, if you do not likewise also carefully press forwards towards that upon which the major skill is based and which embodies everything in this chapter about Handwork. This must occur in many ways, each of which has its own specific terms and manner of speaking. Therefore, it will be necessary to clearly report and inform you about the same, as to what they are and how they should be completed.

|I.17ᵛ| Indeed, the entirety of Handwork is found primarily in the binding (or, that is, remaining): to pursue, slice, strike around, move around, mislead, fly away, set off, counteract, pull, double, invert, snap, miss, circle, round, twist, twist through, change, change through, slice off, crush hands, shift forward, hang, rip out, block, restrict,[52] grip past, run in, etc.[53]

Binding on, remaining, feeling [Anbinden, Bleiben, Fühlen]

This is when the swords touch on one another. There are two types of remaining: the first, when the swords are held on one another in order to see what an opponent wants to fence and where they want to attack their counterpart. Afterward, regarding the second, which occurs with strikes: when you position yourself as if you were going to pull a strike in order to recover but only flick back around obliquely, arriving back in again with the short edge at the point where you had cut with the long edge previously.

---

[49] In order to render the sense in English, the German requires some major rewriting, including changes to the grammatical voice.

[50] While Meyer refers to fencing itself as a 'Kunst = art', he refers to the action of doing so as 'Arbeit = work', not 'Übung = practice'. Arbeit specifically referred to labor with one's hands. He thus labels the central part of a fencing encounter (between the Onset and the Withdrawal) as Handarbeit, or "Handwork".

[51] The techniques in order are: Verführen, Nachreißen, Wechseln, Duplieren, Nachdrucken.

[52] This is Verstillen in the text, which is "restriction".

[53] The sequence of techniques is as follows: Binden oder bleiben, Nachreisen, Schneiden, Umschlagen, Umlaufen, Verführen, Verfliegen, Absetzen, Versetzen, Zucken, Duplieren, Verkehren, Schnappe, Fehlen, Zirkeln, Rinden, Winden, Durchwinden, Wechseln, Durchwechseln, Abschneiden, Hände drucken, Vorschieben, Hängen, Ausreißen, Speren, Verstellen, Übergreifen, Einlaufen.

Remember here the word $\mathfrak{F}$ülen,[54] which means something like 'testing' or 'sensing', so that you can be aware whether they are hard or soft on your sword with their bind, etc.

Pursuit [$\mathfrak{N}$achreisen]

This is an especially good Handwork, and whoever is well practiced in this and knows well how to use it can deservedly be praised as a master. You conduct yourself in pursuit like this: you hurry after your counterfencer to the opening (because your counterfencer cuts with their weapon either too high or too low or too far out), and you thus prevent them from completing their cut. This can properly be used against those who fence with the cuts sweeping wide around themselves. I will explain it to you with this example, so that you can better understand it.

When an opponent fences against you, [you must] perceive which [of the four] part[s] they hold their sword in. If they move in the right Ox (which is in the upper right quarter), immediately or in that moment in which they lift up their sword from there to change to the other side (or only pull up to strike), you should quickly and cunningly cut under the same |I.18| and use those cuts and techniques from which you can simultaneously and immediately have your counteraction. If, however, they fence from the low guards, then take note of whether they fence from the left or right sides so that you cunningly follow them with the long edge under their sword and strike at the closest opening immediately and in that moment in which they lift.

Slicing [$\mathfrak{S}$chneiden]

Among the Handwork, this is also correctly considered a core technique. When you are rushed by your counterpart with quick and fast sequences, you have no better technique to obstruct and hinder than with the slice, which you should keep in mind beforehand in all sequences as a special gem invented for this purpose.

You must complete the slice as follows: after you receive your counterpart's sword with the bind, you should tarry there, to feel whether they want to lift from the bind or strike around. As soon as they strike around, then follow them with the long edge on their arm, push them back from you with the Strong or ricasso and let fly. Cut at the closest opening before they can recover.

Striking around [$\mathfrak{U}$mschlagen]

Striking around from the sword means that when you are bound on from your right toward their left, lift back from the same bind to strike around or flick a strike at the other side.

Dropping down [$\mathfrak{A}$blaufen][55]

[Dropping down] is: from whichever side you bind your opponent on their sword, then in that moment in which the sword touches, invert your hand and let it drop down with the half edge

---

[54] "Feeling".

[55] This is a technical term meaning 'to flow, travel, or roll downstream or downhill'. It's what water and other fluids do. It appears in the list of techniques as *umlaufen*, "to move around".

facing downward, and while that occurs, pull your hilt upwards into the high to strike, and this can be executed on both sides.

|I.18ᵛ| **Misleading [Verführen]**⁵⁶

[Misleading] is completed like this: when you display with gestures as if you wanted to fence at an opening of your counterpart, but you do not execute it and instead hit the strike at another opening (at which you think you can most properly arrive without injury). Yet many other techniques belong to misleading, like the Squinter Cut with the face, the missing, flying away, twisting, dropping down, drawing back, the circle, and others.⁵⁷ This is also due to the fact that misleading does not arise solely from the sword, but instead also arises from quite a variety of body language indicators. These are no longer techniques [with the sword], but they are techniques and characteristics of the fencer because it is directed according to each characteristic and custom in fencing: as one fencer is angry, another demure; this one is quick and spry, that one fences slowly. Misleading also uses the same configuration and is directed the same way in the work.

Flying away [Verfliegen]

[Flying away] must occur like this: when you cut at the opening of your counterpart in the Onset or fully within the work, and *they move against you to receive your strike in the air.* Do not allow them to touch your sword with their blade, but instead forcibly draw this strike back in the air with a sweeping motion to another opening. This work is useful against those who are only desirous to pursue your sword and not injure the body.

Setting off [Absetzen]

Because all fencing sequences require two things (as indicated above), namely cutting and moving aside or counteracting the cut with the sword, take note of this here: this Handwork is the correct way to move aside or counteract, with which you not only receive the strike obliquely and do not cut back, but instead, in that moment in which the set off touches, you simultaneously hit at their opening with steps out to the side. Thus, in the Onset when you arrive in the Change [Guard], *if they cut at you* |I.19| *from above,* rise upward toward their strike with the long edge, and simultaneously step with your right foot toward their left, and set them off. In that moment in which it glances off, immediately turn the short edge and flick toward their head.

---

⁵⁶ "Misleading" as an equivalent for *verführen* combines the "leading" part of *führen* with the prefix *ver-*, indicating the deceptiveness of the action.
⁵⁷ The list of techniques is: *Verfehlen, Verfliegen, Wincken, Ablaufen, Verzucken, Zirkel.* 'Wincken = to slightly move a body part in a way that provides a direction or instruction, such as blinking the eyes, nodding the head, waving the hand'. Thus, "tilting". It isn't described in this section, but *wincken* is used as part of the 'third Squinter' in the Third Part of the Sword Section (I.54). Alternatively, it may just be a typo for '*winden* = twist'.

Catapult [Schlaudern]

[The catapult] is nothing other than: you let a cut fly like a catapult at your opponent's head. Therefore, position yourself into the guard of the Fool and pull your sword back through next to your right [side]. In that moment in which you pull your sword back, step toward them with your right foot, and catapult your cut at their head. This Catapult Cut should then fly straight, like a stone thrown from a sling. Whatever else is necessary for the catapult, you will find described in subsequent plays.

Pulling [Zucken][58]

Pulling is that which is very good Handwork. You can masterfully mislead your counterfencer, and you should execute this as follows: after you have bound your opponent with the long edge or cut in at an opening, then quickly pull back upward (as if you wanted to cut at the other side); however, do not continue, but instead quickly complete the cut with the short edge back at that point from which you started.

Doubling [Duplieren]

[Doubling] is making a cut or sequence twice in this way: cut firstly from your right at their ear. When the swords glance together, push your sword's pommel through under your right arm and simultaneously raise both arms. Hit them with the short edge behind their blade at their head. This Handwork is called doubling because a cut can be doubled (or completed twice) by this means: first with the long edge, then with the half edge.

|I.19ᵛ| ## Inversion [Verkehren]

The inversion is: bind onto your counterpart's sword toward their left, and, in that moment in which it touches, push your pommel[59] through under your right arm and simultaneously pull your head toward your right (well out from the strike). Following this, press their blade or arm away from you and downward with crossed hands so that you force them so they can no longer work. You, however, make your own space to work according to your pleasure.

Snapping around [Umschnappen]

There are two kinds of snapping around. The first: when you use inversion to come at their arm or blade (as just described). Then stiffly hold their blade or arm under yours using the crossguard and let your blade snap around at their head during this.

The second: when they have pressed you down using inversion, then yield to them with your left foot toward their right and simultaneously grip up over their right arm with the pommel. Pull that downward and let [your blade] snap around under their blade with the short edge toward their head, so that your hands end up crossed over one another.

---

58 A *Zuck* is "a strong, violent pull", so *Zucken* is actually "violent pulling".
59 Although the text has *Kopff* instead of *Knopff*, "head" does not make any sense here.

Missing [Sehlen][60]

Anyone can well miss; however, only a well-practiced fencer knows how to do so properly, and to use it usefully at opportune times. Therefore, you will want to execute a missed strike advantageously so that you can recover another out of it. Therefore, pay attention: when you cut at an opening and *your counterpart wants to counteract you*, don't allow the cut to touch, but instead let it drop downward and cut at another opening.

    As an example: move into right Wrath in the Onset. As soon as you can reach them, step and cut at the left ear up to their sword, but in the cut and before it touches, lift the pommel and let the blade drop downward as an intentional miss next to their left, and pull around your head. Cut at their other side from the outside and over their right arm at their head.

|I.20ᵛ| ## Circle [Sirkel][61]

When you stand in front of your opponent in the bind and both you and they move the swords in the air above your heads and neither wants to reveal an opening before the other, then the circle is an exemplary and good work to use which you should form like this: cut through in front of yourself with the half edge and crossed hands from above next to their right side, so that your two hands remain above your head. During the cut, cross your right hand strongly over your left so that you can easily reach or scratch their right ear with the half edge. If they then swipe after the sword with their arms downward, step with your right foot well to the side, out to their right side, or step backwards, and cut a straight Hairline Cut at their head.

Round [Rinde][62]

The round has two forms: single and double. The single round is when you withdraw your sword from your counterpart's blade or opening in one sweeping movement over your head and let it fly around in the air so that you form a round circle.

    The double round is this: when you strongly withdraw from their sword so that it circulates twice in full sweeping movements above your head (once on each side). You will extensively see and learn in the sequences how and which of these rounds (both single and double) would be more serviceable for misleading.

Twisting [Winden]

The little word Winden is "to turn over or back" in good German.[63] This work should be carried

---

[60] '*Treffen* = to hit' and '*fehlen* = to miss' are antonyms; Meyer often uses them to indicate the intention or expectation behind an attack. *Treffen* indicates a cut or thrust that is intended to hit or contact an opponent's body, while *fehlen* is an attack that is expected to miss (often intended to be countered by the opponent). *Fehlen* moves an opponent's body and weapon into a position that allows a subsequent attack to hit. *Fehlen* as a deceptive attack should not be confused with *fehl hauwen*, which is the swing and a miss, literally a cut that misses or goes wrong.
[61] The circle involves vertical forward circles in the wheel plane, parallel to your shoulder lines.
[62] The round involves vertical reverse circles in the wheel plane, parallel to your shoulder lines.
[63] See the Twisting Cut (above) for an explanation of this term.

out like this: when you have bound on your counterpart's sword from your right toward their left, then remain firmly in the bind and turn the front part of your blade back inward toward their head, and back out again, so that |I.21| you always remain rigidly below on their sword with the bind (as mentioned, and as can be seen here in this example).

If an opponent cuts at you from the Day, bind on their sword from your right using a Crosswise Cut, and in that moment in which it glances off, push your pommel through under your right arm and turn the short edge back in one flicking strike inwardly toward their head. Remain with the edge hard on their sword in this entire motion. If they perceive the flicking strike and counteract, or if you feel that they want to drop from above away from the sword at your opening, then pull the pommel back from under your arm upward toward your left and hit them again with the short edge through the Crosswise Cut at their left ear.

Twisting through [Durchwinden]

When you have bound on with a Crosswise Cut, and you have (as mentioned previously) twisted the short edge inward toward your counterpart's head, step through with your right foot toward your opponent's right side (between you and them), and simultaneously twist your hilt through under their blade to your left side and move your pommel outwardly over their right arm. Step back with your right foot and simultaneously rip downward out to your right side. Hit with the long edge at their head. Thus, you have both twisted through and also reached or grabbed over with the pommel.

Changing [Wechsel]

Changing requires a practiced fencer because whoever changes inexpertly and at the incorrect time only delays themselves and exposes themselves without cause. It is skilled work for those who are learned in fencing and know how to use changes, and is suited for fencing against those who work only at the sword and not at the body. Changing, however, has many forms: changing in the Onset from one side to the other, changing before the attack from one stance into another.

Item, changing through in the attack against the cut: that is, cut a straight Wrath or High Cut in the Onset from your right toward your counterpart's left side. |I.21ᵛ| *If they cut toward the sword and not at your body:* let the point, including the cut, swipe down and through with crossed hands; step in and cut long to the other upper opening. Observe that you are careful in this, that they don't use pursuit to catch you or set their sword on you.

Likewise, move into Longpoint in the Onset and extend the same long away from you. *If they cut toward your sword and want to beat it out or twist:* let the point drop downward and through, and work at their other side. *If they swipe after your sword and want to counteract,* change through again until an opening is exposed or you observe otherwise suitable work towards which you can cut.

Slicing off [Abschneiden]

You should execute the slicing off like this: hold the sword with arms extended long in front of you or drop into the guard of the Fool. If your opponent cuts at you with long cuts, then slice the same away from you on both sides using the long edge, as many as it takes until you recognize your advantage (so that you can come to other, more suitable work). Pursuit is also secretly apprehended in this slicing off (which also includes the slice), to such an extent that Liechtenauer also includes them in a mnemonic when he ſtates:

Slice off the hard [ones] / from both dangers/directions[64, 65]

That is, slice the hard ſtrikes away from you from both sides; however, more will be subsequently described about this slicing off in the examples and other weapons.

Crushing hands [Ḣånꝺeꝺrücken]

Crushing the hands is almoſt comparable with a slice at the arm (about which there were entries above) because there is no difference except between a High Cut and a Low Cut. *When an opponent runs over you with great, whacking ſtrikes,*[66] move under their ſtrike with the Crown or another high counteraction (or come under them with hanging) and catch their sword on the flat of your blade. Pay attention as you come under their sword; |I.22| *when they rise back upward away from your weapon with their ſtrike:* follow them with the Strong of your sword and drop the ricasso in front of their fiſts so that you catch both of their hands with the Strong of your edge. Push them upwardly and away from you with the ricasso, and ſtrike long at the opening.

Shifting forward [Verſchíeben]

When you ſtand in the right Wrath and [an opponent] cuts at you, let the blade hang behind you and push or shift the hanging blade over your head and under their blade (so that you receive their ſtrike on your flat, and your thumb reſts below on your ricasso on the wide path [i.e., flat]). You can then carry out twiſting or other suitable work (whatever you consider to be the beſt).

Hanging [Ḣången]

Hanging should be underſtood clearly from the previous [example]. Execute it like this: when you ſtand in Plow and your counterpart cuts at you, lift your hilt upward so that the blade hangs somewhat toward the ground and receive their ſtrike on your flat blade. Then work with twiſting at the closeſt opening.

Ripping out [Außreíſſen][67]

If you bind an opponent from your right, invert your sword in the bind and rip out toward your left side in such a way that you two ſtand next to each other in the bind. Endeavor that you can come between their arms from below with your pommel and rip out upward. Or, if you have

---

[64] If *geferten* comes from *Gefährde*, then it means 'risk, hazard, danger'. If *geferten* comes from *Gefärte*, then it can be the movement of travelers, the travelers themselves, the path or direction traveled, or the movement of chess pieces. Embrace the puns, particularly in the verses.

[65] Verse 93 of Liechtenauer's didactic poem on the long sword.

[66] *Büffelarbeit* indicates rough, unrefined work, in which quantity exceeds quality.

[67] *Reiszen* means 'to carry out a fast, tearing or ripping motion', where *risz* is 'a tear, fracture, crack, or fissure'.

grabbed them from above with the pommel over the arms (or in whatever way this may have occurred), then rip out downward (as you will then hear about later in the sequences).

|I.22ᵛ| ## Blocking [Sperren]

Take note: *if an opponent stands before you in Change [Guard] or in the guard of the Fool,* then drop cunningly down onto their blade with the long edge. When it glances off or touches, cross your hands and block them so that they can't break free. Or, if they strike at you from the front, then drop down on their blade with crossed hands and block them.

Restriction [Verstillen]

You should carry out the Restriction this way: *if a fencer arrives before you, who quickly works at the four openings with any kind of work and wants to quickly execute the same from over their head,* drop down on them with the slice on their arms or their sword, and don't let them move back up. Instead, wherever they want to go, follow them with the slice hard on their arm and restrict their movement so that they can't work. As soon as you see your opportunity, push them away from you with the slice and let fly at the closest opening.

Gripping past [Übergreifen]

The grip past [your crossguard onto your blade] occurs like this: cut from your right at their upper left opening. During the cut, grip with your fingers over or past the crossguard or ricasso (but retain the thumb on your hilt). Raise the pommel with the left hand and hit them on the head with the hanging blade above or behind their counteraction.

Running in [Einlauffen]

To run in means nothing other than to run in under their sword so that both swords touch together. Since this relates to wrestling and throwing, I will save it until then, as it is more suitable for understanding to deal with this in sequences related to those [techniques] and I now want to move to the third part of my Recitation.

|I.23| Up until now, gracious reader, you have heard both about the way you can attack your counterpart with cuts and also by what means you can further approach them in the Handwork without injury to yourself. However, this is not enough if a good Withdrawal is not carried out as a third [action]. Therefore, I will provide correct and clear instruction in the following chapter about the Withdrawal.

About the Withdrawal [Abzug]
Chapter 6

Although it has already been stated that a good Beginning relieves almost half of the effort in all things, a bad End can still destroy and ruin everything that was previously done well, as this can be seen in practice. To avoid this befalling you in fencing—that you, after having attacked well and having pushed back safely, are left covered in shame just at the End—I want to actually explain to you here everything about how the Withdrawal should occur.

Therefore, it should be noted particularly well that after each sequence [that you have] fenced, you must withdraw in these three ways: either as the first before your opponent, or last following them, or also simultaneously with them.

If you want to withdraw before them, then be diligent that you crowd and harass them beforehand with techniques so that they must rise up high for protection. *In that moment in which they want to observe how you want to continue with the work*, immediately strike the cut through with the Withdrawal, and move away before (and as) they become aware of it.

If you want to withdraw after them, then take note that this can occur in two ways: firstly, when you wait for your opponent's Withdrawal (thus, *when they cut away*) such that you cleverly |I.23ᵛ| pursue them above their blade with your Withdrawal. Secondly, when you conduct yourself with body language as if you wanted to withdraw before them, but you retain your cut skillfully and secretly short so that *when they pursue you,* you allow their cut to miss and drop down worthlessly, at which point you can cut above their defense at their opening.

If you want to withdraw simultaneously with them, then position yourself so that you always remain above their blade with your cut while stepping outward, and *when they cut away from their right,* you step well out to their left. *If they cut on their left,* you step out to their right and cut in simultaneously with them.

Because every sequence includes this type of thing, you will be able to understand such things sufficiently at the point where the sequences are dealt with.

An admonition about steps
Chapter 7

It would appear as a marvel and peculiar to those without experience in fencing that I also report about the steps, and [those without experience in fencing] will opine that there is not much importance as to how a person steps, and if there were something there, then (as they say), 'the market would teach it well'. However, there is so much more to this, to the extent that (as experience provides) all fencing would be for naught, regardless of how skillfully it is carried out, if the steps are not used correctly for that purpose. Therefore, the Old Fencers— who were also learned and have well considered everything—also set this in their twelve rules:

Whoever initially steps after the cut / may find little joy in their art.[68]

Therefore, each strike must have its own step, which should occur simultaneously with the strike.

|I.24| If you want to achieve anything with the sequences you use but you step too early or too late, then your sequence is ruined and you yourself bring your cuts to naught. Therefore, learn to execute the steps correctly so that your counterfencer can't carry out their sequence

---

[68] The 4th of Andre Paurenfeindt's twelve rules, which were reprinted in Egenolff (1530s): 4ᴿ. Meyer's attribution to "the Old Fencers" is likely a reference to its title, *Der Allten Fechter gründtliche Kunst* ("The Old Fencer's Foundational Art").

the way they want without moving from position, but instead you steal the ground or space from them (as it were).

Now, take note of this in the attack and position yourself as if you wanted to execute large and broad steps, while in truth, you remain close with the feet. In contrast, when they assume that you want to approach them slowly, then you should be quick towards them with broad steps and attack them. So much depends on this that everyone who has learned to fence and to use the same must acknowledge this.

The step is divided into three exemplary types. First are backward and forward, which does not require much explanation since you merely step toward or away from your opponent. Second are also the steps to the side, which are partitioned by the Triangle in this way: namely, stand on a straight line with your right foot in front of your opponent and step with the left behind your right toward their left. This is the single. The second is executed doubly and is carried out like this: step with the right foot toward their left (as previously), then follow with the left behind the right somewhat out to the side and toward their left, and then, thirdly, with the right back to their left. Third, there are the broken or stealthy steps, which are completed as follows: position yourself as if you wanted to step forward with one foot first, and then before you set it down, step back with it behind you (behind the other foot). Because these actually belong to rapier fencing, I will spare you this until then.

|I.24ᵛ| **About Before [Vor], After [Nach], Simultaneous [Gleichs], and Indes**
Chapter 8

Since the correct principal techniques of all of fencing with the sword have been explained in good order previously—how many there are, what they are called, how they should be carried out and completed—you have been guided, so to speak, to the point of bringing small sequences into action.

Because your counterpart may also likewise have had this instruction which you have grasped and can also oppose you in all things, it is necessary that you know prior to this what opportunities you can have in the approach. Practical experience demonstrates how much depends on opportunity, and this is especially true in fencing. This is because no sequence, regardless of how good, can be usefully fenced if it is not used at the opportune time.

Therefore, pay attention in all three parts of any sequence—namely, in the Onset or attack, in the Middle or Handwork, and following thereafter in the Withdrawal—and to the Before, Simultaneous, After, and Indes, the use of which will be no minor aid to you in fencing.[69]

The Before is named because you attack your opponent first with your cuts, and furthermore drive them away. This is so that they cannot get to their intention or sequence, but instead must be in fear and use counteractions so that they can protect themselves from you.

The Simultaneous is when both you and your counterpart are of one mind, completing your cuts simultaneously with one another (which is also incorporated under the word Indes).

The After is when you would be rushed by your opponent (as indicated above) so that you can't complete your intention. There is thus a constant alternation and exchange between the Before and the After |I.25| because now your counterpart gains it, now you gain the same back.

---

[69] Meaning this will be a major aid. German likes to use the negative of the diminutive or superlative as rhetorical flourishes.

Whoever has the After (that is, whoever is then forced to always have to counteract), that fencer should be mindful of the word **Indes** and not forget the same, because they must overtake the Before through it if they want to withdraw without injury. Whosoever pays no attention to [**Indes**], even if they act vehemently, will never learn any good fencing.

Indes [70]

Many have understood the word **Indes** (and likewise its origin) to be in the Latin word *Intus*,[71] and apprehend therein 'fencing inwardly' (which originates in twisting and work of this kind); however, you will subsequently hear that this is not true.

I will leave the little word *Intus* and what it means to the Latin speakers. However, the little word **Indes** is a good German word and contains inside itself a serious remonstrance for quick and serious consideration that fencers should always and quickly consider with intense focus: for example, in that moment in which you initially strike at the left, you also—simultaneously—observe the opening to the right. Then thirdly, you also perceive—just as you rush to the opening that you saw—where or with which techniques your opponent can approach you so that you do not attack your counterpart's opening too forcibly and become injured.

Thus, the little word **Indes** reminds you that you have a sharp perception which observes and perceives quite a lot, and also that you can sufficiently learn your opponent's body language: what kind of technique they have in mind to use, which openings that same technique exposes, and where they will open up.

All of the art of fencing (as Liechtenauer says) is found in these things, and the little word **Indes** reminds you about all of them. If you do not perceive the same [word], guiding all cuts with consideration and care, you will simply rush in to your detriment, as can be seen in all fencers who thus make a frightful noise and (as they say), 'want [to strike] above or outward but never at'.

[70] In this chapter, Meyer will explain his interpretation of *Indes* to be a small word fraught with meaning: in that moment in which you strike, you should simultaneously observe the other openings that you could also hit, and *also* all of the sequences your opponent can use to strike or counteract your action. It is therefore a point of hyperfocus on what you are doing and what you and your opponent could be doing at the same time and subsequently. Meyer states that he will use *Indes* as a shortened form to call all of this to mind.

Within his text, Meyer tends to use *Indes* to indicate a disruptive action: that is, because he is describing actions sequentially that occur simultaneously or disruptively. Thus, *while* you are performing the above action, execute another given action *at the same time* and *in the same space*. Because *Indes* is an adverb of time in German, it will frequently appear adverbially in the translation, thus *Indesly*.

[71] By the time Meyer is writing, *Indes* is an ENHG word meaning 'in the [temporal] middle'. *Intus* is a Latin word meaning 'within, on the inside, inside', which is closer to the meaning in MHG.

|I.25ᵛ| **Introduction to the sequences, how they can and should be formed from
the techniques explained previously[72]**

Chapter 9

Because these techniques discussed and explained previously are actually nothing more than a beginning and an elemental foundation from which all fencing sequences of the sword can be gathered, it is thus necessary that I first demonstrate in what way this should occur before deriving any sequences from them. If you want to write a word correctly so that the letters that are useful and proper for this flow in an orderly way, one after the other from your pen, then you need to hold all of the letters in your thoughts and memory and fundamentally know the type and property of each one. Likewise, you should grasp the previously-described techniques well and hold them in your mind to such an extent that, whenever you approach another to fence, these appear then in your mind when they are needed. However, because not all letters can be used for each and every word, it is also impossible to want to make use of all of the techniques in each sequence as they have been explained.

For that reason, you should pay attention to how your counterpart positions themselves against you (as events require). Also observe how the person is—whether they are fast or slow, large or small—and know how to use your work accordingly and to oppose them. Each entire fencing sequence is, as indicated above, divided into three parts—namely, into the Onset, Middle, and Withdrawal—and these three pieces have been explained in order and I have indicated what is to be observed in each.

Thus, in the Onset, you must initially use any cut through the stances with which you opine that you can best attack and quickly overwhelm your counterpart. And when you then attack in the first part |I.26| and have approached them, or perhaps have come under their sword, then you must additionally have more techniques so that the entire sequence is completed. You learned these from the second part of this book, namely the Handwork, at which point you forcefully push at all four openings and they are not able to approach you with damaging work or any similar techniques, since you then have a sufficient excess of the same.

And so that you then bring the sequence to a close, you must have even more letters for this word and you must seek them in the third part of the book—how the sequence can be completed and you can Withdraw without injury or leaving your counterfencer with a reversal.

Therefore, see that you push them forcefully in the Middle work so that you can arrive at the Withdrawal before they are aware of it (as will be taught in specific sequences). Or you provoke[73] them to cut down at you so that you can simultaneously cut from above over their sword (while stepping outward) and can properly make your Withdrawal and recover. In order that you can understand this in its entirety, I will present to you an entire sequence made from all three things.

In the Onset:

➤ Move into the right Change [Guard].

---

[72] This would seem to be the beginning of the "second part" of the Sword Section, based on the description in Chapter 1.

[73] *Reitzen* is "to irritate and/or entice to do something", so: "provoke". Meyer discusses this tactic extensively in the Dusack Section, see II.16ᵛ.

➢ Pay attention: *as soon as they raise their sword to strike,* strike through, quickly and up-wardly, before them and cut a Crosswise Cut from your right simultaneously with them.
➢ Step well to their left side in the cut.

If they move straight toward your head with their cut:

➢ You hit them with the Crosswise Cut at their left ear.
➢ If, however, you take note that *they do not cut straight at your head, but instead turn their cut with the long edge toward your Crosswise Cut to counteract it,* quickly cut at their right ear with the long Crosswise Cut before [their cut] touches. Step Indesly with your left foot well around to their right.

You have now attacked from the Change [Guard] with two Crosswise Cuts at both sides, one after the other. This moves you out of the first part with this attack. You subsequently want to move into the Middle work; the second part helps you |I.26ᵛ| in this way:

If they strike from your sword around to the other side:

➢ Pursue them with a slice at their arm.
➢ Push them with the Strong of your blade or with your ricasso in a shove away from you.

In that moment in which they are still stumbling from the impact and have not yet recovered:

➢ Hastily lift your crossed arms and hit them with the short edge, over their right arm and at their head (as this has been mentioned) *before they have recovered from the impact.*

If they have recovered and swipe upward to counteract:

➢ Let your sword fly up again and cut a Crosswise Cut at their left ear, with a with a step backward with your left foot.

Or, if they do not rise up or strike around, but instead remain with the slice or the long edge on your sword:

➢ Turn your sword so that your half edge contacts theirs and rip their sword out to your right side. Let [your sword] snap around in the air Indesly, so that your hands rise, crossed together, high over your head again and then hit them like before.

Before they recover from the rip with the short edge at their head:

➢ Step back following this with your left foot and cut a Middle Cut across with the long edge, from your right at their neck, and, in that moment in which it glances off, pull up to your right with High Strikes.

Thus, you now see how one technique always has to be used in series, according to oppor-tunity and necessity, and must be set together until an entire fencing sequence is created.

And remember this as not least: that an entire sequence can also be completed with only two or three strikes, especially if you hastily attack with the first strike and cut away with the second, or if you carry out three strikes: namely, attack with the first, follow after with the second. In these strikes, you either hit with the first or last, whichever can occur most properly.

Whenever and at whichever opportune time such must occur is not important to discuss here; after you have understood all other fencing sequences in this book and learned them diligently, then (as they say), 'the marketplace will teach you'.

|I.27ᵛ| ## How one should fence at the four openings.
Chapter 10

Because, dear reader, all components that support sword fencing have been explained previously to such an extent that any serious person who assesses such things (in addition to diligent practice) will have sufficient instruction for understanding all subsequent sequences, I now want to describe, using one stance after the other, how one should hold oneself therein and how one should fence with another from each.

Prior to this, and because all of your cuts and techniques should be directed at or toward the four parts of the opponent, you must also be attentive to the same four previously-mentioned parts. At the beginning of fencing from the stances, it is therefore necessary that I actually describe these parts, which I want to present and set down using the following example. And firstly:

➤ In the Onset, when you approach within one Klafter[74] of your opponent, strike up from your right in front of them through their face—once, twice, three times—so that in the third strike upwards in front of them (when you have your left foot in front) you arrive in Longpoint.

➤ From there, let the front part of your blade drop downward toward your left and, in that moment in which your blade has just dropped downward, pull your hilt upward while this is happening, step and cut the first from your right side toward their left ear.

➤ As soon as this strike has hit, quickly pull back in one sweeping motion and cut the second obliquely from below and toward their right arm. With regards to this cut, retain your crossguard high over your head.

➤ Simultaneously with this Low Cut, step with your left foot slightly toward their right outward to the side toward them.

➤ As soon as this has also hit, you should quickly pull your sword back upward and toward your right, and cut from your right at their left lower opening.

➤ Before that has correctly touched or |I.28| hit, pull back around your head again and cut the fourth slantwise toward their right ear.

➤ From there, cut a Crosswise Cut around and withdraw.

These first four cuts should be completed quickly and swiftly, from one opening to the other, including all of their steps.

Additionally, your sword or swordblade can hit and touch in three ways when forming the cuts properly: firstly with the long edge thereof (as was just taught), then with the short edge, and finally with the flat. Therefore, it will also be necessary that you can nimbly guide the short edge to all four openings (just as well as previously with the long edge), then also finally to hit with the flat (just as was mentioned here previously with the half edge), flying freely from one opening to the next—namely, with the inward flat to the right and with the outward (which is the reversed or inverted flat) to their left.

---

[74] In central Europe, a *Klafter* was a unit of length, volume, or area measured according to a person's outstretched arms, and as a unit of length equals approximately 6 feet or 1.8 meters.

In order that you become more practiced in the same, you should always change around with the first strike. Thus, if one time you have cut your first strike at the upper left opening and the second to their lower right opening, then continue as taught above, (as the outermost numbers in this small illustration indicate here). You should afterwards cut the first at their lower left, the second at their upper right, and continue as the second number shows you in the small image mentioned previously. Afterwards, cut the first at their lower right, the second at their upper left, and continue as the third numbers indicate. Finally, you cut your first toward their right, and continue as the inner number indicates. [75]

And learn all of that first with the long edge as mentioned, then with the half edge, and finally direct yourself in the work with the flat. When you can do this well, then the second part follows: namely, that you know to protect the four openings from these cuts as taught, and either hold them off with your swordblade or |I.28ᵛ| you rebuff them from you with counter cuts (which is better). These are the two principal parts in fencing and the origin of all, from which all other aspects flow.

Furthermore, a chance aspect follows as the third, which is actually called the practical application.[76] The practical application is this: that you can now guide your cuts well from the stances to all parts of the opponent, which is the first principal part in fighting and must be brought into the work in the Before. If your counterfencer is equally and quickly finished with the second principal part—the counteraction to remove or to hold up your cut in the After so that you cannot reach your intended goal with these cuts—then the third aspect, which is called the practical application, comes into play. This is the cunning knowledge[77] and teaches you how—when you perceive your strike as being unsuccessful or unproductive at one point—you quickly and immediately pull back from that point before this strike hits or you allow it to pass over as an intentional miss, and then guide it to another opening. If they also want to counteract that, then pull the same back and then let fly vigorously from one opening to the other, as long and as many as needed until you can overwhelm and hit.

In order that this may be better recalled and understood by those learning, I will provide and set forth a few good examples of the same which, in my opinion, easily and clearly teach this. From this, the benevolent reader can receive sufficient report to understand all types of favorable and demonstrated techniques (as discussed previously in the Middle work) like this:

When you have struck upward into the Onset (as taught previously) and have recovered to strike, let the first and the second hit hard (as above), but do not let the third hit. Instead,

---

[75] The descriptions of the third and fourth sequences are reversed compared to the numbers on the diagram: the third should begin on the opponent's upper right and the fourth on the lower right.

[76] *Praktik* is the practical aspect of a subject, or the practicing of that subject, the training/exertion required to gain mastery. Use of "craft" or knowledge in GRIMM is limited to the secretive, evil, deceptive, alchemical, dark arts.

[77] *List* is both knowledge, its clever application, and cleverness/cunning by itself. It appears mostly as a positive noun in ENHG but picks up more negative connotations over time.

quickly pull that one back in one sweeping motion before it hits, so that you can hit with the fourth faster and earlier. Item, if the first hits, quickly pull the second and the third back in one fast sweeping motion, |1.29| and let the fourth hit. Likewise, threaten with the first and the second up to the opening and pull the same back, and guide another in to the second closest opening. When you use pulling [of cuts], you can and should attack with the first, but changing or cycling through (as taught previously with regard to the numbers), namely that you now pull this one back and then the other, and you allow [your] strikes to miss. While doing this, you have the same concerns and pay the same attention as if they were to attack you at any opening so that you would soon be on their sword with a bind from this outward pulling. The dropping down and missing and the like now flow from these pulled cuts.

That is, when you guide a cut at these abovementioned parts of the opponent, and in that moment in which you perceive that *they would counteract such a strike:*
➤ Don't pull back immediately, but instead—*depending on whether they do not perceive your observation*—allow [the strike] to drop down past on this same side as a complete miss.
➤ Cut quickly at another opening *before they correctly conceive of* what you have planned.
Example: when you have recovered to strike using the upward strikes (as taught above):
➤ Step and cut in from your right, high and toward their left ear.
➤ *As soon as they swipe at that,* let your blade drop quickly downward next to their left with the half edge, and use this to jerk your pommel and hilt upward.
➤ Cut quickly with the short edge at their right ear so that your hands cross in this cut.
Item:
➤ Let the first [strike] at their left ear hit hard.
➤ Let the second run quickly past as an intentional miss on their right (as above).
➤ Strike deeply at their left ear.
You can also—if you have cut in hard with the first—let it drop down quickly on both sides afterwards and still attack the closest opening (if it is open). You can do all of this obliquely and cross-wise (as taught previously) and can also direct [the strikes] singly and doubly on opposite sides in the |1.29ᵛ| work, according to your desire and opportunity.

Furthermore, also learn to quickly guide your entire blade in the work, firstly with the long edge, then also with the half edge, and also with the flat toward their sides at the upper and lower openings, together in one complete sweeping motion, as follows.
➤ In the first attack, cut a High Cut long at their left ear.
➤ In that moment in which it glances off, pull both hands upward so that your pommel is shoved through under your right arm while moving upward and cut quickly with the long edge from below also at their left.
➤ Step Indesly toward them with your left foot behind your right and lift your hilt over your head in these cuts.
Conversely to this:
➤ First cut a Low Cut at their lower opening with the long edge with a step forward of your right foot.
➤ Pull quickly back upward to your right and cut the second from above, also at their left, with a step backwards of your left foot behind your right toward them like before, so that you stand protected behind your blade.

Item:

➢ Pull a High Strike from the right at their left with the half edge, crossing your hands in the air to hit at their left ear with the half edge (as you can see in the two upper portraits on the left in the Figure designated with C).

➢ Pull your crossed hands back upward and hit again with a Crosswise Cut from below at their left ear.

Or also conversely:

➢ Cut the Crosswise Cut from below at their left with a ſtep forward.
➢ Pull quickly upward next to your right and thruſt your pommel through under your right arm during this pull upward.
➢ Flick back at their left with crossed hands from high on your right.

In this way, you will hit with the flat below and above on one side; this works on both sides. And take note: when you hit at the lower right opening, whether with the flat, long, or short, your hands will cross. However, when you |I.30| hit at their upper right opening, your hands do not always cross. Take note of this in the following example:

➢ In the Onset, shoot through in front of them and ſtrike from your left at their right ear with the half edge without crossed hands (inſtead, your pommel projeċts out toward your left).
➢ Pull quickly back upward toward your right and cross your hands in the air. Hit them with crossed hands at their lower right opening from your left.

In that moment in which all of this [occurs], ensure that you have ſtepped out twice to their right with your head well behind your blade.

You can also hit with the flat and long edge from below and above toward their right (as I juſt taught you previously) so that you should threaten and wrench the cuts from one opening to the other. That is, you should pull and threaten here at one side and at the lower and upper opening in parallel.

Namely, when you guide a cut at the upper opening, and you take note that *they are not cutting, but instead are moving to counter your sword,* don't allow your cut to hit. Instead guide your blade to the lower opening. *If they come under the cut,* continue to move with your cut against the Strong of their blade.

The twistings on the sword arise from this work, namely:

➢ When you have bound on their sword from your right toward their left, remain strongly on their blade.

➢ Thrust your pommel unexpectedly through under your right arm yet continue to remain on their sword.

➢ Jerk your pommel back out and twist the short edge at their head from outside.

Thus, you have found three options for the edges and flat (and this on both sides): namely, the outward and inward long edge; Item, outward and inward short edge; likewise inward and outward flat.

You now understand that the third aspect in fencing (as mentioned above) is nothing other than |1.30ᵛ| a correct practical application of the two first principal parts in fencing, through which practical application you are taught how—that is, in the first principal part—you allow the stances and cuts to transform, drop downward, change through, fly away, and miss, according to chance opportunity, in order to steal such cuts from the one who would counteract and remove [them].

Likewise, in the second principal part of the counteraction, the practical application teaches you how you suddenly withdraw your counteraction back from them, pursue them, slice them, press them, etc., executing these around their cuts such that they end ineffectually, or at least that they cannot complete them nor end them at their intended target. And that is the sum total of all practical applications: namely and firstly, that you attack your counterfencer in a virile way with cuts through the stances to their disadvantage and without injury [to yourself], using whatever cunning knowledge and quick execution as can occur. After you have then attacked them, you drive them farther away with Handwork that is simultaneous or gains the upper hand,[78] so that you subsequently either safely withdraw at your pleasure as the third aspect or, if they have to yield to you, that you follow them carefully. The extent of these types of practical application and the variety of ways in which the two are used in naming and in fencing were described in greater detail previously in the chapter about Handwork (as you will find). Therefore, I will continue on to describe fencing from the stances.

Fencing from the Stances
Chapter 11

Although much pertains to the stances, I do not want you to tarry long in any one of them because they were not invented nor provided for this reason, but instead |1.31| so that you can know how—when you lift your sword upward to strike or you have been cut at (while you thus draw in your shoulders in the lifting)—you should quickly and immediately guide your sword

---

[78] *Obliegend* means 'to come to rest on top, lie above, be victorious', etc. "Gaining the upper hand" preserves the locative aspect of *ob[en]* and also the victorious meaning of the verb.

toward them from that outermost point (to which you have arrived while lifting your sword), as is treated here in the Guard of the Day, which guard is created by the High Cut. Thus, when you rise for the High Cut (to execute it), the outermost point to which you arrive using this rising is called the Day. If you are not quickly attacked here (in that moment in which you are still rising to strike), then continue with your High Cut.

This is also the reason that experienced fencers also pause occasionally in the same: namely, that you do not undertake any cut or strike thoughtlessly. And also, after you have already lifted for the same previously-considered strike and have recovered there and should now re-lease the strike, that you should tarry for a small while—and virtually only a blink of an eye at that same outermost point—to further consider whether it is useful to complete your previously-considered strike or whether a better opportunity might have occurred or become relevant to which you would convert to another cut (still from that outermost point), and that you complete the motion of lifting to the High Cut with a Crosswise Cut accordingly.

This is the most proper reason for the invention of the stances and is therefore the reason that it is permitted[79] to fencers that they occasionally rest in a stance, to observe what the other [fencer] is undertaking (so that they will catch them in their own sequences more correctly). This type of waiting, merely to observe and to be aware of their [opponent's] undertaking, requires skill and great experience, etc.

In order that you can have additional knowledge (how and in what way you should use your sword from the Day against strikes flying at you from your counterpart), I wanted to provide the following examples, both for *when they cut or don't want to cut.*

|I.31ᵛ| The first sequence

First, when you approach an opponent and have raised your sword up high by striking up or otherwise lifting (into a High Cut), *if they cut toward the left at your head while you are doing this:*

➤ Leap well out from their cut, somewhat forward and toward their left, and hit with the outward flat toward their in-flying[80] strike, so that you hit their sword in the Strong.

Execute this strongly so that the front part of your blade swings in above their sword at their head in this hit. This will certainly connect if you hit simultaneously with them and arrive above their sword with yours.

➤ Whether this cut hits or not: pull your sword back upward and cut obliquely upward from below at their right arm (in contrast to the previous cut).

---

[79] *Nicht gewert* is "not denied to one"; however, the positive, "permitted" allows for more fluent English.

[80] Because German has directional prefixes *hin-* and *her-* to indicate movement away from or toward an object, I have used "in-flying" and "in-coming" as translations of the adjectives *herfliegend* and *herkommend* respectively, which Meyer often uses to describe weapon movements by both parties.

➤ During this cut, step with your left foot well out toward their right and duck your head far behind your sword's blade.
➤ From there: pull quickly upward again and tilt[81] your short edge at their left ear.
➤ If you perceive *that they will swipe in response:* don't let it hit, but instead let it drop down as an intentional miss.
➤ Quickly cross your right hand over the left in the air and hit them with the short edge deeply[82] at their right ear.
➤ Immediately cut a Crosswise Cut around and withdraw.

And take note here: *if they could quickly follow your abovementioned Low Cut and are so close on the Day that you can't drop down,* pay attention:
➤ *In that moment in which they lift up from your sword,* follow them with a slice at their arms. etc.

The second sequence

If they cut toward your left from below:
➤ Step out toward their left once again and cut on top of the Strong of their sword with the long edge.
➤ As soon as your sword touches or glances on theirs, pull your sword back upward in the air and hit with the short edge in a downward flick deeply at their left ear, while stepping farther around toward their left.
➤ *They will want to counteract that in haste and rise upward against it.*
➤ Therefore, cut quickly with the long edge at their right ear in turn, stepping well toward their right (as before) during this strike around and simultaneously keeping the crossguard high over your |I.32| head.
➤ And take note: *as soon as they strike around,* drop on their arms with a slice once again.
➤ *If they don't want to suffer that but instead want to work free,* follow them, remaining on their arms.
➤ *When they misperceive[83] in the slightest,* then let fly at another opening and cut away from them.

The third sequence

If they cut at your right after you have arrived in the High Guard:
➤ Step quickly with your left foot toward their right out from their cut and simultaneously drop on top of the Strong of their sword with your long edge.
➤ In that moment in which you drop onto their sword, thrust your pommel through under your right arm so that you hit high with crossed hands and the short edge over or next to their sword at their head.

---

[81] *Winken* is a small movement, but is restricted to body parts quite early, particularly the eyes. *Wenken* [*weichen*] is "to slip/yield to the side, bend to the side, list (as in ships)" particularly with directions.
[82] *Tief* means the opposite of high, thus both "low" and "deep". It can also mean 'strongly or extensively', reflecting the depth of an emotion.
[83] *Versehen* = positive meaning of perception, negative meaning of misperception.

If they lift their sword upward and toward their right:
- ➤ Let the half edge drop down next to the same.
- ➤ During that, ſtep well out to the side and toward their left.
- ➤ Cut with the long edge ſtraight from above at their head.
- ➤ Pull quickly back upward and hit with a Crosswise Cut from below at their left ear with a ſtep backwards of your left foot.
- ➤ Cut away from them, etc.

The fourth sequence

Take note. When you arrive in the Onset with your sword high in the Guard of the Day and you perceive *that they will not cut at you so haſtily,* such that you can begin your sequence well in the Before:
- ➤ Cross your right hand over your left above your head so that it appears as if you wanted to thruſt at their face.
- ➤ While doing this, ſtep toward them with your right foot and simultaneously pull your sword around your head and toward your left, cutting powerfully through Crosswise (from your right) at their left ear with the short edge.
- ➤ Pull quickly back and threaten them with a Crosswise Cut long and toward their lower right opening.
- ➤ Don't let it touch, but inſtead yank your sword upward in the same sweeping motion and let your short edge drop down deeply at their left ear for the third cut.
- ➤ Hit deeply at their right ear with the short edge and crossed hands.
- ➤ As soon as this hits, ſtep back with your left foot and |I.33| cut with the long edge from below toward their left arm. (You ſtand as shown by the portrait on the left in the small upper tenants on the right side in Figure G.)

Take note here. If, when you are ſtepping backwards, *a Low Cut of this type is cut at your lower left opening:*
- ➤ Step toward them with your left foot and drop onto their sword with crossed hands and the short edge.

This fixes their Low Cut in place, as this can be seen in the other portrait on the right in the abovementioned tenants [Figure G].

> And also take note: *in that moment in which they pull their sword back upward*, jerk your sword completely to your left with crossed hands.
> *In that moment in which they hit in again*, take out the in-flying cut with your outward flat (from your left and toward their right), strongly across so that your sword flies around above your head in one complete sweeping motion and your hands cross over one another again in the air while your sword flies around above.
> Step well toward their right, simultaneously keeping your hands high, and let the half edge drop down through a circle[84] next to their right ear (regardless of whether the same hits or scratches).
> Cut long with a step backwards.

I have actually described this sequence in this way because many good sequences can be learned and fenced from this. Therefore, you should not merely learn it well, but also consider it diligently. In this way, I will set before you another sequence of this type with a different beginning.

Another

In the Onset, when you arrive into the Day or High Guard, let the blade sink downward (as before) in front and toward your left side.

> Pull around your head, step, and cut a Middle Cut across with the long edge toward their left at their neck or temple.
> As soon as it touches, pull back around your head and cut the second (also a Middle Cut) across from your left toward their right, also at the neck.
> As soon as it glances off, cut the third: a High Strike with the long edge straight from above.

These three cuts should move quickly in series in a single sweeping movement.

If you would like to create more space, then raise your pommel upward toward your left and pull around your head.

---

[84] *Zürck* = *zirken*, "the round that a guard walks, the circular motion of a needle sewing cloth, circular movement". The motion described is also defined above under *Zirkel*, see I.20ᵛ. FORGENG assumes it is a type-setting error for *Zwerch*, and RASMUSSON assumes that it is a type-setting error for *Zuck*.

➢ Take [their blade] out by ripping upward next to your left with the flat or short edge and from below (through their right and toward your right) so that your blade flies around again in the air.

➢ |I.33ᵛ| Cut down from above with the half edge and with crossed hands, next to their right ear and past it as an intentional miss.

In addition, if you can then reach them with the short edge while moving past, let it hit.

➢ Cut a strong Wrath Cut at their left side afterwards.

➢ Subsequently, cut away from them.

This is indeed a very serious and strong sequence which no one will be able to defend against as long as you have the Before.

Counter for the Stance or Guard of the Day

If you perceive that *an opponent prefers to cut up into the Guard of the Day, after striking up over their head:*

➢ Move into the guard of the Key in the Onset.

➢ From there, raise your crossed hands over your head and simultaneously step toward them with your right foot.

➢ During the step, strike strongly and upwardly with the short edge next to your right thigh (through your opponent's hair line from below) such that the sword flies above your head and back again from your left to their right as a Low Cut. After this, remain with your hands high in the counterposture.

➢ In that moment in which it touches, step quickly with your right foot toward their left and cut with the short edge deeply at their left ear in one sweeping movement.

➢ From there, cut two Low Cuts in one movement.

➢ Subsequently, hit at their right ear with a Crosswise Cut.

➢ Simultaneously, step Indesly backwards with your right foot behind your left.

This allows the Crosswise to go deeper.

➢ When that has then occurred, you can quickly withdraw from them with cuts.

Or, *if an opponent who approaches you rises up early,* then pay attention. *In that moment in which they rise from the lower guards:*

➢ Follow them quickly with two strong Low Cuts from both sides—cut quickly and immediately from below from whichever guards or stances you want.

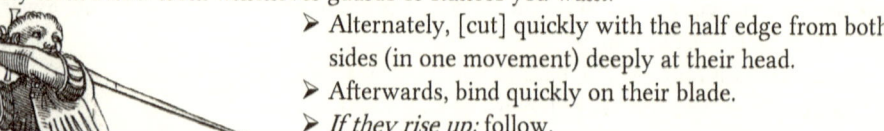

➢ Alternately, [cut] quickly with the half edge from both sides (in one movement) deeply at their head.

➢ Afterwards, bind quickly on their blade.

➢ *If they rise up:* follow.

➢ *If they remain:* twist, rip out, and whatever work you can carry out most proximately.

|I.34ᵛ| Wrath Guard

If you move into the Wrath Guard in the Onset:

➢ Step as soon as you can reach them and cut a swift Wrath Cut at their left ear, *which they will then have to defend against.*

➢ Afterwards, quickly cut an opposite Low Cut (that is, at their lower right opening). You have now attacked.

➢ During this, *before they have recovered to work and pulled their arms in to strike,* drop down on their arm with your sword and prevent them in their movement so that they can't work.

➢ *Before they truly perceive this,* thrust them away from you with an unpredicted shove, *so that they stagger as if they would fall.*

➢ During this, hit at the closest opening that you are certain of.

➢ *If they recover and cut at you:* you should respond with the set off or slice and drop on their blade and against their strike.

➢ *If they move away from your blade:* slice on their arms once again.

➢ *If they remain on your sword:* thrust their sword out to the side with your ricasso and let your sword fly quickly at their closest opening, and from there quickly back on their sword.

➢ *If they don't want to allow their sword to be caught:* follow with your sword on their arms so that you force them according to your will.

In all sequences, you should follow from the sword to the body, and from the body to the sword. *If they want to pull or fly away,* always use the slice to assist (because whoever can't slice also can't fence usefully). If you can carry it out correctly, then you will force them as you want unless they themselves can counter the slice. However, you will find few of them, and those who can't guide the slice correctly, their slice is countered quickly.

If you stand in the right Wrath Stance and *your counterpart cuts at you from their right toward your left:*

➢ Move, with a step forward of your right foot, while you push [your sword] forward over your head and under their blade, catching their cut on your flat such that your thumb is below and the blade hangs down next to your left side somewhat toward the ground.

➢ In that moment in which it glances off, step with your left foot to their right side and twist the short edge under their sword inward at their head (as the small center tenants show in Figure L).

➤ As you have now twisted, hold your sword with the short [edge] on theirs and subsequently rip upward with toward your right your sword (as the central small scene teaches you in Figure F) |I.35| so that at the end of this rip with crossed hands, your hands cross in the air.

➤ Hit at their lower right opening with the inward flat (while your hands remain up high).

➤ *As soon as they swipe to counteract in response,* don't let it touch, but instead pull back upward and cut a Glancing Cut at their left ear.

In this strike, let the blade swing in deeply over your hand and fence quickly away from them.

If your counterpart cuts at you from above:

➤ Step and cut a Middle Cut across from your right through their in-flying strike, and with the long edge away from you so that your blade flies around with the half edge toward their left ear.

➤ Let it drop down next to the same, and immediately pull from your right toward your left around your head.

➤ Step and hit them with the inverted flat from your left to their right ear across through the middle line (as can be seen in the large portrait on the [left][85] side in Figure A).

In case they don't want to cut:

➤ Position yourself in the right Wrath [Guard] and lean forward over your front thigh. Remain standing on your left foot and cut slantwise from your right over your left leg into the left Change [Guard].

➤ From there, rip back upward and outward with the short edge (through the strike line through which you have cut from above) so that your sword returns to your right shoulder.

➤ Execute this once or three times and finally, when you observe your opportunity, lift your short edge in a jerk upwards into the air from your high left and let it snap around over your head in the air into a Low Cut at their lower right opening while advancing twice.

---

[85] The text says "right", but the action described fits the left fencer better, especially if the fencer is mirrored as it is here (the original is left-handed and striking from right to left with inverted flat).

➤ Before this correctly touches, hit again over your hands deeply at their left ear with the short edge. (Let your pommel snap far upward during this; this will move it deeper.)
➤ Pull back around and threaten a cut on their right; however, quickly cut a Crosswise Cut again at their left, with a step backwards and withdraw yourself.

In the Onset, cut up into right Wrath [Guard] and, *as soon as your counterfencer lifts up:*
➤ Raise your hands high over your head and let the front end[86] shoot toward their face, as if you wanted to thrust; however, pull back and hit with inverted hands or the outward flat from your lower right at their left ear or arm, together with a step backwards.
If they cut in from above simultaneously with you:
➤ After the swords have contacted, pull quickly around and hit obliquely and deeply at their upper right opening with the inward flat (so that your hands cross).
➤ Immediately pull back as if you wanted to cut at their left, |I.35ᵛ| but don't do that; instead, pull back without hitting.
➤ Cut with the short edge in a circle at their right ear so that the short edge scratches the skin on their ear and your hands remain high over your head during this.
➤ In that moment in which the circle moves around, step backwards and cut a straight Hairline Cut at their head.
➤ Pull quickly upward with the crossguard horizontal, that is, move with the Crown over your head.
➤ From there, cut a Crosswise Cut at both sides—the first at the right with the long edge, the second at the left with the short edge (such that your thumb always remains below on your ricasso)—and withdraw.

Rule

When you stand in the right or left Wrath [Guard] *and an opponent cuts at you from below at either your right or left opening:* cut at that [cut] from above with the long edge and, in that moment in which the weapons meet, shoot the point on top of their sword at their face. Lift your hands Indesly and work at the closest opening using the sequences mentioned above or below.

Left Wrath Guard

If you move into the left Wrath Guard in the Onset:
➤ Drive one cut, two, or three over your right thigh (as previously over the left).
➤ Step and cut strongly through from your lower left and up through their right, so that your sword flies around in the air back to a Low Cut toward their right.
➤ Immediately pull around your head and cut a strong Crosswise Cut at their left ear.
➤ Furthermore, flick cross-wise and across at all four openings: with the palm up (that is, with inverted or reversed palm) on their left, regardless of whether lower or upper; and with the inward flat on their right (that is, with the palm down).

[86] Meyer uses the terms *vorderen ort* and *hinteren ort* as synonyms for the tip and the pommel, and to specify them in systems in which both ends are used to strike the opponent.

|I.36|

The Ox plus[87]

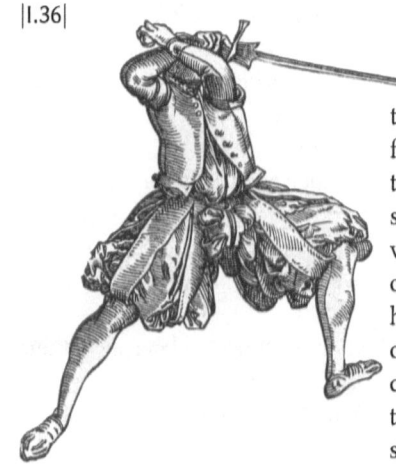

I hope that you have enough instruction from the sequences taught previously and have gained the information [as to] how you set up your cuts and sequences toward the opponent's four openings, and how you should sometimes let fly with a twist, slice, likewise a drop downward, circle, or fly away. This should not be understood as occurring solely from those guards in which such things have been explained, but also for fencing in the majority of all other stances. Therefore, because the Ox is an especially good stance for attacking your counterpart, I want to provide a short teaching and rule here [about] how you should attack or overwhelm your opponent in the Before and force them to counteract you.

And first of all, take note: you have four primary attacks from each side according to the direction of the four main lines (as such was presented for you to see visually at the beginning of the chapter). These lines are the correct paths for all cuts which can be guided and cut by you toward your counterfencer.

Therefore, when you approach with the Plummet Cut toward your opponent, if you hold and maintain the point toward your opponent, this Plummet is called the Ox (as taught above) because of the thrust that it indicates. You can attack out of this stance as soon as you can reach your counterfencer—both from below and above, slantwise or across, as the lines presently show. Regardless of whichever line you [follow to] attack from one side, you should also cut opposite to that same line—across or obliquely—whether with the long or short edge or with the flat.

Thus, if you complete it quickly and strongly in the Before, you force them to allow you other work against their will because, *even if they are already working against this Onset,* you are already at their throat, following this with pursuit, slicing, pressing, and the like. In this way, you do not allow them to complete any work. Therefore, sequences taught previously are also used for this reason, directed both at the attack and in the fencing afterwards. An example:

When you move in the Onset through the Plummet into the guard of the Ox:

➤ As soon as you can reach them, cut a powerful Wrath Cut slantwise from your right toward their left ear with a large step forward of your right foot.
➤ As soon as the cut touches or hits, immediately pull back around and cut |I.36ᵛ| opposite this at their [right][88] arm, also with the long edge.

---

[87] *Mit* here is assumed to be short for *mit anderen/allen,* "with (all) other things".

[88] The text has *lincken arm*; however, the context is the arm opposite the left ear, and the following action takes place on the opponent's right side.

For this cut, step well toward their right with your left [foot], and remove your head well to the side behind your blade.

In that moment, they will perhaps be ready either to cut or otherwise to extend their sword in front to counteract.

➤ Therefore, let your blade hang down behind you and away from their right arm, and simultaneously pull your hilt around your head toward your right.

➤ *If they do guide their blade to cut or extend it to counteract,* take out their blade powerfully and strongly with your long edge or flat (across from your right toward their left) so that you completely break through with your blade.

➤ Let your blade fly (in this movement to take out their blade) with a Crosswise Cut in one sweeping motion back up high around your head toward their left ear.

➤ From there, pull your sword back around your head and cut a strong strike, sweeping in from outside at their right ear with the outward flat. (You see this flat strike in the above-mentioned Figure K in the larger portrait the right side.)

Also, take diligent note: step with the left foot well outward to their right side in this strike.

➤ From this flat strike or Rebound Cut: pull your sword high over your head (maintaining your hands at this height) and let the blade fly around with the long edge at their right arm.

➤ Don't let it touch, but instead, quickly cut a Crosswise Cut toward their left ear with a step backwards of your right foot and withdraw.

If this sequence has already been used against you, then you still have the slice in your repertoire (as taught above) with which you can create space again, to either complete the sequence or to undertake another sequence.

Item, when you can reach your opponent from the Ox in the Onset (as just taught):

➤ Pull your sword around your head and hit strongly with the outward flat from your right (directed across at their left ear).

➤ From there, pull quickly around your head and hit them again with the outward flat from the other side (also across).

➤ After these two cuts, fence whatever appears good to you according to opportunity.

Thus, you can always attack cross-wise and opposite, and also fence farther outward.

In the meantime and when the opportunity arises, you can also attack slantwise from one side, across from the other, and on one side perhaps with the long [edge] and from the other with the short [edge] or flat.

Finally, also take note of this: *if an opponent can quickly overwhelm you in this guard*, so that you cannot move in the Before with any sequence, shoot the forward point into their face with a step forward into Longpoint. During this push forward, turn the long edge toward their in-flying cut. As soon as you have caught it, twist on their sword at the closest opening.

|I.37ᵛ|

Unicorn

Item, in the Onset:
> Move with your left foot forward and strike with the short edge upward and through their face from your right—once, twice—and the third time, remain in Longpoint with your sword extended in front of you.
> Turn the long edge upward toward your right so that your pommel moves through under your right arm and your hands cross over one another. Lift your crossed hands and you stand in Unicorn (as described previously).

> From there, cut an additional two Low Cuts together (with your left foot remaining always in front); the first [cut] upward from your right, the second from your left, both equally close next to your body so that your hands arrive cross-wise again (as before) with the second Low Cut.
> Lift quickly back into Unicorn. Raise your left foot somewhat during this lifting but set it down again quickly.

By using this body language and these formal, meaningless movements,[89] you provoke them *so that they prefer to cut at your left opening.*

In that moment in which they cut in:
> Let your blade drop down in front of you and simultaneously pull your sword around your head to cut across with the long edge (from your right toward their in-coming strike), with a step forward of the same foot, so that you catch their strike horizontally on the strong of your blade.
> As soon as the blades glance together, leap with your right foot even farther around toward their left, and swiftly raise your sword slightly up from their blade.
> While you are lifting slightly (as mentioned), thrust your sword pommel through under your right arm so that your hands cross.

---

[89] *Ceremonie*, GRIMM, vol. 31, column 670: "Since the Reformation, considered to be empty and external, therefore superfluous customs, and simultaneously directed at the body language and gestures, employed to deceive the unlettered people." *Seit der reformation als leere und äuszerliche, daherüber-flüssige gebräuche angesehen und zugleich das augenmerk mehr auf die gebärden und gesten gerichtet: dasunverständig volck zuo betriegen.*

➤ Flick quickly (with the inner flat or short edge) behind their sword and at their head (during the step outward toward their left mentioned previously). This can be seen in Figure C (in the smaller tenants to the left).

➤ In this way, you expose your left side and *they will rush to the same*.
➤ You then do nothing more than pull your pommel back from under your right arm and turn your sword into Longpoint (so that the long edge is turned toward their blade).

Thus you stand in the Simple Brace[90] (as is taught by the other smaller scene to the right in the aforementioned Figure [C]).

---

[90] The translation of "Simple Brace" for *gerade Versatzung* comes from the diagonal braces used in half-timbered building construction, which are called *Versatz/Versatzung* in German and "brace" in English. A *gerader Versatz[ung]* has no additional offsetting wedge to affect its angulation, thus it is "simple" as opposed to complex. (See GRIMM, vol. 25, col.1047.) In fencing with the sword, the Simple Brace appears similar to Longpoint; however, it is aimed slightly higher.

Or,

- ➤ If you have turned the half edge inward and toward their head with crossed hands so that you expose your left side, *they will rush to fence toward the same (as mentioned above)*.
- ➤ Keep your hands crossed and remove your head well toward your right and shoot your blade well over theirs (the closer to their ricasso the |I.38| better).
- ➤ Rip their blade out toward your left (as you see in the smaller scene on the right in the Figure designated with D).

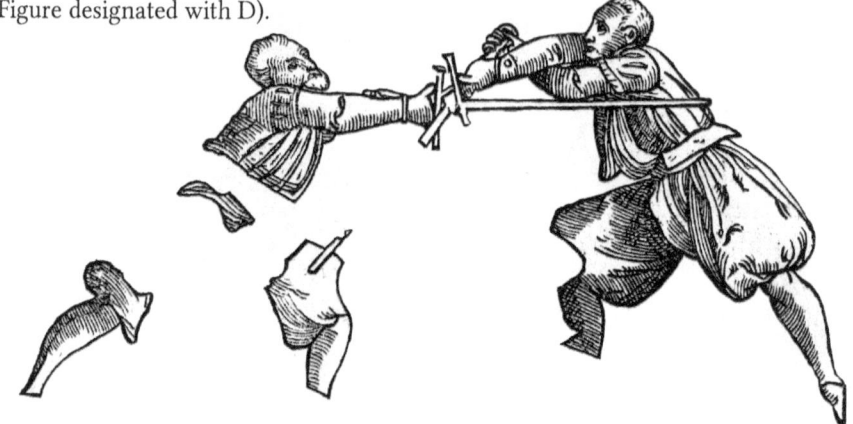

- ➤ As you approach your left [side] during this outward rip, lift your hands and hit back with the half edge (deeply at their left ear) with your palms upward.
- ➤ After this, move quickly back on their sword with the long edge so that you ſtand in Longpoint.
- ➤ Afterwards, withdraw as you please.

Or, as before, when you have moved into the Unicorn in front of your counterpart, then take note; *in that moment in which they cut in from above:*

- ➤ Let your blade move around your head and bind on their sword across from your right toward their left.

- ➤ *As soon as they lift up from the same*, let your blade snap back around so that your right hand moves over the left with crossed hands and drop in front on their arms with the short edge *while they are still lifting up* (as you can see in the outermoſt small scene toward the right side in the Figure designated with I).
- ➤ Thrust with your ricasso powerfully away from you, toward their left and out to the side.

> *While they stumble*, cut quickly at the closest opening or hold them by following after until you can have the advantage.

Item, once you have winged your elbows out and up on both sides, and have arrived high in Unicorn, *if they, your counterpart, then cut from their right toward your left at your head:*
> Step once again with your right foot toward their left (and well out from their strike).
> Drop on them from above with the short edge on the strong of their sword (keeping your hands crossed).

This attack, together with the aforementioned outward step, should be completed simultaneously with one another and toward their in-flying blade.
> In that moment in which the swords glance off together in this way, immediately let the short edge snap back around (away from their sword) and hit with the same [edge] at their head with your palms up.
> Or break through by dropping downward and toward your left.
> Subsequently, pull your hilt back upward around your head and cut with the long edge and with a step outward, etc.

From this Unicorn, you can properly attack with both Low Cuts and the Crosswise Cut as well, and also fence many good sequences (as always), about which you yourself should contemplate further.

|I.38ᵛ| **Key**

This guard is therefore called the Key, because all other sequences and stances can be countered from this stance. Even though such can also occur from others, you must use more force for such purpose than in this one. And the same way a key—a small instrument—opens a large, strong lock without special effort (where you would otherwise have to have exert great force), all other sequences can be skillfully and artfully countered without special effort from this weak stance (as it is viewed). This occurs without risk in this way:

Position yourself into this guard in the Onset (and that even in the same way as you have fenced as mentioned previously in the Unicorn).
   If your counterpart rests to the right or left in the high or low guards:
> Thrust from the Key straight in front of you at their face and into the Longpoint; *they must defend against this thrust (if they don't want to be hit).*
> *From whichever side they strike out at you*, let your blade move deliberately along *the path which they have indicated with their strike outward* and around your head, to cut in at them at precisely the same side *from which they have hit out at you.*
> *If they swipe after your cut* so that you can't hit, let it fly to another opening instead.
> Cut away from them toward another opening *before they observe this.*

Secondly, *if your counterpart does not rest [in a stance], but instead crowds you with cuts—if they cut from above or from below, from the right or from the left:*

➤ Take note: *in that moment in which they cut in*, shoot the Longpoint in front of you toward their face once again, and simultaneously with this push forward, turn the long edge toward their in-flying cut.

➤ When you have received their cut on your long edge in the strong, remain hard on their blade and twist quickly inward or outward at their head.

➤ *If they lift quickly up from your blade to cut toward the other side*, cut or flick at their head or arms *(while they are still moving their sword around)*.

➤ Afterwards, hurry quickly back with the bind on their sword.

➤ Consider pursuit, slicing, ripping outward, or misleading at all times.

From this sequence, which is mentioned previously as a counter from this Key [Guard] against the High Guard, you can gather how you should fence without risk from this guard in the Before and how you should use it to attack.

|I.39ᵛ|

Hanging Point

In the Onset, strike powerfully upward and through, from your left toward their face—once, twice in a wheel,[91] and a third time—always letting your sword sweep forward into the correct position in front of your face or turn into the Hanging Point (as the portrait teaches you on the right side in the Figure with the F on the opposite page). And execute that once or a few times, until you perceive your opportunity to attack with a sequence.

If they, your counterpart, cut at you during this (while you stand in Hanging Point) *from above, across, from below, or even at your fingers, or toward the left at your head:*

➤ Step quickly with your left foot behind your right out toward their left and, *in that moment in which they cut*, simultaneously pull your hanging sword upward toward your right shoulder.

➤ From the same, cut simultaneously with them at their head (in the abovementioned step toward their left).

In this cut, jerk your pommel very strongly to the tendons on your inward arm so that your blade swings at their head more powerfully.

---

[91] *Raht* is somewhat problematic, as it can indicate both *rat* and *rad*. The former meaning 'advice' (from a suggestion to a legally binding decision) and also 'store, supply of a thing', and the latter meaning 'wheel, gear, and possibly rotational movement'. I have no idea where RASMUSSON gets 'sweep'. This seems to be the rotational motion similar to the '*Rund* = round'.

> Keep your pommel firmly on your arm and rip out upward toward your left with an extended blade.
> In this outward rip, let it fly around your head and cut a Crosswise Cut strongly in toward their left.

If they cut toward your right from above:
> Catch their strike on your flat blade and step out toward their right, or remain with your blade on theirs (in that moment in which the swords touch together) and twist the short edge inward at their head.
> Turn quickly with the sword out of the twist into Longpoint, so that in response you rebuff their work away from you with the long edge.

If they fence under your blade at your right ear, however that happens:
> Turn your sword again into Longpoint (the long edge downward), so that you set their blade off. While you set off, step at the same time with your left foot quickly toward their right and thrust your pommel through under your right arm (during this step and in that moment in which the sword to be set off still touches).
> Lift your sword high up with crossed hands and hit quickly with the half edge back down at their right ear.
> *If they counteract this:* let the blades drop down next to your right and step backwards with your left foot back, and while you are stepping backwards, cut a powerful Middle Cut across at their left ear or arm and subsequently withdraw.

It is easy to gather from this how to further fence from this [guard].

|1.40| **Iron Gate**

This Iron Gate is actually the Crossed Guard (as mentioned above), which is fenced from like this.

If an opponent cuts at you from above:
> Lift your crossed hands and catch their cut on the strong of your blade.
> *In that moment in which they remove their sword back from your blade (from the aforementioned cut),* cut at their arms with powerful Low Cuts (*while they pull their arms upward*).
> *As soon as they drop down,* fence at their head.

Item:
> Counteract their High Cut as before and, in that moment in which the swords glance on one another, quickly twist the short edge inward at their right ear.
> Afterwards, quickly twist your pommel back through from below up toward their left side and cut with a step backwards long toward their left at their head.

If they were to fence at you from below:
> Drop from above with the long edge onto their sword in Longpoint.

This Iron Gate or Crossed guard is countered by the Key: namely, thrust toward their face so that you force them upward, and then fence at them from below (*in that moment in which they rise*).

Side Guard

You should principally fence from the Side Guard using the Crooked Cut.
> When you hold yourself in the right Side Guard *and an opponent cuts at your opening:*
> Step with a leap with your right foot toward their left (well out from their cut) and cut with crossed hands above and behind their blade at their head.
> If you don't want to rip out toward your left, pull quickly upward with crossed hands and hit strongly from below at their left ear with the outward flat.

If they don't want to cut, then fence in the way you will subsequently be taught relating to the Middle Guard.

|I.40ᵛ|

Middle Guard [Mittelhut]

You will be instructed about this Middle Guard later in the Dusack [Section]. Therefore, as you will conduct yourself the same with one hand there, you should accustom yourself to using both hands here, because even though I was not initially inclined to place this here, I have not been able to avoid it (as there is no other guard from which the Roses can be properly taught).

Take note *when an opponent approaches you who holds their sword extended before them in Longpoint or otherwise in Simple Brace:*

> Move your blade from the Middle Guard around in a circle (completely around their [blade]) so that you arrive with your blade almost back into the first Middle Guard.
> From there, powerfully sweep the Weak out over their arms at their head.

Or, if they were to attack downward from above at your opening (in that moment in which you were moving around their blade using the Roses):
> Ward off their blade with the half edge (namely, when you have arrived for the second time in the Middle Guard), because *no matter how quickly they rush unexpectedly at your opening,* you will arrive Indesly with the Roses so that you have sufficient time for the aforementioned removal.
> After you have warded them off, allow [your sword] to circulate in the air over your head (in order to mislead them) in a round through a circle at the closest opening.

Or, if you have cut into the left Middle Guard in the Onset and *they, your counterpart, cut from above during your cut*:

➤ Step well out and away from their cut toward their right side and throw your short edge upward or outside of their right arm at their head, and allow your blade to shoot well in with this throw (either at their head or over both their arms).

➤ Afterwards, pull your sword quickly back upward and cut strongly from your left (with the long edge upward toward their right arm).

➤ From there, fence further with them from the preceding and following sequences according to your desire.

And because the Roses can also be properly fenced in Longpoint, I will describe the remaining sequences that I wanted to set out in Longpoint.

|I.41| **Longpoint**

Longpoint was actually called the Breaking Window by the Old Ones, because all other sequences can be countered[92] from it. While it can be sufficiently gathered from sequences taught previously how you should fence from this stance (based on a similar one), I want to demonstrate something about this stance using examples because Longpoint is, after all, the end of all binds.

In the Onset, bind your opponent on their sword using the High Cut, and take note:

➤ *As soon as they lift their sword back upward*, cut at them from below at their chin (between both their arms) *in that moment in which they lift their arms.*

Regarding this sequence, see the two small portraits on the upper left side in the Figure designated with the letter I.

---

[92] *Brechen*, which also means 'to break'.

Item:

➢ Bind on them as before.
➢ As soon as the swords touch together in the bind, break through from below using the Roses between you and them and throw the short edge in at their head at the other side.

Or, after you have broken the bind using the Roses from below:

➢ Rip their sword outward and to one side using the short edge from the other side (so that your hands cross over one another high in the air).
➢ Hit them with the short edge from above (deeply at the head).

Item:

➢ Bind on them toward their in-flying cut.
➢ As soon as the blades touch together, thrust your pommel through under your right arm (also stepping well out toward their left side during this) and rise upward using crossed hands.
➢ Cut through the Roses from below and from both sides with the long edge, behind their arms and at their head.

Item, take note in that moment in which you bind onto an opponent:

➢ Release your left hand from the pommel and quickly grab their blade in the bind onto yours with [your left hand].
➢ Afterwards, move your right hand (including the hilt) through below.
➢ Then beat upward toward your right to take their sword (as you can see in the two smaller tenants on the right side in Figure H).

|I.42| Item, *if an opponent binds onto your sword with an extended sword*, then take note:

➢ In that moment in which the bind touches together, change through quickly below and flick your Weak obliquely[93] at their ear from the other side.

---

[93] *Flächling* (m. subst) means 'flat head'; *flächlingen* (adv) means 'in a flat plane, horizontal'; *flächlings* (adv) means 'oblique'.

You will learn to carry out many lovely techniques from this change through, so you should consider this diligently.

Item, if you take note that *an opponent wants to bind onto you or cut:*
> Guide your sword toward them as if you also wanted to bind on.
> Take note: in that moment in which the blades should just barely touch together, thrust your pommel quickly upward and turn the blade up through the Roses from below.
> Catch their cut in this way on your long edge (as the small scene on the right side in Figure N indicates).

After you have received their cut in this way (as mentioned), you can complete the same sequence in two ways:

Firstly, in that moment in which the swords have touched together:
> Move your blade through completely underneath.
> Rip their [blade] out toward your right side.
> Let your hands snap around in the air (or cross-wise over each other) and cut strongly at their head with the short edge.

This is a masterful progression that will not fail you (*even if they cut from above in a different way*).

Secondly, if you have received their sword, then in that moment in which the swords clash together:
> Step well to their left side and cut back with the long edge from outside over their left arm at their head.

This last cut approaches unexpectedly, swiftly, safely, and strongly.

Change [Guard]

Although it is not necessary to set down specifics for the Change [Guard] because all sequences fenced from another [stance] can be fenced properly from this [stance], I did not want to omit setting a few sequences in this, from

which you can also observe some particularly quick and cunning points.[94] Thus:

If you find an opponent in the Simple Brace or Longpoint (as stated before):

➢ Strike strongly through and with the half edge from the right Change [Guard]—once, twice, a third time.

➢ Strike through under their sword and step toward them with your right foot.

➢ Hit powerfully and high with the flat or short edge at their left ear (as the small scene to the left in the Figure designated with F shows you).

|1.42ᵛ| You thereby force them *to move steeply upward.*

➢ *As soon as they do that:* release your left hand from the pommel and let your blade snap around from below (in one hand) and upward toward their right.

➢ Set the front end on their chest and Indesly grab the pommel again (as you can see the same in the smaller scene on the right side [in the] previous [Figure] with the F).

---

[94] *Geschwindigkeit* means 'velocity, speed', yet also 'quickness of thought, knowledge' as positives, and 'trickery, deceits' as negatives. "Cunning" seems to bridge both the positive and negatives of the use of knowledge.

> Thrust them away from you with inverted hands.
> Quickly release your pommel again.
> Guide your sword around your head and cut long in response while grabbing the pommel.
You should use this same sequence against those who like to run in.

Item, if you have taken note *that your counterfencer likes to and will soon lift up high:*
> Strike powerfully up in front of them.
> As soon as you perceive that *they will lift upward,* cut at them across from your right toward their left arm (*while they are still lifting*).
> As soon as this hits, immediately hit again in one sweeping motion with the inward flat (deeply at their right ear).

Breaking Window

This is actually allocated to the high guards. It is only used in the bind after you have come under your opponent's sword. You should refrain from moving into this guard for as long as you see your opponent's point and blade still in front of you, as you will not be safe from anything [in front of you] in this [guard]. However, as soon as you come under their sword then this is one of the most exemplary guards, and you use it like this:

When you have now come under your opponent's sword, it means that you must guide your sword over your head in the aforementioned Breaking Window such that you expose both your arms and fingers. Therefore, *as soon as they cut at your fingers from above:*
> Step to one side, well out and away from their strike (this applies regardless of which side it is).
> Cut a Crosswise Cut toward their in-flying strike.
> You will both catch their cut on the strong of your blade (close to your ricasso) and you will also simultaneously hit their head with the outer part of your blade *if they otherwise complete their cut without pulling back.*
If they cut at one of your arms:
> Cut a Crosswise Cut upward from below toward their in-flying cut once again, and from the same side as the arm *toward which they want to cut.*
> Take diligent note: *in that moment in which they pull their sword* |1.43| *back up,* during that moment, cut Crosswise Cuts upward or across into their arms.
> *As soon as they drop back down,* catch their blade on your horizontal blade or ricasso.
If, however, they cut a Crosswise Cut, or if they cut from below:
> Cut at them from above and onto their blade.
> Or, observe, so that you cut a Crosswise Cut under their blade before they do.
If they were able to cut a Crosswise Cut under your weapon:
> Let your blade hang well above theirs and press it downward away from you with your ricasso.
> You may thus reach their head behind their blade with the half edge.
Or,
> After you have pushed their blade downward and away from you, you can strike around, etc.
Or you may grip past [your cross-guard onto your blade], like this. *If an opponent cuts at you from above:*

➤ Cut opposite to the same from your lower left with the long edge.
➤ *In that moment in which they remove their blade back up from yours (and thus lift upward with their arms)*, grip past your cross-guard with your fingers onto the blade during that moment and lift with your blade over both arms (as you see the same in the two lower portraits [to the left] in the Figure following in the text [M]).

➤ Rip out to your right side.
  If you want, you can throw them like this:
➤ Step with your right [foot] behind their right and grab them by the neck with your short edge.
➤ Indesly thrust your pommel toward your right and above your right arm and away from you.
➤ Throw them over your right leg toward your left and onto their back.

Item, *if an opponent cuts Crosswise Cuts at your left ear*:
➤ Drop from above with the long edge onto the middle of their blade.
➤ In that moment in which you drop, grip past your crossguard onto your blade (with your fingers upward).
➤ Place your short edge behind their blade on their head (regarding this, view the larger scene in the Figure [M] mentioned previously).

➢ Thrust your pommel, on which you hold them, away from you so that they cannot get free
before you have then wounded them.

If they jerk back out from under your blade:
➢ Quickly follow them and grip [past your crossguard onto your blade] once again, and over
both their arms as before.
Or:
➢ *In that moment in which they cut the Crosswise Cut:*
➢ Catch their Crosswise Cut on your hanging blade.
➢ Reach with your left hand onto their ricasso and crossguard.
➢ Twist them out away from you (as the smaller upper tenants also show you in the afore-
mentioned Figure [M] on the [upper] right side).

And finally, take note: as often as you bind on with a Crosswise Cut against a High Cut, twist
the short edge inward at their head.

Conversely, *if an opponent were to twist in at you*, then take note indeed *in that moment in
which they twist the short edge inward toward your left ear from the Crosswise Cut:*
➢ Twist the long edge upward toward their blade.
➢ You then stand in the old Squinter Cut (mentioned above) and you hit them on their head.

This works |I.44| on both sides, like the other sequences, because *if they twist in toward your
right or left, they expose their other side*; therefore, you can easily hit their head by counter
twisting. *When they twist inward*, you twist outward; therefore, you hit and *they miss*.

Take note, when you twist in at an opponent and you enter into the potential danger that *they
will want to twist toward you* (as recently taught):
➢ Twist equally as far [as them], and in this twist, rip out with your half edge facing the
side toward which you have twisted inward.
➢ Let your blade snap around.
➢ Or, fence another technique.

These twistings are eightfold, which will be touched upon sufficiently now and again in techniques. More about the aforementioned twisting will be treated further at another point.

You have now been properly instructed in this first and second part of fencing with the sword. Both about the division of the opponent and also of the sword, followed by the Onset, Middle work, and Withdrawal, in addition to other necessary techniques and teaching including the examples in the second part drawn from the first part. Whatever else additionally concerns necessary techniques, you will find sufficient information in the following book about fencing with the sword, as much as I intend to write at this time.

|I.44ᵛ| The third part of the Sword [Section], in which the following didactic
          poem[95] is explained with many nice and swift techniques, with which a
          fencer who loves the art can usefully read and train themselves in the same.

                Didactic poem for Free Fencers [Frei Fechter Zedel]

    1   Take note: if you want to learn skillful fencing
            You should listen diligently to the Recitation.
        A fencer should hold themselves to high standards:
            Not be a braggart, gambler, or drunk;
    5   [They should] also not blaspheme nor swear,
            And not be ashamed to learn.
        [They should be] God-fearing, modest, and also calm,
            Especially on the day on which they will fence.
        [They should] be moderate, honor their elders
    10   And women. Listen further:
        You should strive at all times for
            All virtues, honor, and virile strength.[96]
        So that you can serve
            [The] emperor, king, princes, and lords with honor,[97]
    15 And also be useful to the fatherland
            And not be a disgrace to the noble art [of fencing].
        Also take diligent note of the word Indes, also Weak and Strong,
            The Before and the After[98]
        Learn feeling to check for soft and hard
    20   Step with the strike whether it is close or far
        Keep the division [of the opponent] well in mind
            Also guard yourself against great wrath.
        Observe the guards and the cuts
            So that the counters are obvious to you.
    25 High, Wrath, Middle, and Low:

[95] Historically, a *Zedel* was a written instrument, in MHG it was specifically a legal instrument, a record or complaint/pleading. ENHG both reduces and expands this original meaning. In the judicial
sense, it is reduced to information presented as a list that was less informative than a legal brief.
However, the concept is also expanded to include an abbreviated or abridged way of communicating
that is contextually comprehensible (the implication being that it is incomprehensible without this
context) and written using terms that are limited to specific locations or to specific technical language.
  The fencing *Zedel* (recitations, didactic poems) meet this definition in that they often contain lists,
they are highly abbreviated or abridged, they are contextually comprehensible, and use highly specified terminology.
[96] This and the preceding line are inverted from the German order because English is an SVO language.
[97] This and the preceding line are combined for English fluency.
[98] This and the preceding line are combined for English fluency.

From them execute all of your marvelous [techniques]:[99]

Squinter, Hairline, Crooked, and Crosswise

    And many more techniques according to your desire.

[100]Ensure that you are the firſt at the field.

30   Before they [can] adopt a ſtance, attack them.

Pay attention to Indes; underſtand me correctly:

    Hit them, before they position themselves in a ſtance.

Whatever ſtance appears good to you.

    Hit them in the After with bold courage.

35 Guide your cuts powerfully from your body;

    Direct your work at the four openings.

If you cut a Crooked Cut, lift quickly.

    With crossed hands, throw the point on their hands.

Let the circle touch on the right,

40   Hold your hands high if you want to mislead them.

|I.45| When you cut a Crooked Cut at their ſtrong,

    Take note: turn through, overrun with this.

You should contemplate misleading with the pommel

    You will annoy them with Tagging contacts, flicks.

45 Step well with the Crooked [cut] if you want to counteract,

    Crossing over will injure [the opponent].

[Cut a] Crooked Cut at the flat if you want to ſtrengthen yourself.

    You should take diligent note of how to weaken them.

As soon as it touches above and glances off,

50   Pull up to the opening if you want to enrage them.

Also, if you want to shoot through correctly,

    Cut Crooked, Short, [or] change through on their ricasso.

Take note: if they want to confuse you with a Crooked Cut,

    Remain correctly on the sword and execute the War,

55 With twiſts, slices, and whatever else.

    Do not let yourself fly too far aſtray.

Also flick with the Weak at the right.

    Flick twice and guard yourself with the ricasso,

And twiſt ſtrongly around your opponent's ricasso.

60   Shove off Indesly and ſtrike swiftly.

---

[99] *Wunder* in ENHG is only attested to mean the miraculous or that which invokes awe. Attempts to link it to a person who inflicts wounds [*Wunde*], have failed to find evidence in the literature (GRIMM, vol. 30, col. 1782, etymological discussion). MHG traces *wunder* (neuter) to the wondrous, and *wunder* (masculine) to *verwunder* which is linked to the wonderous or miraculous, or someone who creates something wonderful or miraculous. There are precious few instances cited in the LEXER (vol. 3, col. 1022) in which 'wunt = wound' offers *wunder* as a plural form: *süezer wunder âne swert* and *ieglîcher wunder von dem bette gie* are attributed to Gottfried von Strassburg and Heinrich von Friberg (who completed Gottfried's unfinished *Tristan*). Instead of the *drei wunder*, here it has has become '*deine wunder* = your marvels or wonders' and is followed by the crooked cuts.

[100] The gloss begins here.

You should carry out the Squinter Cut wisely.

 You can also double it by twisting.

You should also consider the Crosswise Cut as worthy,

 Your skill with the sword becomes compete with it.

65 The Crosswise Cut can counteract

 Everything that comes from the Day.[101]

Execute the Crosswise Cut strongly in the attack;

 Also take note of reversals and misses in addition to it.

Be quick to the Plow and Ox,

70 Threaten them fast with cuts at contrary ends.

Take note of which type of Crosswise Cut is executed with a leap;

 You also miss with it; it touches according to your desire.

You should doubly execute the Misser,

 Likewise double the step and slice.

75 From the sword to the body, reverse

 Twice with it or slice into the weapon.

Pursuit is exceptionally good;

 Protect yourself with slicing, twisting

In the case of a second [slice] or in the middle [of a technique]

80 Let it fly away. Begin with this.

Execute hits towards all four ends;

 Learn pulling if you want to play them false.

Bring the slice off and catapult [to your fencing] as well.

 Rebuff hard and hostile [actions] with the slice

85 Do not rely too much on the Crown,

 Otherwise you will receive injury and insult from it.

Strike powerfully through with Longpoint;

 This holds off all hard and hostile actions.

See that you counter all cuts and techniques correctly

90 If you want to avenge yourself for your part.[102]

You should use the hangings deliberately

 If you want to wrestle; do not grab at the wrong time.[103]

|I.45ᵛ| If you also want to know the core [knowledge] of the Masters,

 Learn to step correctly in all sequences.

95 If you don't counteract too often, you will be much freer;

 Joachim Meyer warns you against doing that.

Introduction to the Third Book

Dear Reader to whom I have granted my didactic fencing poem which, based on the Authorities, I have composed, improved, and correctly organized: I also wanted to explain it to some

---

[101] This couplet appears in reverse order for English fluency.

[102] This couplet is not included in the gloss.

[103] The remaining lines are not included in the gloss.

extent, using many beautiful and clever sequences and examples (so that many can gain more use from it) and to provide a small introduction in order to understand it. Because it is so rich in techniques and all types of cleverness, the longer you ponder this introduction, the more techniques you can gain from it. The fact that rhymes without an explanation are not very useful is clear from other preceding fencing treatises. And you should know that the first part of these rhymes has been explained sufficiently in my previously-taught fencing; therefore, I will take the initiative with this one:

> [104]Ensure that you are the first at the field;[105]
> Before they [can] adopt a stance, attack them.

That is, when you want to fence with an opponent then ensure that you are the first at the location so that you can position yourself in your prepared sequence in good time. Then, you should hold them up with cuts and steps to such an extent that *they do not have the time or space to position themselves in a stance or to execute sequences according to their desire.* And you should thus overwhelm them quickly with clandestine steps before they perceive them. You can gather how this is brought to pass sufficiently from the following couplets.

> Pay attention to Indes, understand me correctly:
> Hit them before they position themselves in a stance.[106]

You should understand this in this way: when you are in the Onset, pay attention and *when they act as if they want to position themselves into a stance,* |1.46| then do not let them move there in peace, but always thrust continuously through in front of them. Pay attention and observe, and *in that moment in which they want [to position themselves in] their stance,* attack straight at the closest opening. Act as if you wanted to cut strongly, but let it miss or fly past and attack another opening. As soon as you are halfway there with your blade (or have arrived on their sword) then don't rest, but instead cut a Crosswise Cut, or strike around, or rip out, or slice, or twist, or execute whatever work seems to come most properly to hand for you.

> No stance should appear so good to you;[107]
> Hit them in the After with bold courage.

One could ask here how this is to be understood since there are so many good stances, and now and again many sequences have been shown and taught from the same. Regarding this,

---

[104] Beginning here, the text is based on JML: 10^R–40^R.

[105] Meyer later glosses *plan* as the '*platz* = location'. *Plan* sort of equals "plain, field", if you squint. This is one of the points where you can see him forcing the rhymes to support the meaning.

[106] Compare Egenolff (1530s): 4^V, which has *In des hab acht* rather than *Indes nim war*, but the meaning is the same.

[107] This line is different in the actual verse, '*Es kom dir für was Leger gut* = Whatever stance appears good to you'. Both are different from the version on Egenolff (1530s): 7^R, '*Keins Legers ich dir werd sein gut* = No stance will be good for you'; Egenolff's version of the couplet was part of a taunt directed at the opponent, not advice for the fencer.

you should take note that there are many good stances and that beautiful and good sequences can be fenced from some stances, as some have been shown to you and contained herein.

From these rhymes, however, you are taught that it is always better to avoid completely assuming a stance because your counterpart can easily steal [information] from the stances about what kind of sequences you have in mind to fence, which cannot be predicted from the cuts.

In addition, you should learn from this how you should hit a fencer or come at an opening (when an opponent stands before you in a stance), and you should also understand everything that can be brought to fruition in the After.

When your counterfencer stands in a stance, cut opposite [that stance] to the other opening. *As soon as they move from their stance (to counteract your cut)*, when it just barely touches (or it is better if it doesn't touch at all), pull around your head and hit in at the part or quarter *which they have abandoned in order to oppose you.* However, so that this is even easier [to understand], I will set you an example.

Namely, as soon as you take note in the Onset *that they have positioned themselves in the Wrath Guard:*
  ➤ Cut through from your left toward their right in front of them. Ensure that you are not too close to them.
  ➤ In the cut through, let your sword shoot around in the air as if you wanted to thrust from right Ox.
  ➤ Before you have completely indicated that and acted this out with body language: step hurriedly and cut from below at their left while keeping your hands high.
In this way, you force them [to act]: *they either must counteract you out from their stance, or they must cut in simultaneously with a step backward.*

If they cut, then take note:
  ➤ As soon as the swords touch together, pull around your head and cut at their right ear with the short edge. Execute this in one sweeping motion and with crossed hands |I.47| (as this Figure [G] indicates).

In this attack or fencing, which is quite an ambitious technique, you should observe the step in particular, and allow your body to move well after the cuts.

When you threaten to strike at one point, you can mislead them quite well. Therefore, you must secretly steal the ground from them in this attack and in the Onset, act as if you were going to step close and narrow. *Then, before they perceive it,* step broadly to attack.

Or, in contrast, allow yourself to be noticed at the Beginning with large steps *so that your counterpart also pays attention to that and will want to oppose you fast, and to seriously arrive before you.* Then, restrain your steps and act with moderation *until they make useless movements for no reason.*

In that moment in which you perceive your advantage—as soon as you observe your advantage—then you will be next to them with fast, broad steps.

Guide your cuts powerfully from your body;
Direct your work at the four openings.

In these rhymes, you are taught how you should guide your cuts to all four ends powerfully and long—that is, freely flying at all four openings along with the body, which you should move fully with the cuts (as mentioned above).[108]

If you cut a Crooked Cut, lift quickly;
With crossed hands, throw the point on their hands.[109]

The crooked cuts are fenced in many ways because all cuts that are cut with crossed or crossing hands are called "crooked cuts". Therefore, a Squinter is numbered among the crooked cuts. This applies regardless of whether they occur with the half or whole edge whenever you hold your hands crossed.

Firstly, take note:

When a fencer cuts from their right and from above, straight at your head:
➤ Step with your right foot well out of their strike to their left so that you evade their strike in one leap to their left.
➤ Cut with crossed hands from your right toward their cut.
➤ You thus arrive with your sword blade between their head and sword on their half edge, which faces toward them.
➤ In that moment in which it touches, step again with your right foot toward their left side and counteract with your blade or displace your blade from their blade to their arm (between their head and sword).
➤ Press their arm downward in one jerk with your hands still crossed.

During this, you will certainly find an opening at which you can cut |I.47^V| according to opportunity, and you should not delay as soon as you see the opening.

Furthermore, when you approach your counterpart in the Onset, be observant, and *in that moment in which they lift their arms to strike:*
➤ Cross your hands in the air and throw the point (that is, the Weak or outermost part of your blade) on their hands or arm.

---

[108] JML: 11^V adds an extra paragraph here.
[109] Compare verse 42 in Liechtenauer's didactic poem.

Take note that this should take place (as mentioned) *in that moment in which they lift their arms to strike.*

And before they have completed that, you should be back on their blade with a Crosswise Cut because these sequences should move swiftly and fluidly.

Let the circle touch on the right;
Keep your hands high if you want to mislead them.

[The] circle also arises from the crooked cuts and is a particularly good technique for misleading (better than others) because it does not drop down meaninglessly, vainly, or without touching like other misleading techniques (dropping down, flying away, and the like) because if one uses it correctly, the circle hits very hard with the half edge as it moves forward.

Perform this technique like this: when you stand in front of an opponent in a bind (after you have come under their sword with the attack) and you hold your sword up high over your head, then *as soon as they allow you space and don't bind onto your sword, but instead also hold their sword with the point upward:*

- ➤ Cross your hands in the air and cut with the short edge (with your hands thus crossed) from above down at their right ear[110] so that, regardless of whether your blade hits it or not, it moves around in a circle next to their right arm. Keep your hands high over your head during this.
- ➤ *As soon as they swipe after the circle:* step with your left foot well out toward their right side.
- ➤ Cut with the long edge behind their blade and over their right arm at their head.
- ➤ Along with the step, move your body (including your head) well outward to their left side out from their strike.

A good sequence from the circle

When you stand in front of an opponent in the same work (as mentioned previously), pay attention:
- ➤ Whenever you have the opportunity, step with your left foot out toward your left side.
- ➤ Simultaneously with the step, cut a circle past their right, however, such that when [the cut] passes by on their right, it grazes or hits.
- ➤ Simultaneously with the circle, step through with your right foot between yourself and them to their right side.

---

[110] JML: 12ᵛ illustrates this.

> During this step through, cut a Crosswise Cut from your right toward their left in front at their face (as you can see in the upper tenants in the Figure designated with K).[111]
> Jump Indesly well out to their right and cut with the Weak at their sword.

|I.48ᵛ| **Step well with the Crooked [Cut] if you want to counteract; Crossing over will injure [the opponent].**[112]

This should be understood in this way: when you cut in with a Crooked Cut, step well out of their strike simultaneously with your cut (so that you withdraw your head out of their strike behind your blade).

Secondly, when you have bound onto their sword with a Crooked Cut (if you have an opportunity to do this), you should quickly cross over. Subsequently, snap around or twist a flick at their head, or rip out, or let it overrun, or the like.

A fine sequence from the Reversal

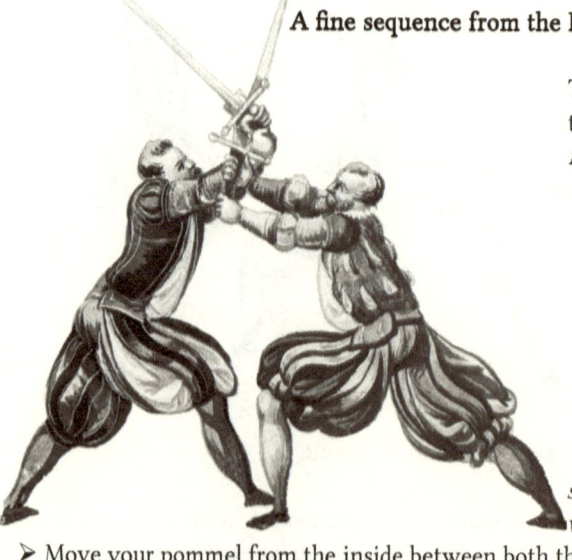

Take note in the Onset and pay attention *when your counterpart rises up in front of you:*
> Step and cut from your left with the short edge and crossed hands at or over their right arm.[113]
> However, step well toward them in this Crooked Cut.
> Reverse your sword and rip downward and out to your right side.

If they work above with their arms, so that you cannot force them downward:
> Move your pommel from the inside between both their arms.
> Release your left hand from the hilt and grab your sword blade.

---

[111] JML: 14ᴿ illustrates this.
[112] Compare verse 43 in Liechtenauer's didactic poem.
[113] JML: 15ᴿ illustrates this.

> Rip upward (as is shown in the Figure with the O).

The counter to this:
> Release your left hand, *so that they rip outward in vain.*
> Follow their rip upward Inde₅؛ly with a slice at their arm.

Do not allow them to come to any other work, nor allow them to be free [of you]. When you have observed your advantage, let fly at the closeſt opening.

Cut Crooked at the flat if you want to ſtrengthen yourself;
You should take diligent note of how to weaken them. [114]

This is a teaching about how you should weaken your opponent's in-coming ſtrike. You should execute it like this:

In the Onset, pay attention *when your counterfencer cuts at you from their right:*
> Step well out of their ſtrike and cut with crossed hands and the long edge on the Strong of their blade in the flat.

You weaken |I.49ᵛ| them with this so that they can scarcely recover to another [ſtrike].
> Before they recover, you can be at their head with turns or flicks.

You should also carry out the counter; if you take note that a fencer will oppose you with a Crooked Cut on your in-coming ſtrike to weaken you:
> Quickly change through under their blade and work at the side from which they originated their Crooked Cut.

As soon as it touches above and glances off,
Pull up to the opening if you want to enrage[115] **them.** [116]

These verses are quite necessary to note because they seriously remind you to pay attention to the openings as they occur. It is certain here that if you pursue the matter correctly, as often as it touches above or as often as two ſtrikes glance together above, you can rush to a lower opening. You will not lack in this; however, in order that you can underſtand it better, take note of this through the following example and sequence.

Thus, in the Onset, as soon as you truſt that you can reach your opponent:
> Step and cut a powerful High Cut at them from your right.

---

[114] Compare verse 44 in Liechtenauer's didactic poem.
[115] *Toben* means 'to rage, be furious'. The *be-* prefix makes the verb transitive, thus, enrage. It may also be a pun on '*betäuben* = deafen, stun' with the emphasis on strikes to the ear.
[116] Compare verse 45 in Liechtenauer's didactic poem.

➤ In that moment in which it glances off, strike quickly around back at their left ear.
➤ During this, step with your left foot behind your right.

You thus hit twice equally [with them] or you complete two strikes on one side before they complete one.

Likewise, *if an opponent cuts at you from above* (as before):
➤ Cut from your left and from below toward their strike so that you catch their High Cut high in the air on your sword with crossed hands.
➤ As soon as it glances off, cut with the front short point, that is with crossed hands, away from their sword in a circle at their right ear.

This should occur quite speedily so that, in that moment in which the blades touch together, [you] also simultaneously strike down from above with the half edge.

Furthermore: act in the Onset with body language as if you wanted to cut from above.
➤ As soon as you observe *that they will swipe upward to counter your cut,* twist your High Cut into a Low Cut before it actually touches above.

This is a true Misser, and you will hit their left ear before they perceive it.

Finally, *if your counterpart cuts from below:*
➤ Drop onto it with your long edge.
➤ In that moment in which it glances off, strike quickly at the closest opening in a sweeping motion or strike around away from their sword with the flat in a twisting flick at the closest opening.

|I.50|

**When you cut a Crooked Cut at their strong,
Take note: turn through, overrun with this.** [117]

When you cut a Crooked Cut at an opponent *and they resist strongly* so that you cannot achieve anything with crossed [hands] or other work from above:
➤ Turn through below with the pommel.
➤ Reach for the other side with the same [pommel] from outside and over their blade or arm.
➤ Rip downward.
➤ In this [downward] rip, hit them on the head with the long edge.
➤ Or reach with the pommel between their two hands (as can be seen in the Figure [O] printed previously, in the two tenants on the left side). [118]

---

[117] This couplet and the following one appear in a different order in the gloss than in the poem.
[118] JML: 16^V illustrates this.

Counter if an opponent cuts a Low Cut at you

➢ Cut with the long edge from above on top of the strong of their blade so that your hands are crossed or so that you cut a Crooked Cut.
➢ In that moment in which it glances off, shove the blade directly in front of you, twisting your short edge in a flick around at their face or at their head (during this shove).
If they rise up to defend against your flick:
➢ Rise up also, pull around your head and strike at their lower opening.

You should contemplate misleading with the pommel;
You will annoy them with Tagging contacts, flicks.

That is to say, when you cut a Crooked Cut at their right *and they resist or counteract high:*
➢ Twist through below with the pommel and (with body language) act as if you wanted to grip past with the pommel (as taught previously).
➢ *Before or when they notice that:* flick back in with the short edge at whichever side you originally cut the Crooked Cut at.

Item, in the Onset, attack your counterpart strongly with a powerful Middle Cut across at the left ear.
➢ Quickly pull your pommel around your head and threaten them with it (as if you wanted to thrust at the other side with the pommel).
➢ *In that moment in which they swipe at you to interrupt the thrust:* flick with the short edge back at their left ear.
➢ Step with your left foot back behind your right during the flick.
➢ Cut away from them.

|I.50ᵛ| ### Also, if you want to shoot through correctly,
Cut Crooked, Short, [or] change through on their ricasso. [119]

This is truly a small Master Sequence. When you are in the Onset, position yourself in right Wrath.
➢ *As soon as they bring their sword [up] into the air:* cut a free High Cut at them but do not complete it.
➢ Instead, cross your hands in the air (so that your right moves over the left), and cunningly cut a Crooked Cut through their cut with the short edge.
➢ Step out Indesly with a doubled step to their right.
➢ Cut with the long edge at their right ear or use a change through to arrive on their ricasso toward their right.
Work with twisting, slices, and whatever kind of work you can.

A sequence from the shooting through

Pay attention in the Onset. *As soon as your counterfencer lifts their sword to strike:*

---

[119] Compare verse 46 in Liechtenauer's didactic poem.

➢ Cut a Crooked Cut quickly and cunningly through during this (as mentioned above) so that you arrive on their right and on their ricasso from outside.

➢ When it just touches, swiftly twist the short edge inward at their head.

➢ Yank your pommel well upward in this twist so that the short edge drives even deeper.

If they lift to counteract this:

➢ Let your blade snap back around so that your right hand moves back over the left.

➢ Flick back under their right ear during this snap around, stepping well out to their right with your left foot during this.

➢ Then cut a Crosswise Cut deeply at their left ear with a step back.

➢ Twist your short edge on their sword back out from below and at their left ear.

➢ Afterwards, cut away from them.

Everything should be completed quickly using twisting.

**Take note: if they want to confuse you with a Crooked Cut,
Remain correctly on the sword and execute the War
With twists, slices, and whatever else.
Do not let yourself fly too far astray.** [120]

|I.51| With these lines, you learn how you should behave against an opponent *who binds on your sword with a Crooked Cut.* Two techniques will be mentioned here that are of service—namely, remaining and the War—which function like this:

If a fencer binds on your sword with a Crooked Cut, don't pull away quickly. Instead, remain on the sword to feel what kind of work will be necessary here. *So that when they move away,* you pursue, *or if they remain,* you twist.

Twists, slices, reversals, and outward rips are called the War, because one [fencer] always counters the other [fencer's] sequences this way, and one counter follows from another because *if they defend against one [counter] then they cause or assist you in the second [counter].* Therefore, both [fencers] War around the Before.

You should also pay attention when *a fencer wants to fence at you with Crooked Cuts,* so that you do not fly astray and *let them move from one opening to the next.* As soon as you move away from their Crooked Cut, you are completely exposed to them if they have a little knowledge about how to send in [their sword] during this.

Now, take note of a good sequence arising from this.

If an opponent cuts a Crooked Cut from their right at you:

➢ Set their cut away and upward using the long edge.

➢ In that moment in which it glances off: remain on their blade with the bind, twisting your pommel indesly and upward toward your left, and twisting the blade downward toward their left, [which twists] the short edge at their left ear.

This should occur simultaneously and in one step so that you hit with certainty.

If they were skilled enough that they could also turn the Crooked Cut into Longpoint:

➢ Twist the short edge to flick at their head from inside.

➢ Afterwards, twist the pommel quickly back from underneath and down to your left side.

---

[120] Compare verse 47–48 in Liechtenauer's didactic poem.

➤ Reach with the pommel over their blade or arm and rip it outward.

Or, if you have completed that, take in hand another sequence that you view to be most proper here.

Flick with the Weak fast at the right; [121]
Flick twice and guard yourself with the ricasso.

Take note: in the Onset, move into the right Change [Guard].

➤ From there, strike up through their face so that your sword runs around over your head in a round.
➤ Step with your left [foot] well to their right and strike across at their ear with the outward flat, from your left toward your right. Move your head well out of the way with that (as mentioned previously). [122]

➤ In that moment in which it glances off, thrust your pommel quickly through under your right arm and flick with |1.51ᵛ| the inward flat in a flick (from below and back up) at their right ear.
➤ In this twist, remain hard on their ricasso with your sword, and simultaneously push hard away from you.
➤ *If they resist:* then let your sword move slightly outward and pull around your head.
➤ Strike a strong Glancing Cut with the outward flat at their left with your palm upward (so that your pommel rises well upward—this [allows] the cut to go deeper).
➤ Twist the pommel back through under your arm and flick from inside and behind their blade at their head.
➤ Remain on their ricasso constantly and hard and twist hastily back outward.
➤ You then stand back in the Glancing Cut (as previously).

Work whatever you want further according to the four openings.

---

[121] This line is slightly different here than the actual verse. "Also flick with the Weak" becomes "flick fast with the Weak" in the gloss. Not a major difference in meaning.
[122] JML: 18ᵛ illustrates this.

Item, *if a fencer cuts a High Cut at you from their right:*

➢ Cut a High Cut toward theirs and simultaneously with them as well.

➢ In that moment in which it glances off, thrust your pommel hastily through under your arm and flick back inward at their head.

➢ Before that actually touches, pull both of your arms (which are crossed) upward and toward your left and rip upward and around on their blade.

➢ Flick back with the outward or reversed flat at their left ear from below, as just [completed] above.

This double flick should occur quickly, and because it is an especially quick sequence, I have presented it to you in detail because it is sure [to hit] when you bind from one side on their sword, remain hard thereon, and twist thus inward and outward in a double flick at one side (at the upper and lower part of their head).

If they swiftly counteract the flick, you have an opening on the other side that you can touch with a circle, or you can flick around in one motion.

In my opinion, you can take good note of and learn twisting from this.

Also, twist strongly around their ricasso;
Shove off Indesly and strike swiftly.

That is, *if an opponent wants to defend themselves from the double flicks and counteracts you:*

➢ Catch their ricasso with yours (as the small portrait on the right in the Figure designated with I, who has caught the arms of the other [fencer]). Yet do so in such a manner that you don't release your left hand from the hilt.

➢ Jerk their sword outward and away from you to the side in one thrust.

➢ Let the short edge snap around Indesly [to strike] deeply at their other opposite opening.

|I.52ᵛ| **Another**

If a powerful and coarse fencer[123] *cuts at you* so that you cannot approach them with such finely detailed work:

➢ Cut the first simultaneously with them.

---

[123] A '*büffel/püffel* = buffalo' is someone who is unlearned, unsophisticated, and able to execute only the crudest work. A strong, dumb ox.

Take note at that point:

> *In that moment in which they pull their arms back toward themselves,* move your horizontal blade onto both their arms from below.

> During this run underneath, release your left hand from the pommel and grab your blade in the middle (as the smaller tenants show on the left side in the previous Figure N).[124]

> Rip both their arms outward to the side with your ricasso and crossguard, releasing your left hand in the thrust or rip.

> Cut quickly afterwards, either short or long.

Therefore, take note: *when an opponent arrives before you who attacks crudely from above,* observe how you can counteract one strike or two until you perceive the opportunity to strike most optimally. *When they have lifted,* move quickly under their arms and step well under them *so that they strike their own arms onto your blade.*

You should carry out the Squinter Cut wisely;
You can also double it by twisting.

There are three types of Squinters: namely, two Squinter Cuts—one from the right, the other from the left (which is not dissimilar to the Crooked Cut with crossed hands, as is mentioned above in relation to the crooked cuts)—and the third is a Squinter with the face, in that I look at a point as if I wanted to strike there (including all of the body language), then I don't do that but instead strike anywhere else.

You have been taught about these Squinter Cuts previously in the first part [of the Sword Section], and because they were additionally mentioned here and there in the sequences, it is unnecessary to treat them further here. Therefore, I will only say something about various counters and the like that arise from them.[125]

Counters against the Squinter

Take note: *as often as a fencer carries out a Squinter Cut against your long cuts, they expose their right side.* Therefore, don't allow them to arrive on your sword; instead, change through below and (after this move through) cut long at their right from your left.

---

[124] JML: 21^R illustrates this.

[125] Instead of this paragraph, JML: 21^V–22^R describes the first two Squinters. One is a short edge, cross-armed cut against the opponent's sword, similar to a crooked cut. The other, which he labels the "Old Squinter", matches the one in Chapter 3 and references the illustration on the next page of this book (compare with the large pair in Figure G).

|I.53| **Counter.**

Item, *if an opponent changes through under your Squinter Cut to your right side:*
> Remain steady with the point directly in front of their face and turn the long edge toward their blade.
> Let your pommel move through Indes under your right arm and step with your left foot well out to their right side.

Then they will have changed through in vain because you will arrive at their head using the other Squinter Cut and crossed hands.
> Let that run down immediately next to their right in a circle and Crosswise Cut at their left ear.

Counter to the Plow

Take note: *when a fencer approaches you in the guard of the Plow*, feel free to attack them with the Squinter Cut. *As soon as they rise upward*, work at their lower openings and furthermore at all four ends.

Counter to Longpoint

Item, *if an opponent stands before you in Longpoint:*
> Present yourself with body language as if you wanted to cut a long High Cut at their left ear.
> Don't do that, but instead turn in the air and cut a strong Squinter Cut on top of their sword.
> In that moment in which it glances off, shove the point forward at their face.
> *They must counter this. In that moment in which they rise upward,* pull your sword in one sweeping movement around your head and cut across at their right ear with crossed hands and the short edge (I call this the second Squinter).
> Let your left hand move well upward and under your right arm—this moves the short edge deeper.
> Pull back around your head and rip their blade out with the flat across from your right toward their left (so that your sword flies back over your head) and let the short edge shoot in deeply at their left ear.
> Immediately cut two Low Cuts at their right and left.
> Cut away Indes.

|I.53ᵛ| **Another**

Item, *if a fencer arrives before you and they like to bind on you long from above, or they like to execute their first strike long from the Day:*
> When you then approach them, strike through before them (upward and toward your left) so that your sword's blade shoots around over your head toward your left in the Plummet.
> Threaten them as if you wanted to strike toward their left. *They will doubtlessly be ready for this and cut towards that.*
> Therefore, let your sword snap back around Indes over your head with the right hand

over the left and (simultaneously with their strike) strike in with the short edge at their right ear (as taught above).[126]

If you do this correctly and step well in addition, you will certainly hit.

If they counteract and rise upward (as they must rise up if they want to counteract):

➤ Immediately pull around your head and cut with the long edge, across and from below, at their left forearm (close under their pommel at the bone[127]).

You will certainly hit one of these two openings: either the right ear or the forearm.

➤ Pull your hilt back upward and around your head and cut a strong Long Cut at their left and at their head.

In this third strike, step well around to their left side with both feet in a double step. This allows the cut to proceed well.

This is a good, serious sequence, if you want to fight in your opponent's space (bring it home to them).

Another Sequence from the Squinter Cut

When you are just approaching them in the Onset: present yourself as if you want to carry out a long, strong High Cut.

➤ *In that moment in which they rise up to oppose you:* turn the short edge in the air from your right toward their left and pull your pommel upward.

➤ Strike over their arm or hands with the short edge.

➤ Step well to their left side and let [your sword] move around and past in a circle.

➤ Afterwards, cut long at their closest opening or fence at them from the Low Cuts.

|I.54| ### To double the Squinter

Item, in the approach, cut a Squinter Cut from your right on top of their sword against their cut.

➤ In that moment in which it glances off, turn your sword on their blade and slide on their blade out toward your left side.

---

[126] JML: 22ᵛ illustrates this.

[127] *Knochel* is a typo for something. *Knochen* are bones. *Knichel/Knickel* is the projecting bone on the outside of the wrist. *Knöchel* are knuckle bones. Because of the subsequent reference to the forearm, '*Knichel* = projecting wrist bone' is the most likely meaning.

➢ Also step with your right foot farther toward their left.
➢ Let your blade move around your head and cut the second Squinter Cut (also from your right) deeply in from above and behind their blade.

[Cut] the one like the other, quickly one after the other, twice at their left, with a doubled step. This is a swift sequence against slow fencers who hold their arms too far out from themselves.

Or, once you have completed the first Squinter and want to also carry out the second:
➢ Push your pommel quickly through under your right arm (while you are guiding your sword through the air).
➢ Cut the second with crossed hands (also at their left, as previously) and pull your head well toward your right.

The third Squinter is misleading using your face

In the Onset, strike up into the guard of the Day.
➢ As soon as you can reach them, immediately turn the short edge toward them while in the air. Use your face to act as if you wanted to cut a Squinter Cut at their left.
➢ Don't do it, but instead let the Squinter run past next to their left as a deliberate miss and work at their right.

Or flick your eye[lids] at their right and strike quickly back in at their left, using your entire body. It is lovely and fast work which is not easy to describe, but [can be] shown with a living body.

Take note: a fast sequence from the Squinter

In that moment in which you just barely approach them: wing your elbows out in front of them so that you arrive into the Unicorn with crossed hands.
➢ During this elbow movement, lift your left foot somewhat upward (following the change in weight that you achieve by raising the crossed hands high and coming through the Unicorn). You thus stand as if you wanted to shoot through.
➢ *As soon as they extend their sword out:*
➢ |1.54ᵛ| Cut down with crossed hands and the short edge from above and toward their right, up to the outermost point of their sword blade.

➢ Before it touches, turn your short edge around and strike a Squinter Cut.

That is, with the short edge from your right at their left ear, arm, or face, with a step forward of your right foot toward their left.

You thus hit them (like the large portrait teaches on the left side in Figure G).

- ➤ When it hits, let your blade run out a bit to the side and away from their left and simultaneously thrust your pommel through under your right arm.
- ➤ Cross your hands so that your half edge snaps back around in front of them toward their left and over their head or arm.
- ➤ Rip their sword outward with crossed hands (from your right toward your left) or cross [your sword] over both of their arms.
- ➤ *If they resist* so that you can't rip out or cross over, let your pommel pass through underneath and [use it to] grip over their right arm.
- ➤ Use wrestling against them.

About the change through

The change through is appropriate against those who fence using the Squinters or Crooked Cuts.[128] Take note like this: *if they don't extend their hands far in front of them in their cuts, but instead keep them close by in their fencing,* this is one that you can happily [use the] change through at a distance.

Item, *if they fence out of the twisting, reversing, Crooked Cut, Squinter Cuts, or any other type of technique in which they shorten their strikes, or they can't fence long in front of themselves (as occurs using those techniques), then before they complete their technique to the halfway point,* you should change through against them to the other side (which they expose due to this shortening [of the strike]). By doing this, you force them to counteract and they must let the Before pass to you.

Item, *if a fencer already fences wide and long in of themselves with the long edge, but more at your sword than at your body,* then you should change through at their closest opening and let them completely miss with their cuts.

Therefore, be diligent in what you fence at (whether with the long or short edge) so that you properly cut at the opening (that is, at their body). And even though it isn't always possible that you can cut close to the body, you should drop in after their sword at the opening as soon as they change through.

Take note of this rule in all cuts, so that when you touch or contact their blade with the strong of your sword in the bind, immediately and simultaneously and in that moment in which it glances off, you should simultaneously cut in with the Weak of your blade (that is,

---

[128] JML: 25ᴿ illustrates this.

with the outer part) at their body or |I.55| closest opening so that your sword simultaneously hits their blade and body. Or, as soon as your Strong touches their sword (while [the swords] still mutually glance together) you should turn the Weak to the next opening and snap and twist with flicks.

Furthermore, keep this instruction [in mind] if you want to fence these techniques against those who know how to change through against you (namely, while you guide your sword in the air to a Squinter Cut or Crooked Cut).

Item, [when you use] crossed hands, missing, and the like: *as soon as you perceive that they want to change through,* drop out of such work into the long slice (that is, into Longpoint at the opening that they will offer to you during the change through) *because as often as they change through, they expose themselves.* In that moment in which you rip through at the opening, guard against their sword with your long edge (so that you turn the Strong toward them) *wherever they want to arrive next,* and simultaneously remain at their opening with the short edge. As soon as you have touched, do not tarry any longer, but instead let it fly quickly away from one opening to another.

You should also consider the Crosswise Cut as worthy; Your skill with the sword becomes compete with it.

The Crosswise Cut is also one of the proper Master Techniques with the sword. You should know that if the Crosswise Cut did not exist (as it is currently used at the present time), only half of fencing would be possible. [This would be] especially true when you are inside and under your opponent's sword, as you cannot fence through the Cross with long cuts. Even though I have written a lot about the Crosswise Cuts previously—enough that if one can fence, they could gain sufficient understanding from there—yet, because much depends on the Crosswise Cut (as stated), I have undertaken to write more, both for great masters and even more for students. Therefore, I will not merely repeat the Crosswise Cut, but instead I will write more fully about it to teach those who love this art.

Pay attention in the Onset as to *whether your opponent wants to attack you from the Day (that is, from above).*
- ➤ Strike up from the right Change [Guard] toward your opponent's face.
- ➤ *In that moment in which they want to strike or cut:* let your blade move next to your left and around your head (so that your flat faces upward and your thumb is under on the ricasso, or the ricasso lies on your thumb).
- ➤ Step with your right foot well around their left side toward them.
- ➤ Simultaneously with |I.55ᵛ| this step, cut from your right side toward their left ear with the half edge, so that your hilt (including your thumb below) stands high over your head to counteract.

[This is] so that, *wherever they strike,* you receive their strike on the strong of your blade.
- ➤ Simultaneously with this, you will hit their left ear with the Crosswise Cut with the outer short edge from below.[129]
- ➤ As soon as the swords touch together or glance off, strike deeply and obliquely at their right ear with the long Crosswise Cut (so that your thumb remains underneath).

[129] JML refers back to the illustration on folio 14ᴿ.

Secondly, take note: when you simultaneously cut in or bind on with a Crosswise Strike, immediately seek the opening (above and below) on the same side by reversing and snapping around, or [with] Crosswise Cuts, crossing over, pursuit, slicing, crushing hands, and ripping outward. *As soon as an opponent cuts at your from above,* counteract with the Crosswise Cut. In that moment in which the swords glance off or touch together: reverse, cross over, seek[130] the opening, and fence whatever work you can first execute that has been mentioned here.

Therefore, Liechtenauer states correctly in his enigmatic verses:

> *The Crosswise Cut takes away / that which comes from above.*
> *Item, Crosswise Cut with the strong, / take note of your work with it.*[131]

That is: counteract all strikes from the Day with the Crosswise Cut, or as I have set it here in my [own] rhyming lines:

The Crosswise Cut can counteract
Everything that comes from the Day.
Execute the Crosswise Cut strongly in the attack
Also take note of reversals and misses in addition to it.

If an opponent cuts at you from the Day:
➢ Cut a Crosswise Cut strongly against their strike.
You force them [to act] with this so that *they must drop more deeply downward with their cut.*
➢ In that moment in which it glances off, thrust your pommel through and under your right arm, reverse, and press downward.
➢ Let the blade immediately snap back around with the short edge into their face, but such that you remain with the edge on their arms in the reverse and snap around.

This sequence proceeds well when you execute it with speed.

If they evade you [by moving] their arms upward too quickly:
➢ Let your blade move around your head so that the long edge arrives in front and across on their arms through a Low Cut (as the small scene indicates on the left side in Figure [N]; however, don't release your left hand from your hilt, but instead thrust them away from you with crossed hands).[132]

---

[130] Literally '*schrenck ubersich die Blöß* = cross upward to the opening'; the translation above is based on JML, which has *schrenck vber such die blöss.*
[131] Verses 49–50 of Liechtenauer's didactic poem on the long sword.
[132] The left scene in Figure I doesn't show this technique; Meyer either meant the *right* scene in Figure I or the left scene in *Figure N*. The right scene in Figure I looks like the action being described, but the clarification not to grab the blade only makes sense if the left scene in Figure N is intended.

|1.56ᵛ|
> **Be quick to the Plow and Ox,**
> **Threaten them fast with cuts at contrary ends.** [133]

The rhyming lines are also quite clear in themselves (like the others as well). Namely, that you should quickly cut a Crosswise Cut at the Ox and Plow (that is, at the lower and upper openings), to the left and right, cross-wise and obliquely, to all four quarters as with other cuts (as has been taught previously and extensively in the section on the four openings). [134]

> **Take note of which type of Crosswise Cut is executed with a leap;**
> **You also miss with it, as it touches according to your desire.** [135]

Take note in the Onset if you want to cut a Crosswise Cut at the upper left opening: leap well toward the same, and also let the pommel move well upward. This allows the Crosswise Cut to move deeply at the head, especially if you can misrepresent the body language—that is, leap unexpectedly and simultaneously with the Crosswise Cut, *so that they don't observe the leap* until it has occurred and the Crosswise Cut has hit.

If they perceive it and defend or counteract the same: step swiftly toward their right side with the left foot and, with your body bent forward, cut a Crosswise Cut from your left to their lower right opening. You will certainly hit this if they have counteracted the first Crosswise Cut.

Item, if you leap and strike high and deeply with the Crosswise Cut (or another type of flat [cut] at the opponent), don't let it hit but instead let it drop down as a deliberate miss next to the left. Strike at another opening hurriedly with the Crosswise Cut, then you will hit according to your desire because *before they think to counteract the Crosswise Cut,* you have hit them at another location—as long as you execute this with your whole body (that is, that you use the body language for this technique).

> **You should doubly execute the Misser;**
> **Likewise, double the step and slice.** [136]

The Misser is a good technique against fencers who like to counteract (like the previous techniques from the Crosswise Cut). When you cut at an opening, and you take note that they [will] counteract the cut in response, then let the cut run past as a deliberate miss |1.57| and strike at another opening.

Double misses are a skillful technique and require a well-practiced fencer. However, I will place here and describe some techniques as double and single, from which you can well learn all kinds of Missers.

Position yourself in the Onset into the Wrath Guard on the right.

---

[133] Compare verse 51 in Liechtenauer's didactic poem.

[134] JML: 27ᴿ adds two extra paragraphs here.

[135] Compare verses 52–53 in Liechtenauer's didactic poem.

[136] Compare verses 56–57 in Liechtenauer's didactic poem.

➤ *As soon as they lift their sword into the air,* cut through from your right and around your head (with the long edge and extended arms) at their right side to deliberately miss, so that your Crosswise Cut flies powerfully around in the air toward their left ear.

➤ Don't let it touch, but instead pull back around your head and cut with the long edge, so that the flat sweeps in powerfully at their right ear.

➤ Now reverse, snap around, and let it fly, and carry out whatever work you can.

Item,

➤ In the Onset, cut a long High Cut at their upper left opening.

➤ When you have scarcely touched their blade above in the air, change your High Cut into a Crosswise Cut and strike with the Crosswise Cut from below at their left ear or arms.

These are the correct fencing techniques from which many sequences are fenced.

A Misser with a deceptive step

In the Onset, cut a lofty High Cut and when your blade barely touches on their blade:

➤ Immediately change the High Cut into a Crosswise Cut.

➤ Simultaneously with the Crosswise Cut, step through with your right foot between you and them and out to the side toward their right side.

➤ During this same step, cut a Crosswise Cut at their muzzle[137] such that the point is between their arms (as you can see in the small upper scene in the subsequently printed Figure [K]).

➤ Immediately let it snap around again and strike with the short edge and crossed hands back around at their right ear, or cut with the long edge in response.

➤ Regardless, leap well out to their right to the side with this strike.

Two-fold or double miss

Item, in the Onset, before you actually arrive before them:

➤ Cut through next to your right so that your weapon shoots over into the Plummet.

---

[137] *Maul* is the mouth that animals have. When used for people, it functions to dehumanize them.

➤ Step forward toward them with your right foot.

➤ Let your sword move around your head and pull a |I.58| High Strike from the Day into the air. However, cross your hands and threaten to hit them with the short edge. [138]

If they swipe in response and want to counteract:

➤ Turn your hands back around and change the curved edge[139] into a Crosswise Cut.

➤ Don't let the Crosswise Cut touch either, but instead let it run forward in a deliberate miss.

➤ Strike the second at their right side; that is doubly missed.

These two Missers are completed together in the air in one sweeping motion as you twist around their blade. However, you can break off in the middle into a counterposture whenever you want, or you can turn into [a counterposture] if they were to reach you (such that you could not get to them with your sequences). However, when you have forced them to counteract you, then the double Misser is very good, and it goes very fast.

Item, this is also called a double miss because one allows it to drop down doubly (or twice) in order to mislead their opponent.

Another from the double Misser

In the Onset, bring a High Strike from your right high into the air.

➤ Before it touches, turn the short edge toward them as if you wanted to cut a Squinter Cut.

➤ However, don't let the short edge touch either, but instead let it also run quickly down as a deliberate miss and swing your Weak at their right ear with crossed arms.

➤ Let that fly quickly upward again and attack them with your sword at the closest opening (or on their sword).

➤ From there, [attack] at their body and on top of their arms.

Counter to the Crosswise Cut

Take note: when you bind with a fencer from above or cut in simultaneously with them, observe whether *they want to strike around with the Crosswise Cut.*

---

[138] JML: 28^R illustrates this.
[139] A term used primarily for the back edge of a dusack.

➤ *In that moment in which they strike around:* strike your Crosswise Cut before [theirs] and under their blade at their neck (as the large scene shows on the left side of Figure L).

Item, *if they cut a Crosswise Cut from below* so that you cannot come underneath [their blade]:

➤ Catch their Crosswise Cut on your ricasso by pushing forward.

➤ Thrust your pommel above your right arm and well away from you, turning the long edge over their blade (up from below and from outside) toward their head (as the larger scene indicates on the right side in Figure N).

|I.59| **About stepping**

Much importance is placed on steps. Therefore, observe that you give each strike its [corresponding] step, because if you cut at their opening and don't step with your foot from the side from which you have cut, your cut will be useless. When you are only threatening to cut but not completely cutting, you should also not fully step but instead use body language to act as if you had stepped. Practice will teach you this better.

Execute a double step like this: when you are stepping with your right [foot] to their left but your technique requires that you must step even farther around, then step quickly with your left following your right (either behind your right, outside, or past it). Scarcely after you have set your left down, you can step forward with the right.

To double the slice,[140] observe the following sequence.

If an opponent cuts at you from their right:

➤ Cut also from your right toward their strike, yet with the short edge and crossed hands.[141]

➤ In that moment in which the swords touch, step in a doubled step with the right foot (quickly and farther around toward them on their left) and drop away from their sword with the long edge onto their arm.

➤ Now cross your hands over.

➤ *If they rise up and will not suffer your slice:* follow them with a slice from below into their arms.[142]

➤ Thrust them away from you with your crossguard and ricasso before they recover.

➤ Cut in response.

That is the correct, classic slice, and is the purview of a master.

Item, when you have sliced an opponent on the arms from above, you can pull the sharp edge through their muzzle.

From the sword to the body, reverse
Twice with it, or slice into the weapon.[143]

This is the correct gloss about the previous rhymes which say: twist twice or slice into the weapon. Understand it like this:

[140] While *Schnit* is probably a typo for *trit* and presents a logical reading, the following sequence does concern slicing with a doubled step, and Meyer usually uses the verbal forms for stepping.
[141] JML: 29ᵛ illustrates this.
[142] JML: 28ᴿ illustrates this.
[143] Compare verse 75 in Liechtenauer's didactic poem.

> When you slice from the sword onto their arm, you should reverse immediately.
> *If they escape upward:* you should twist your pommel back out from under your arm, which reverses your sword again.

Slice into the weapon means:

> If the double reverse has missed, you should pursue twice [while] remaining with the slice on their arms.
> *If they defend against that:* drop on their blade with the slice and ensure that you don't let them move away out of your advantage, but instead always pursue [them].

|I.59ᵛ| **Pursuit is exceptionally good;**
                  Protect yourself with slicing, twisting

Pursuit is varied and diverse and should be executed with great caution against those fencers *who fence around themselves with long cuts without any skill.*

Execute it like this. When you approach them in the Onset with your left foot forward and you hold your sword in the Day, *if they cut long at your head from above:*

> Do not counteract them, but instead observe how you can slip through with your head and sword under their blade toward the other side *while it is still in the air flying toward you (so that their cut doesn't touch [you]).*
> Let them miss in this way: *in that moment in which they are still dropping with their sword and cutting downward toward the ground,* cut quickly and cunningly at their head from above while you step through (as mentioned above) *before they recover or lift back upward.*

If they lift upward so quickly that they counteract you:

> Remain hard on their sword and feel whether they push strongly further upward.
> [If that is the case], let your blade escape quickly and easily upward.
> Step and strike a Crosswise Cut around at their right.

Item, *if a fencer binds onto you from their right:* pay attention to *when they strike around.*

> In response, follow them with a slice on their arm at their right.[144]
> Or, if you stand in the guard and *they drop their sword onto yours before you lift up:*
> Remain below on their sword and lift upward.
> Feel Indesly whether they want to carry out a cut or twist from the counterposture.
> Do not let them move away from your sword, but instead follow them and work Indesly at the closest opening.

Also take note about what pursuit is: *when an opponent lifts too high upward* so that you pursue them from below with cuts or slices (in that moment in which they pull up to strike). The same [holds] if they move astray or too far out to the sides: pursue their weapon to the opening from above. However, pay attention in all pursuit; *if they evade you:* turn your long edge toward their weapon.

Also consider the slice, because you can use it to force them out of their work.

---

144 JML: 31ᵛ inserts an illustration here, but it seems to have no relation to the text.

> In the case of a second [slice] or in the middle [of a technique],
> Let it fly away. Begin with this.

|I.60| That is, you should pay attention so that when you have set upon them with the slice, you do not immediately release them, but instead follow them once or twice with the slice in order to hinder them from their work and sequences in this way. *When they least expect it,* you should slyly and unobtrusively let your sword slip away to another of the closest openings before they become aware of this. This is a true Master Technique; thus, it is a Beginning.

> Execute hits towards all four ends;
> Learn pulling if you want to play them false. [145]

You must be well taught in the four openings if you want to fight with some certainty because then you can fence whatever cuts and sequences you want, as well as you want. However, if you do not know how to break off in each quarter, to transform the intended technique and change it into another, more suitable work (depending on the person who fences against you and counters your sequences), it may occur that you intend [to use] a sequence at one opening but *they move against this such that you might have another, closer opening.* Yet this opportunity will escape you if you continue to fence your intended sequence without taking note of other opportunities of chance. Therefore, be diligent so that you can quickly conceive of and can fence at all four openings in a free-flying manner, because you only have three ways to cut and to strike (that is, with the long and short edges and with the flat), from which all fencing is composed together in harmony with the four parts of the opponent, from which all other techniques of opportunity arise (that is, pulling, doubling, dropping down, which have been sufficiently dealt with previously). [146]

> Bring the slice off and catapult [to your fencing] as well;
> Rebuff hard and hostile [147] [actions] with the slice.

As you now allow your sequences to run to all four openings (as taught previously), you should also simultaneously pay attention to their movements (that is, to their sequences) so that you stop [148] them and slice them off according to opportunity. Therefore, hinder them and slice off their sequences long until you observe your opportunity for other work. The |I.60ᵛ| two hostile actions are the strikes from both sides. When you slice them off, ensure that you do not overlook any opportunity and also do not slice too far away from their body (so that they don't move through).

Pay attention to the catapult. *As soon as they expose one side,* rush up from below with the flat to their ears. Slice quickly back down at the opening. (With regard to more information on the catapult, see the first part in the chapter about Handwork.) [149]

---

[145] Compare verses 74 and 90 in Liechtenauer's didactic poem.

[146] JML: 33ᵛ–36ᵛ adds four extra paragraphs and two illustrations here.

[147] *Gefehrt* can mean both '*gefährde* = dangerous' or '*gefährte* = that which moves'. Embrace the pun.

[148] *Stecken* means 'to cause to remain fixed, to stick in place'. In this case, something like spiking a wheel.

[149] See I.19.

Do not rely too much on the Crown,
Otherwise you will receive injury and insult from it.

Take note: when you counteract with your crossguard horizontal, high over your head, that is called the Crown.

When you take note that *your opponent wants to run under your High Cut with the Crown:*

➤ Do not let your High Cut touch at all, but instead pull the cut so that they lift upward in vain.

➤ Cut a Middle Cut across with the long edge at their arms or forearm (if you want to lame them).

Therefore, as often as you take note that *your opponent likes to rise up high to counteract,* present yourself with body language as if you wanted to cut high. Do not do it, but instead strike quickly around at a lower opening with a Crosswise Cut, the flat, or the long edge.

In summary: *whoever wants to use the Crown against you,* execute a Misser against them.[150]

Strike powerfully through with Longpoint;
This holds off all hard and hostile actions.

Stand with your left foot forward and strike through your opponent's face from your right (so that the half edge leads)—once or four times—quickly, one after the other. As soon as you drive them upward, attack from below, either with a Crosswise Cut or the long edge.

Take note, when you thus strike up toward them, observe:

When they cut at you from your right and from above:

➤ Turn your long edge toward their blade while striking upward and catch their cut in the air at the Strong of your sword (so that your blade extends somewhat horizontally, with your point upward and out toward their left).

➤ Step quickly with your left foot to their left and thrust your pommel **Indesly** |I.61| through under your right arm.

➤ Strike using the short edge by sliding down the back of their blade to the pommel (as the tenant teaches you on the left side of the Figure designated with B).

➤ Simultaneously step with your right foot well toward their left.

---

[150] JML: 37ᵛ–38ᵛ adds two extra paragraphs and an illustration here.

➤ Quickly jerk the pommel back out from under so that you stand with your sword in the Crosswise Cut or in Hanging Point.[151]

Item:

➤ Strike upward in front of them and let your sword fly around above next to your left.
➤ Step and cut powerfully from below and from your right at their arm.
➤ *If they counteract,* twist through below with your hilt and use your pommel to catch over their arm from outside.
➤ Release the hilt with your left hand and grab your blade with it [your left hand] to assist your right.
➤ Strike with the long edge at their head (as the tenants show you on the right side in Figure O).

You should let the Obscured Cut rebound;
Throw the Crosswise Cut around, be diligent about the flick.[152]

➤ Strike the Longpoint into their face.
➤ Step and pull your sword around your head and strike with the inside flat through the middle line (from your right at their left ear).
➤ Move your head with you and quickly twist back on their sword and around the outside flat at their left.[153]

You should use the hangings deliberately
If you want to wrestle; do not grab at the wrong time.

This is when you have both run in toward one another. You should make certain of the attack when you want to wrestle with the other, because with one grab you can jeopardize yourself (if you miss), and you will scarcely be able to rescue yourself from this without injury.

---

[151] JML: 39ᴿ adds two extra paragraphs here.

[152] This couplet is not present in the poem at the beginning of the third part. The first line, *Den Blendthauw soltu lassen bröllen,* is similar to a verse from Martin Syber's didactic poem: '*Denn Plinthaw las Prellenn* = Let the Obscured Cut rebound' (JMR: 40ᵛ).

[153] JML: 39ᵛ–40ᵛ adds two extra paragraphs and an illustration here. This is the end of its Sword Section.

About this grappling and wreſtling: take note of these following examples, of which the firſt originates from the High Cut.

In the Onset, hold your sword with the blade extended in front of you so that the front end faces out toward their face.

➢ As soon as you can reach them, let your blade drop downward from their face toward your left.

➢ |I.61ᵛ| Pull your sword around your head with the blade hanging.

➢ Cut a high and powerful ſtrike toward their left ear with a leap forward.

➢ As soon as the cut hits, immediately pull your blade back upward and remain with your hands high over your head.

➢ However, let the blade drop downward next to your left side into a Low Cut toward their right arm.

➢ During this, bend downward swiftly with your upper body and release your left hand from the hilt. Grab the middle of your sword's blade with that hand and quickly lift it up from below onto both of their arms (*while they are high with their arms to counteraƈt the firſt ſtrike*).

➢ When you have thus captured their arms between your two hands with the long edge of your sword: twiſt powerfully with the front end over both their arms from outside.

If they are too ſtrong and force you upward:

➢ Remain with your blade equally hard on their arms and twiſt your pommel through below.

➢ Use the [pommel] to catch over their left arm from outside and rip powerfully downward and outward.

➢ In that moment in which you rip outward and downward and toward yourself with your pommel: use your left hand to set your front end over their left arm and inside of their right [arm] at their face.

If they want to work further with the sword:

➢ Catch their right arm with the front part of your blade as well.

➢ Rip downward and hit them in the face with your pommel.

Counter

When you perceive that *your opponent wants to twiſt over both your arms from outside with their blade:*

➢ Release your left hand from the pommel and grip your sword's blade in the middle with [your left hand].

➢ During this same time, move the pommel between both their arms and use the [pommel] to catch over their right arm from the inside.

➢ Rip towards you with the pommel and use the blade to push away from you.

➢ Thus, you take their sword.

In this way, one counter always follows from the other.

Another

In the Onset, hold your sword with extended arms (extended[154] long and in front of yourself). Take diligent note of the opportunity. When you observe it:

> ➤ Once again,[155] cut powerfully (with your sword pulled around) |I.62| from above toward their left ear.
> ➤ In that moment in which the cut hits, lift both arms up and thrust your pommel through under your right arm during this same time.
> ➤ Swiftly release your left hand from the hilt and grab back on your sword's pommel with the same [left hand] over your right arm.
> ➤ Thrust behind their sword toward their face.

Or, if you have a short[ened] sword:

> ➤ Thrust in from above (between both their arms) and set the front end on their chest.
> ➤ Step back with your right foot and shove them away from you with the front end.
> ➤ Let your sword move back Indesly and cut completely through from your right (across at their left ear) with a step backward of your left foot.
> ➤ Cut the last from your left through their right to withdraw.

Or, when you have thrust in at an opponent from above between their arms:

> ➤ Release your left hand from the pommel and reach with the [hand] underneath their right arm to grab your blade.
> ➤ Turn [your blade] upward in front of their face and toward their left, and rip downward with your pommel or with your right hand.

Thus, you take their sword.

Although you can learn many counters from the preceding sequences through deliberation, it is also useful that I set one here as well.

If an opponent moves high with their hilt or both hands against you (as taught):

> ➤ Move high with the Crown as well, so that they cannot set their point on you.
> ➤ Also, approach them even closer.
> ➤ Release your right hand from your hilt, reverse it, and reach between both their arms for their hilt.
> ➤ Jerk toward your right side with your reversed hand.

Thus, you take their sword.

In the Onset:

> ➤ Strike upward from your right through their face (strongly and powerfully) with your extended sword.
> ➤ Immediately step and cut across from above with the long edge (from your right toward their neck) with a step forward of your right foot.
> ➤ As soon as the same hits, cut the second from your left toward their right (also across from below).

---

[154] *Ausgestreckt* is repeated in the original text; ergo, "extended" appears twice.
[155] This first cut is meant to be a repeat of the first technique in this section: hold your sword in Long-point, let your blade drop down and pull the hanging blade around your head to attack their left ear.

➢ In that moment in which you execute this Low Cut, release your left hand from your hilt and grab your blade in the middle.

➢ Twist the front part of your blade over their right arm and at their face.

➢ *If they raise their arms and want to counteract:* remain high with your right hand, (including the crossguard) and change downward with your left hand (including the front end), down from above and around next to their right arm.

➢ Move the front |1.62ᵛ| point in between both their arms and catch over their right arm and from the inside with the [front end].

➢ Shove away from you with the pommel and rip toward yourself with the blade.

Thus, you take their sword again.

Counter

Take note: in that moment *in which the Low Cut toward your right hits* and you have counteracted the same:

➢ Release your right hand from the sword and reach for their sword's pommel (over their left arm from outside).

➢ Rip it toward you and onto your right side.

Thus, you take their sword.

Another technique for running in

When you perceive in your opponent that *they want to rush you with a High Strike and run in [at you] with this:*

➢ Raise your hilt with your hands open and facing upward and catch their cut on your hilt (as this is illustrated by the upper tenants on the left side in Figure O).

➢ As soon as the cut hits your hilt: move your crossguard from above between both their arms and step backward.

➢ Rip downward with the crossguard toward yourself and out.

➢ Use this pull to cut at their head.

Throws

When a fencer has run under you with their sword so that both of you have approached quite closely:

➤ Throw your sword away behind you.

➤ Bend down quickly in front of them and grab both of their legs with both hands.

➤ Pull towards you so that they fall on their back (regarding this, see the upper and smaller tenants on the left side in Figure D).

Or, if you both have approached too closely and they hold their sword above their head so that their blade hangs down a little bit behind them:

➤ Release your right hand from your sword and move the same over their left arm from the outside.

➤ Grab their right hand close to the joint.

➤ In that moment in which you grab them by the hand, simultaneously step with your right foot behind their left.

➤ Pull your right hand backwards, upwards, and away from you and shove with your left hand (including the sword) in front on their chest.

➤ You will drop them onto their back over your extended leg.

|I.63| When both of you approach closely together:

➤ Step with your left foot between their two legs.

➤ Release your left hand from your sword and turn your back a little. During this, reach through with your left hand (released from the sword) under both their arms, and with your right hand (including the sword—or throw it away from you) over their left arm from the outside, so that you grab your left hand again with your right above both of their arms.

➤ Squeeze both of their arms together toward your left shoulder.

➤ In the meantime, step forward with your left foot in front of both of their feet.

➤ Swing yourself toward your right side, bending yourself forward and downward in that swing, to throw them down in front of you.

Another

Or, in that moment in which both of you have approached closely together by running in so that you must grapple:

➤ Step with your right foot between their two legs.

➤ Release your left hand from the sword and reach through with your right hand (including the sword) under their right arm, from outside and around their body.

➤ With the left hand, reach through to their right thigh from the inside (the lower the better), and ensure that you have stepped well through between their two legs with your right foot so that you can assist the throw with the same by blocking and jerking outward.

➢ Lift them in one sweeping motion and throw them behind you onto their head.

Another

If an opponent runs under your sword with a Crosswise Cut, take note:
➢ In that moment in which they cut the Crosswise Cut toward your left, counteract this Crosswise Cut with a hanging blade and reach through with your left hand (which you should release form the sword) under your blade and theirs and over their right arm from outside.
➢ Attack their throat with the same reversed left hand.
➢ During this [movement of your left hand], step with your left foot behind their right.
➢ Throw them on their back once again.

In order that you have a foundation for wrestling and grappling, I will explain and present the hangings and twistings somewhat broadly: the fighting sequences for the short sword originate and flow from these techniques. The hangings are properly used in the work in two ways. Namely, they are first used |I.63ᵛ| to catch the cuts, to let the same slide off on top of the flat of the blade, and to fence in response. Or to use this [blade] hanging in front of an opponent to go under their sword and to twist in under it. These are generally brought into use and completed from the Plow or from one of the Low Cuts. Secondly, they are used to hang inward or over, which are originated from the High Cut and from the Ox. The two types of hangings are understood like this:

How you should twist inward and outward out of the Low Hangings

In the Onset, approach with your right foot forward.
➢ Strike upward toward their sword with arms extended such that the half edge faces upward and precedes [the sword] in this strike upward.
➢ Afterwards, when your front end arrives equal [in height] to the belt (namely, of your opponent) in this upward strike, *and while they are striking in:* thrust your sword's pommel swiftly through and under your right arm, and lift upward towards their left with hands crossed so that your blade hangs a little toward their right side.
During this rise upward, ensure that the flat of your blade rests on your thumb.
➢ Catch their cut on your horizontally extending flat blade, so that both swords touch together at their Strong in this upward catch and movement underneath, and that your head is covered and well defended under your sword.
➢ Immediately when their cut touches on your blade, jerk your pommel back out from under your right arm toward your left side, using this to turn your half edge inward behind their blade and across over their head.
➢ In this twisting inward, remain with your sword hard on theirs (so that you can feel when they will want to move away).
➢ You should also withdraw your head well away from their blade toward your left side during this twisting inward.
If, however, you perceive indees that *they will want to rush down from above* (because you have somewhat exposed yourself with this twist inward):

➤ Step with your left foot quickly out toward their right side, thrust your pommel back through under your right arm, lift with both arms, and turn the half edge back outward at their right ear.

In this way, you can twist inward and outward on the one side (with the blade under their sword) and you can reach over with the pommel on the other side.

Thus, when you have caught their sword in the previous way through hanging in front, in that moment in which the swords still touch together:

➤ Step quickly with your right foot |I.64| toward their left side and reach in with the pommel from outside and from above (over their right arm) and rip out toward yourself.

Or, after you have twisted inwardly and outwardly *and they have raised their arms up high:*

➤ Twist the pommel inwardly between both their arms and reach with the same [pommel] from inside and over their right arm. rip it outward and toward your left.

➤ If you want, you can release your left hand from the hilt and grab the middle of your blade with the same [hand].

➤ Afterwards, as you pull with your right hand toward your left side and toward yourself, shove away from you with your left hand toward their left.

➤ You can also shove under and counteract with a hanging blade from all stances, and thus twist inwardly and outwardly on their sword from both sides (as has been taught).

How you should direct the Inward Hangings and twist from above in the work

In the Onset: execute a powerful cut from above toward their left ear.

➤ In that moment in which your cut touches or glances off their sword, thrust your sword's pommel through under your right arm again and hang the blade inward (with crossed arms) well behind their blade and at their head.

Or:

➤ After you have twisted inward from above, hang your blade well over both their arms and rip out toward your left with your sword.

➤ In this twist inward, as soon as you feel that *they want to rush down at your opening from above:* yank your sword's pommel back out once again, lift the same [pommel] back high upward, and hang the half edge back inward at their head from the outside.

➤ *If they defend themselves against this:* twist quickly through and underneath with the pommel, and inward over their right arm from outside.

➤ Rip out downward toward yourself again.

Or:

➤ Reach inward and between their arms. Catch one of the same from inside with your pommel and rip it toward you again.

Item, when you guide a High Cut in on top of your counterfencer and *they counteract you with a Crosswise Cut,* then take note:

➤ As soon as your cut touches on their blade, thrust your sword's pommel high upward (regardless of whether it is under your right arm or not) and hang the blade over theirs in at their head.

➤ You can remain hard on their blade and twist inwardly and outwardly (according to opportunity).

➤ *If they defend against these twists and rise up:* |I.64ᵛ| twist your pommel around their arm (*in that moment in which they lift up*).

You can thus twist powerfully inward and outward (from all sides) when you have bound on to them.

Finally, you should also always consider three things every time you want to twist inward:

Namely, the first is the cut; secondly, the twist inward in and of itself with which you should hit; thirdly, the slice (because you must expose yourself with the twist). Therefore, you should remain with your sword hard on theirs, *depending on if they want to pursue or move away* (while you are twisting). You pursue them with the slice and remain on their sword and guide it outward and deflect it with the slice.

Or *if they move away*, you attack from inside at the closest opening with the abovementioned slice.

A good counter for all High Cuts

In the Onset, position yourself with your sword in the left Change [Guard] and take note. *In that moment in which they cut in toward you from above:*

➤ Step with your left foot toward their right (or well out to the side on your left).

➤ Simultaneously with this step outward, cut upward toward their in-flying cut with the long edge (such that you thrust your sword's pommel through under your right arm in this upward cut and you catch[156] their strike at the Strong on your long edge with crossed hands high in the air).

➤ In that moment in which the swords now touch together and glance off, jerk your sword's pommel back out and downward from under your right arm so that, due to this outward rip, the front part of your blade moves back out behind their sword and sweeps inward over their right arm and slantwise through their face.

➤ In that moment in which you twist the Low Cut back out (through a High Cut and toward their face)—precisely during this time—step with the left foot further out toward your left side and back behind you (as you have an illustration of the same in the larger scene on the right side in Figure H).

---

[156] The original has a participial form here, '*aufgefangen habest* = have caught'; however, that disrupts the fluency of the instructions and the order of operations indicated by verb tenses.

➤ Cut through their face in this step back.

When you have quickly executed and completed this Low Cut, including the step out and twist outward in one cutting motion, then it will succeed certainly and well.

Many other lovely sequences are derived and brought to completion from this Twisting Cut.

End of the Sword [Section]

[154]Organized description and instruction in fencing with the dusack, including many virile and swift sequences collected in a good order and presented one after the other, by means of which youthful students can be instructed to quickness and can subsequently fence so much the better in the Rapier [Section].

1   Reach wide and long with this weapon;
        Hang forward and over in response to the cut.
    Step forward with your body,
        Guide your cut inward and powerfully around them,
5   Let it fly to all four ends.[155]
        You can deceive them with body language [and] pulling.
    You should counteract with the Strong,
        [And] simultaneously injure them with the Weak.
    You shouldn't approach closer
10   Than the point from which you can reach them with one step.
    If they want to swiftly run in at you,
        The front point will drive them away from you.
    If they were to have run in at you,
        You should be the first with grappling [and] wrestling.
15   Observe the Strong and the Weak,
        Indes makes the openings clear.
    In addition, step correctly in the Before and After,
        Take diligent note of the correct time,
        And don't allow yourself to be quickly frightened.[156]

---

[154] This is on an unnumbered page between I.64 and II.1, and has a special frame.

[155] The four parts of the opponent. The four openings.

[156] This poem also appears in Rösener (1589) as a section titled *Vnderrichtunge auch nützliche anweisung des Fechtens, sampt dem gantzen* Fundament *im Dussacken* ("Instructions and useful advice in fencing as well as the whole foundation in the dusack"). It's unclear if Rösener copied this poem from Meyer, or if both men were presenting an older didactic poem known by fencers.

Contents of fencing with the dusack
and the order in which this type of fencing will be described
Chapter 1

After the foundation was laid previously in the [section on] fencing with the sword, the dusack now follows, which takes its origin from the sword as the correct source of all fencing (both that directed with one hand and that with two). Therefore, I want to place this here—not only because it is the most commonly-used [weapon] for us Germans (following the sword) but also because it is a beginning and foundation for all weapons that are used in one hand—and subsequently to explain and consider all of its circumstances and associated sequences in order.

While the dusack is closely related to the sword—so that the majority of sequences which are fenced with a sword used in two hands may also be fenced with a dusack in one hand with very little change—I still want to follow the same order here in the description of [the dusack] that I previously maintained for the sword. And therefore, because enough was reported previously in the Sword [Section] about the division of the opponent[157] and how this is connected and about the division of the weapon,[158] I will first take up the stances or guards and also discuss how many there are and how they should be directed in this work.

Afterwards and secondly, I will set and treat the cuts and how they should be completed (in order).

Thirdly, I want to discuss the division of the opponent with respect to how the cuts currently under discussion should be directed (namely, about their usefulness and use), which |II.1ᵛ| previously passed over in the Sword Section as being more appropriate here, so that nothing about this topic is omitted.

Fourthly, I will append a necessarily-useful teaching and admonishment about counteractions and how one should use the cuts for three different purposes.

And lastly, after all of the necessary appropriate subjects have been dealt with regarding fencing with the dusack, I will consider one stance after the other, and how you should fully fence from the same or arrive into the same through cuts, and present them using several examples, and also describe how to correctly set the previously-taught techniques together so that a complete fencing sequence is constructed from them.

About the stances or guards and their use
Chapter 2

Although the stances which are commonly used for the dusack have their origin in the sword and are also considered to be not dissimilar, they still have a different use after one has changed the weapon—both in name and in fencing from them.

Therefore, I considered it necessary to set those stances here no less than as previously with the sword, in the same order, and to bring to light how they are to be used specifically and

---

157 "Opponent" is used as an equivalent to *Mann/man* throughout the translation. The inclusive "they/them" will be also used instead of the masculine, singular pronouns from the original. In addition, Meyer uses several other terms to refer to the opponent, which will be reflected in the translation.
158 Division of the opponent, I.3ᵛ; of the sword, I.4ᵛ.

beneficially. Firstly, there are five stances: namely, the Wrath Guard, the Bull, the Middle Guard, the Boar, and the Change [Guard].[159] As you position yourself in them on the right, you should also direct them in the work toward the left. Furthermore, you also have five straight down in front of you through the upright hair line: namely, first the Watchtower, secondly the Slice (which is the counteraction from above), the Longpoint, the Bulwark of two types.[160] In rising again, you have the fifth, namely the Arch, which is the other counteraction from below, as you will hear in the chapter on counteractions.

Because the stances will be depicted subsequently in Figures in their assigned sequences, I considered it good to save the lecture about the stances (how you should position yourself in them) to be described most appropriately while viewing the Figures. Each stance is then marked with its specific sign, which can be found in the subsequent register.

Therefore, I now want to continue to describe their use, and how they can be of service. And for the first, I really do not want you to wait in the stances for your opponent's attack. Instead, as soon as you can reach your opponent and quickly overwhelm them, you should attack them with your sequences according to your advantage and fence from the same. However, it occurs rather often that you cannot begin your sequence in the Before, nor even carry it out usefully, without receiving injury from an opponent who fences against you. Therefore, it necessarily follows that you position yourself, cautiously and yet with grace and serious demeanor, into that stance in which they cannot cut at you without their own injury and disadvantage. Secure in your stance, you can thus observe and attack them to your advantage or wait for their cuts.

On the other hand, they are also serviceable in that you can divide all of your sequences in an orderly way using the stances, so that as often as you strike down into a stance or draw yourself up into one, you will quickly know and consider what |II.2ᵛ| kind of sequences you should most properly fence against them from that point and stance, such that you are not delayed by long contemplation, because you always move from one stance into another in all cuts and sequences.

Thirdly, you should learn to recognize your counterpart's fencing from their stances, and what types of sequences they will likely fence against you, and accordingly easily derive[161] approximately what sequences they will fence against you when they have brought their weapon high or low into a stance. Furthermore, because one is generally taught about stances, as above, one can usually trace or take note of their opponent's fencing.

And for the fourth, you should know that you should not tarry motionless in any stance, but instead always change from one stance into another, and not solely because you mislead them in this way, but rather to confuse them [to the extent] that they cannot know what kind of sequence will be fenced at them or [what kind of sequence] they should fence at you.

Take a look in the Rapier [Section] about the use of the guards. I will now subsequently continue to the cuts, which are the correct primary subject in fencing.

---

[159] The guards on the sides are, in order: *Zornhut, Stier, Mittelhut, Eber,* and *Wechsel.*
[160] The guards along the centerline are: *Wacht, Schnitt, Lang ort,* and *Bastei.*
[161] This is an odd use of *abnehmen,* which generally means "to remove, take away", but here seems to follow the meaning of "derive a rule from an analogy", *die regel aus der analogie abnehmen,* GRIMM, vol. 1, col. 81.

About the four cuts, with four good rules about how one should learn and execute them correctly, including numerous appended examples

Chapter 3

Following the previous explanation of the stances or guards and their respective uses, I now arrive at the major element of fencing and that is, indeed, the cuts, which arise from the stances. There are, however, no more than four [cuts] from which the others have their origin and provenance (as will be subsequently demonstrated). Because these four cuts form the correct foundation of all fencing, I want to explain the same in an organized fashion as a service to the good-natured reader—and not just merely how they are cut, in and of themselves, but also how they can be taught and executed in many useful ways. This is so that you can see and recognize how one technique always develops from another and how the originating one offers assistance to the other [or derivative] one. However, I have considered that the four cuts cannot actually and properly be taught—much less be understood by students—without prior knowledge and understanding of a few lines which serve to assist in the instruction in the cuts. Therefore, it will necessarily follow that I will first teach those lines: their extent, their shapes, and their names. Since the cuts don't number more than four, the routes or |II.3ᵛ| lines through which they are cut are also four.

Thus, about the first [line], it is the upright line through which the High Cut is guided and cut and is therefore named the hair line, because the opponent is divided into left and right by this line. The second [line], the slanted or hanging line through which the Wrath Cut is guided, is called the wrath line (after the Wrath Cut), otherwise also called the strike[162] line. The Middle Cut is completed through the third [line], the crosswise or middle line. The fourth, upwardly slanted line indicates the path for the Low Cut, just as it shows the path for the Wrath Cut from the other side from the top downward. Thus, the Low Cut is also guided upward through the same line through which the Wrath Cut is cut downward (slantwise from above).

That point where you imagine the four lines intersecting is the same as the chin. Thus, the crosswise or middle line crosses the opponent across both shoulders. This locates the intersection correctly, and you can then cut, not only the four major cuts, but you can also use this to direct the other cuts surely and well. However, you should not opine that you are not permitted to guide the cuts no lower or higher than the lines indicate, but instead you should understand that you merely learn to guide

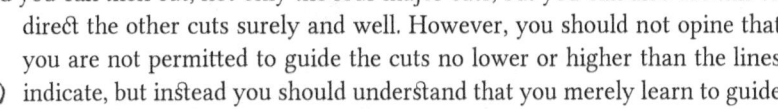

---

[162] *Strich* in this case is probably a typo for *streich*, thus the "strike line" in the translation.

and cut the cuts this way [through the lines] in the beginning.

Secondly, you should also take note: *when your counterpart cuts at you,* and you want to cut equally at them, you must guide your cut exactly at the height of their line; otherwise, you don't counteract nor are you guarded by your cuts. If you guide your cut toward your [counter]part in the Before and they are not ready to oppose your cut with any strike, you can then cut under or over their dusack at their body regardless of where the lines indicate (as will be sufficiently taught later in the sequences).

In the fourth chapter, after the lengthy discussion about several methods for learning to cut the cuts through the abovementioned lines, I will continue about the use and usefulness of the cuts, and how or which one counters another, as you will require no small measure of support in order to understand and correctly cut the sequences.

|II.4ᵛ| **The first rule about how you should guide the four cuts, cut by cut, from one side, each through its associated line, firstly halfway, that is up to Longpoint; secondly, completely through the line**

Stand with your left foot forward, hold your dusack in Bull (or as the larger portrait on the left shows in the Figure [B] printed previously). Step and cut straight from above down through the upright line up to the point at which the lines cross over one another. Then you stand with an arm extended in Longpoint. Regarding this, examine the larger portrait (also on the left) in the subsequently-printed Figure [C]. From there, let the forward part of your dusack sink and drop downward toward your left. In that moment in which the forward part drops downward, simultaneously pull your hilt (with a hanging blade) upward and around your head to a strike. During this and while you pull

your dusack up for another cut, simultaneously pull your rearmost foot up to the front right until you again have a complete step forward with your right foot for your prepared cut. Cut again through the upright line (as previously), but not farther than to the intersection of the lines (into Longpoint). From there, remind yourself not to cut farther (namely, as before), and execute one cut three or four times. As many times as you have now cut forwards, you should also cut and step backwards. As you previously moved the rearmost foot up to the frontmost (so that you could step farther forward with the front), when you want to step backwards in the cut, you must yield with the front [foot] back to the rearmost [foot] in that moment in which you pull up to strike. As you previously advanced with the right foot for the strike, you must now step back with the rearmost and left [foot] during the cut, identically to how you previously cut the High Cut through the upright line up to the point. Thus and in this way, you should also complete the other cuts through their respective lines.

Namely, the Wrath Cut through the slanted hanging line, the Middle Cut through the crossing [line], and the Low Cut through the upward slanting line. And always not farther than up to the center point, again letting it drop down toward your left and recovering by pulling your hilt upward for another identical cut. These cuts should serve you, in that you learn to stop all of your cuts at the halfway point (before they are then completed) and turn into a counterposture, so that you can receive your counterpart's cut in the complete course [of the cut] with identical cuts.

|II.5ᵛ| Secondly, you should cut completely through the abovementioned lines. Therefore, position yourself to stand with the dusack as taught above. From there, cut completely through the upright line with extended arm and turn your right side well into the cut toward your left, so that your dusack passes back by your left in this cut. In that moment in which your dusack moves back through next to your left, simultaneously pull your hilt upward next to your left, around your head, and up into the Watchtower again to strike. Then you stand as the portrait shows on the right in the Figure designated with the letter B printed previously.

In that moment in which you pull your dusack up to strike (as mentioned), you should simultaneously draw your rear foot up to the front [foot] so that you can step forward with the right [foot] for this cut (as was taught just prior to this). Step farther forward with the right foot and cut from the Watchtower again straight from above (as previously), completely through the upright line. Perform one cut three or four times, backwards and forwards, until you are well practiced.

As you now complete the High Cut through the upright line, you should also cut the other three completely through their associated lines in this way. So that you can better understand this (because it is hard for an inexperienced [person] to do), I want to set down here the Low Cut: in which way it is to be completely cut through the line.

Thus, stand with the left foot forward, hold your dusack in the previously-mentioned Bull, step and cut upward from the bottom of the rising line through their face with the long edge.

In this cut, turn your hand in the air so that your thumb faces your left and your short edge rises up to your left shoulder in the completion of the cut. At that point, turn your right side (as mentioned above) following your cut well toward your left as soon as you have arrived close to the left shoulder with this cut. Pull the hilt quickly back upward around your head for another stroke and afterwards draw your left foot up to the right.

Then, cut again from your right from the bottom of the slanted line upward through their face (with a step forward of your right foot) so that you come to your left shoulder as before. From there, gather yourself again for another cut.

Execute then another cut like this, three or four times in a row, backwards and forwards, as you have done with the previous half-cuts. As you have previously been taught to catch the ſtroke of your counterpart with the half-cuts, now learn to rebuff and cut away their cuts by cutting through with these.

|II.6| The second rule, how you should execute the cuts through a line counter to one another

For the second, after you can now cut the four cuts from your right (each one halfway and completely through its line as was juſt taught) you should then also learn to cut each line from both sides counter to one another. Thus, ſtand with the right foot forward (yet with the feet not too far apart from each other) so that you can ſtep forward into the firſt cut, and then ſtep and cut from your right from above completely through the upright line with extended arm, so far that—in that moment in which your dusack runs back through next to your left—it shoots into left Bull. Following this, cut ſtrongly and powerfully from below through the upright line (with the long edge upward) so that your dusack shoots around over your head, through the Plummet,[163] [and] back into right Bull.

In these cuts, you should always keep your right foot forward when ſtanding and ſtepping so that you can ſtep forward for each ſtrike and recover for a further ſtep (as taught above). In this way, ſtrongly execute the High Cut from your right from above together with the Low Cut from your left from below through your opponent's face. And allow both the High Cut from above to plummet headlong from above and the Low Cut to shoot around upward from below such that the front end[164] always points toward their face after shooting forward.

As you have now cut through the upright line from below and above (opposite one another) you should also cut through the other two lines—namely, the slanted hanging line and the

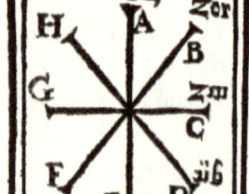

crossing line—always from both locations opposite one another.

However, when you want to cut through the upward rising line from your right (which is designated with the two letters D and H opposite one another), you shouldn't allow it to shoot through but simply drive upward over your forward right thigh (from your right from below) up into the left Wrath or to your left shoulder, slanted through the abovementioned rising line up and down counter to one another. You have already been taught at length about this driving and why it is appropriate and useful in the Sword [Seĉtion].

|II.6ᵛ| The third rule about the beginning of the misleading and how one should pull the cuts back and change into another

As you have now learned to move through each line opposite one another, you should furthermore also learn to pull the cuts back, namely like this:

---

[163] The *Sturzhauw* is described on II.9.
[164] The terms *vorderen ort* and *hinteren ort* come from the polearm section, where they refer to the two ends of the staff, but occasionally Meyer also uses them as synonyms for the tip and the pommel of other weapons.

After you have positioned yourself into a stance in front of your opponent according to opportunity, step and cut from above with arms extended and the long edge along the upright line to the head, and **Indesly**[165] take note whether they want to oppose your cut by counteracting.

As soon as you perceive this [to be the case], don't allow your cut to touch or to hit their counterposture, but instead pull your cut quickly back before it meets with their counterposture, and cut strongly and powerfully next to your left [side] from below upward through the same upright line (as the portrait on the right[166] in this Figure [D] indicates).

Then, in turn, cut the first next to your left side from below almost up to their counteraction and, in that moment in which it should now just touch, pull it quickly back upward around your head and cut from above through the upright line (that is, through their face).

In this way, cut from your right toward their left following the middle line up to their counteraction and also don't let it touch, but instead, in that moment in which it should just hit, pull back around the head and cut from the other side completely through the same middle line. You should learn to do this through all four lines, just like the previous two cuts. Pulling them back like this is the beginning of all misleading.

The fourth rule, how one should cycle[167] through the cuts, one after another

Fourthly, it is also necessary and useful that you can cycle through the cuts fluidly and through another. This can occur in three different ways. First, I want to show you the reasons for this

[165] On I.25 of the Sword Section, Meyer explains his interpretation of *Indes* to be a small word fraught with meaning: in that moment in which you strike, you should simultaneously observe the other openings that you could also hit, and also all of the techniques your opponent can use to strike or counteract your action. It is therefore a point of hyperfocus on what you are doing and what you and your opponent could be doing at the same time and subsequently. Meyer states that he will use *Indes* as a shortened form to call all of this to mind.

Within his text, Meyer tends to use *Indes* to indicate a disruptive action, that is, because he is describing actions sequentially that occur simultaneously or disruptively. Thus, *while* you are performing the above action, execute the following action *at the same time* and *in the same space*. Because *Indes* is an adverb of time in German, it will frequently appear adverbially in the translation, thus *Indesly*.

[166] The scene depends on how one interprets the instruction. The fencer on the right is preparing to start the upward cut. The fencer on the left has completed the upward cut.

[167] *Wechseln* is commonly translated as "change", which is one of its meanings. However, "exchange, replace, or alternate" are also more common interpretations in this time period, and "change" is considered a derivative of "exchange/replace". In this particular situation, Meyer is not using *wechsel* in its transformative meaning (see II.64ᵛ), but instead to describe cutting through one line after another after another in a semi-repetitive cycle. This interpretation is supported by GRIMM, vol. 27, col. 2743. For cycles that only have 2 repeated actions, I will use '*wechsel* = alternate'.

cycling through the three lines: namely, through the two slanted [lines] (as the hanging and rising) and across through the middle line. Afterwards, [I will be] appending some examples.

When you now cut slantwise through your opponent from your right (from above or below) |II.7ᵛ| so that you arrive with your weapon on your left, cut the second quickly again from your left across through their right (following the middle line through your opponent). If you have cut from your left through the slanted line (regardless of whether through the rising or hanging line) so that you arrive through toward your right with the cut, immediately cut from there also through the middle line—as before from your left through their right, but now from your right through their left. As often as you cut through a slanted line from one side, immediately also cut across through the middle line from the other [side].

Furthermore, take note: when you have cut the first through from your right (slantwise from above) and also the Middle Cut from the other opposite side so that you have arrived back on the right side, then don't cut back from above through the slanted hanging line, but instead cut upward from below through the rising line and thus upward through the second rising line.

Afterwards, cut a Middle Cut again from your right through to your left, so that you can again cut the High Cut slanted through their right. And this forms the foundation for changing all cuts between the two lines as in the slanted and crossing. As often as you cut across from one side, [that is how] often you should cut around again from the other slanted line. Therefore, if you cut from these slanted lines, whether from above or below, cut from the other side across. In order that you can better understand this alternation, I want to set three useful examples here.

Example with six cuts

➤ Therefore, step and cut the first from your right, a Wrath Cut toward their left through the hanging line which is designated with the letters B and F.
➤ Cut the second from your left toward their right across through the middle line.
➤ Cut the third through from your right toward their left through the upward rising line (strongly with a Low Cut upward) so that your dusack hangs down behind your left shoulder at the end of the cut.

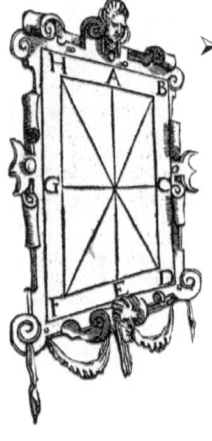

➤ From there, cut another Low Cut powerfully slanted upwards through their right [side].
➤ Afterwards, cut the fifth (a Middle Cut) from your right toward their left through their crossing line.

> Cut the sixth, however, straight from above |II.8| (following the hair line at the head or through their face) with another step forward.

These six cuts should quickly follow one after the other. For these cuts, always remain with the right foot forward and as you want to step farther forward for the cuts (as you should have one step for each individual cut). Always recover yourself with the rear foot somewhat to the front [foot] so that you can have another advance with the right [foot].

Cycling the Cross [Cuts] by means of the Middle Cut

Cutting the Cross [Cuts] long and alternating from one side to the other by means of the Middle Cut is very good, in which you should always cut the Middle Cut with arms extended strongly in front of you, the other two slanted lines through the cross, and perform it like this:
> Namely, cut the first through the slanted line from above, toward their left.
> Cut the second through also along the slanted line from above, from your left toward their right.
> Cut the third across through the middle line, from your right toward their left, so that you arrive with your dusack on your left in the Middle Guard.
> From there, begin again, and cut the first from your left slanted through their right.
> The second slanted through their left (both from above).
> The third again a Middle Cut toward their right from your left, and additionally cut back again from your right.

Execute this one [sequence of] cut[s] six or seven [times], one after the other. Always remain with the right foot forward for these cuts.

Cross alternation

Perform the Cross alternation like this:
> Stand with your right [foot] forward (as always) and cut the first from your right through their left Wrath line with a broad step forward. Thus you arrive in the left Change [Guard].
> From there, rip back upward through the abovementioned Wrath line (through which you have come downward with the Wrath Cut) with the short edge, upward toward your right shoulder. Let it move around in the air above your head.
> Cut the second through their right Wrath line slantwise over your |II.8^V| forward right thigh, so that your dusack arrives next to your right with the point on the ground.
> From there, rip upward again with the short edge toward your left shoulder, exactly through the line through which you have cut from above. Let the dusack move around above your head again.
> Cut again from your right through their left, so that you arrive again in left Change [Guard].
> From there, rip upward again through as before, and continue like this.

Powerfully execute one cut [or sequence of cuts] three or four [times], as you please, through your opponent's face. Thus, you now have the four Primary Cuts, and how you should cut them in many ways. I have treated this so extensively because all fencing is contained in these four cuts, as mentioned above. And it is assured that, if you know well that you can cut the four cuts as mentioned above, all sequences will be easy for you to fence. The additional secondary cuts now follow.

About the secondary cuts, which have their origin in the four primary cuts, and how one should use them in the work
Chapter 4

Because the four primary cuts have been laid as a foundation for all other cuts, I also want to set down as much as is necessary for you about the secondary cuts (which grow out of the same), and firstly the reason for their existence and to show how they differ from the others.

Namely and firstly, you should know that the cuts explained above are not only cut straight (as taught previously) but are also cut curved. This was also mentioned in the Sword [Section], as then the hands are turned or reversed in cutting so that you don't hit with the (front) long edge but with the (rear) short edge or with the flat. Therefore, as the cuts are cut differently in this reversal [of the hands], they are also named differently, regardless of whether they are cut from above, slantwise, across, or from below, as this will be seen in the Crooked Cut, which alone of them is also called this in dusack, because the curved edge[168] leads in the reversal and it hits with the same. Thus, this reversal is the first and most common reason for the many names of the cuts.

|II.9| However, some cuts are not reversed when cut and are still named differently. This is the reason and it arises from this: firstly, according to the intent of the one who cuts with it, as occurs with the Missed Cut and the Ramming Cut. Even if they are High Cuts, they are named like this because my intention with the one is to miss and with the other is to ram (if an opening were exposed to me).

Some names are linked to the external movement of one's body language, like the Wrath or Rage Cut.

Some also receive their names from the shape, which is viewed as identical to their cutting action, like the Rose Cut. Likewise, some are named for the limbs at which they are aimed, as you will see in the Rapier [Section] in the Hand Cuts and the like.

From these reasons listed above, you can easily understand the subsequently listed cuts, and notice how far they differ from the others. The secondary cuts are fifteen in number, namely:

Plummet Cut [Sturzhauw] is brought into motion from the High Cut and Wrath Cut

[The] Plummet Cut: this is most often used in the Onset. Thus, in the Onset, cut a High Cut from your right (including the step) back through next to your left, so that your dusack shoots or plummets around back over your head so that the front end faces toward your opponent's face at the end of the plummet around. [This is] not dissimilar from the left Bull, except that you must thrust the point forward and longer away from you and toward your opponent's face. It has its name from this plummet over, otherwise, it is itself merely a High Cut, except that you allow the front end to drop downward over the hand toward your right.

[Then] Indesly pull your hilt upward around the head and simultaneously lift your left foot (stepping forward with the same) in that moment in which you pull upward, and cut from your left back through next to your right, |II.9ᵛ| so that your front end plummets around over your head so that the front end faces toward your opponent's face (just as previously). Then, let it

---

[168] The curved edge is another term for the short or half edge, used primarily in dusack.

drop down again toward your left, and cut from one side to the other until you can arrive at your appropriate placement.

Crooked Cut [ƙrumpḫauw]

[The] Crooked Cut is executed like this: rotate the grip of your dusack so that the curved edge leads and hits in cutting. Then cut from above or below with the curved edge according to opportunity.

Short Cut [ƙurʒḫauw]

[The] Short Cut: in the Onset, when you perceive that *they want to cut from above*, pay attention. *In that moment in which they lift their dusack for the cut:*
➢ Pull your weapon toward your left shoulder during that moment.
➢ From there, cut across simultaneously with them (through with the short edge) above their arm at their face, so that your [palm] faces upward in the cut.

Thus, you take their ſtrike from them and simultaneously hit, etc. [The] Short Cut is also carried out at the sides and short under their weapon. You were taught about this in the Sword Section.

|II.10| ### Overwhelming Cut [ʒwíngerḫauw] [169]

[The] Overwhelming Cut is completed in two ways. Firſtly, when you ſtand in Middle Guard on the left and from there cut your counterpart's cut in front of you away with the long edge. Regarding this, see the Middle Guard. Secondly, it is now executed like this:
➢ Namely, ſtand with the right foot forward, hold your dusack with arm extended in front of you in the Slice or Simple Brace.
If your counterfencer cuts at you from their right and guides their cut high:
➢ Let your point drop downward and simultaneously pull through with your hilt toward your left under their dusack, *so that they cut in vain above your dusack.*
➢ Cut quickly from the outside over their right arm (*in that moment in which their dusack ſtill falls downward toward the ground*) at their head.

Ensure that you have withdrawn your head under your counterpoſture [170] by bending your body, so that they cannot reach you as they pass through.

---

[169] *Zwingen*, in a figurative sense, means "to overwhelm, overcome with force", in particular "to bring an opponent down through the force of weapons or with the strength of one's arms". GRIMM, vol. 32, col. 1240, meaning B.1)a).

[170] Meyer uses the term *Versatzung* to refer to a previously-mentioned stance, or as the noun equivalent of '*versetzen* = counteract'. In LF, *versatzung* (*obsessio* = barricade/obstruction) is the response to the *hut* (*custo* = guard). In the Latin translation of Lew's gloss in PMM, *Versatzung* is translated with '*permunitio* = completely walled, defensive position' or '*castrorum* = defended military encampment'. Therefore, *Versatzung* by itself, as a reference to a specified or unspecified stance, is translated here as a defended position in response to a particular action by the opponent, thus a counterposture.

Roaring Cut [Brummerhauw][171]

[The] Roaring Cut therefore has this name because it moves in one sweeping motion so quickly that it has an effect like a blustering wind. Execute it like this:

➤ Observe how you [can] drive your opponent high with their counterposture.
➤ Indesly rotate your grip so that you hold your dusack curved [edge outward].
➤ Pull your hilt with hanging dusack around your head and cut next to your right from below across with the curved edge—with a step forward of your right foot toward their right—under their dusack at the bone of the forearm[172] or inwardly at the tendons, |II.11| depending on *how high they have lifted.*
➤ Let your dusack shoot back in front of your face to the counterposture.

You will be taught more about this Roaring Cut in subsequent sequences, because it is used and fenced in many ways.

Awakening Cut [Weckerhauw][173]

Perform the Awakening Cut like this in the Onset:

➤ Cut in at them with a powerful High Cut.
➤ *If they counteract the cut,* take note: in that moment in which it glances off or touches on their counteraction: turn the cut into a thrust, pushing your dusack around onto theirs at their face (as the two tenants on the right side in this Figure [P] teach you).

---

[171] *Brummen* is a rather dull noise, distinct from *brüllen* which is rather sharp, screaming noise. *Brummen* is associated with lions, oxen, bears. Larger flies and bees *brummen*, otherwise *summen* is used. Thunder *brüllt* if nearby, but *brummt* from a distance. *Brummen als per ein lew* [roaring like a lion] is a catchphrase to describe fury and wrath. Latin equivalents are *rugire* and *fremere*. GRIMM, vol. 2, col. 428.

[172] *Ulna brevior.*

[173] A *Wecker* is a person who wakes others (prior to alarm clocks), also something that brings the dead back to life. The result of this cut is definitely startling and would absolutely function to "awaken" the person on the receiving end.

➢ *If they rise up:* cut upward through their arm with the curved edge (as you can see in the larger tenants [in Figure P]).[174]

Rose Cut [Ꝛosenhauw]

Rose Cut; *if you find an opponent tarrying in Arch:*
➢ Act as if you wanted to cut from above at their head.
➢ Don't let the cut touch, but instead move through and around from outside and under their right arm so that you move in a circle around their dusack.
➢ Let it drop down again in the air next to their right side and cut at their face.

Thus, you can also move in a circle around their counteraction on the other side and cut in, wherever you find them open.

|II.11ᵛ| **Dangerous Cut [Gefehrhauw]**

Perform the Dangerous Cut like this: when you approach your opponent in the Onset, pay attention. As soon as you trust that you can reach them, take diligent note of *when they want to cut. In that moment in which they pull their dusack up to strike:* cut at them from above next to their hilt at their face or chest (in that moment in which *they have lifted their dusack up high*).

This Dangerous Cut must be carried out with caution, otherwise it is dangerous; therefore, it is named the Dangerous Cut.

Rage Cut [Entrüsthauw]

[The] Rage Cut is completed in many ways, [and] also named with two names: as the Armored Cut[175] because it catches the High Cut in the air and holds it like a lance rest on armor;[176] and

---

[174] The center pair of fencers in Figure P appear to show this technique, while the larger pair of fencers in front demonstrate the Roaring Cut at the tendons.

[175] *Rüsthauw.*

[176] *Gerüst*, GRIMM, vol. 5, col. 3781: *insbesondere eine vorrichtung an der rüstung, der am plattenharnisch angebrachte eiserne haken, in welchen beim angriff der speer eingelegt wurde.* "In particular a device on the armor, the iron hook applied to plate harness in which the spear is held during an attack." This is a *Gerüst* in German, and an *Arrête* in French.

it is named Rage because it comes so seriously and unpredictably, as if you had become unpredictably enraged [and filled] with wrath toward them. Execute it like this to oppose their cuts: *if an opponent cuts at you from above,* take note. *In that moment in which they pull their dusack into the air to strike:*

> Pull quickly around your head **Indesly** and cut across a little from below and upward toward their cut, so that you catch their cut still high in the air with the long edge and with the dusacks horizontal (so that your dusack faces horizontally between you and them, as the larger portrait to the right shows in this Figure [G]).

In that moment in which the dusacks thus touch together, you can carry out many lovely sequences, etc.

Furthermore, it is also performed in this way:

> Stand with your left foot forward and hold your dusack in Boar (as you will subsequently learn).

> *If they cut at you from above:* raise both arms to the counteraction, so that your dusack comes to lie with its back on your left arm and leap in under their strike.

> In that moment in which it glances off, thrust outside of their right arm at their face, step away, and cut across toward their |II.12ᵛ| left and through their face (as you can see in the Figure designated with the letter O, in the small tenants on the left).

Missed Cut [Feþlþauw]¹⁷⁷

[The] Missed Cut: in the Onset, if you find *your counterfencer in Arch or Simple Brace:*
➤ Step and cut outwardly at their right arm from above.
➤ In that moment in which you notice that *they want to counteract:* let your short edge drop strongly downward in front of their arm and simultaneously pull your hilt back upward so that your cut doesn't hit.
➤ Instead let it run in front of their arm and miss, *so that they vainly move to counteract.*
➤ Step quickly outward to their left and cut at them straight through their face.
    You can thus let this cut miss near any opening of those who move against the cut to counteract it.

Obscured Cut [Blendþauw]¹⁷⁸

[The] Obscured Cut is carried out in many ways; however, take note of this technique.
➤ *If an opponent cuts from above:* catch their strike high over your head in the air on your long edge (with your dusack horizontal so that your point projects toward their left).
➤ As taught above about the Rage Cut, as soon as it beats or touches: turn the short edge inward in a flick in at their face.
➤ Immediately after the flick: turn your dusack with the hilt upward toward their left.
➤ Pull quickly back toward your right and cut with the long edge forward at their face.
➤ In addition to this cut, step well around toward their left in this strike and away from¹⁷⁹ their strike.

|II.13ᵛ| **Flicking Cut [Sþnellþauw]¹⁸⁰**

[The] Flicking Cut: take note when you stand in Arch in front of an opponent. *If they don't want to cut:*
➤ Pull upward into the Watchtower as if you wanted to cut from above; however, don't do that.
➤ Instead, turn in the air and cut with the long edge up from below at their right arm in a flick and twist the dusack back toward your left shoulder.
➤ From there, cut a Defensive strike through their right (either above or below their arms) through their face.

---

¹⁷⁷ A *Fehlhauw/Fehlhieb* is a cut that is meant to deceive your opponent. In strategy, it is a cut that deliberately misses in order to provoke your opponent to a specific action. In contrast to a successful deception, *fehl hawen* is to cut in vain, ergo one that misses (by mistake).
¹⁷⁸ A *Blende* is "a cover or means for obscuring eyesight". *Blenden* means "to obscure one's eyes, or to obscure defensive aspects from a besieging army". This is an attack at a target you can't necessarily see clearly, and an attack whose initiation is difficult to clearly see.
¹⁷⁹ The text has *auff sein streich*, which is probably a typo for the usual wording of *aus sein streich*.
¹⁸⁰ This is tricky, because *schnellen* is both extremely fast movement and also extremely destructive, meaning "dash to the floor, to fling belongings out of a house, to deceive through speed, and to snap one's fingers".

Item, *if an opponent stands before you in Arch and doesn't want to work:* flick at them above their dusack at their head with the short edge. Or *if they hold their dusack high in a counter-posture:* flick under their counterposture at their face.

Twisting Cut [Windthauw]

[The] Twisting Cut is a cut that twists [and] is nothing other than you cut in (regardless of side) and you move back out toward that same side.

Thus, if you cut in from your high left, pull and turn the dusack downward toward your left and outward, all in one smooth motion (as if you wanted to cut a piece from their side equal to half a man) and this works on both sides.

Ramming Cut [Bochhauw] [181]

Perform the Ramming Cut like this: when you two stand in front of each other in a high counterposture in Arch and neither wants to cut Before the other, but instead waits for the other's strike: |II.14| drop down before them and cut straight in front of you at their dusack so that you touch their chest with your hilt so strongly that their dusack rams against their face once or twice. You thus force them to work.

And take note, *as soon as they lift up:* step out to their left side and cut in next to their hilt at their face, or undertake other appropriate sequences after you have thus driven them upward.

Change Cut [Wechselhauw]

[The] Change Cut is among the five cuts initially taught to students [and] which fundamentally does not differ (in itself) from changing through the cuts from one side to the other. However, because the alternation of the cuts and sequences is often considered in this [Dusack Section], it is unnecessary to introduce this here.

Cross Cut [Kreutzhauw]

[The] Cross Cut is (in itself) two Wrath Cuts from both sides which are completed through the two slanted and hanging lines; thus, strike through your opponent slantwise from both sides and cross the strikes over one another in this way.

[Execute] the cut like this: stand with the right foot forward and cut the first from your right through their left and the second from your left through their right (both slantwise through their face).

Practice this cut four, five, or six times (backwards and forwards), yet so that you always remain with the right foot forward |II.14ᵛ| so that, when you want to step, you recover with the

---

[181] *Bocken* is "the beating of your heart". *Bock* is "goat"—rams are sheep, but 'ramming' is what goats do with their horns/heads. *Pochen* is "to beat/knock on a door with a knocker". "Ramming" preserves the pun.

rear foot so that you can step forward with the right—because you should always take at least one step for both cuts (thus cut through the cross from both sides).

You should learn to cut this Cross Cut, including the abovementioned four cuts, with extended arms in a free-flying, powerful, and quick way. In the cuts, don't keep your arms in the chest (as they say), that is close to yourself, because that person who fences short and keeps their arms close to themselves is easy to mislead and to touch. However, the extension must remain within bounds according to the opportunity at that time.

Therefore, I want to especially remind you, so that even if you wanted to make use of this book in no other way, then before everything else, you would want to learn to cut the cuts long, free, and well. And if you can cut the Principal Cuts[182] well, the others will all be easy for you. Because when you fence whatever sequences you want, regardless of how good [the sequences] are, if you can't cut these cuts well individually (in and of themselves and each according to their art) and execute the sequences correctly, you will not perform much of use, because as is often said, all fencing is based on the cuts.

Finally, take note of how the cuts counter one another, thus a rule in brief: namely and for the first, the High Cut (if you guide it against your opponent's dusack at the Strong toward their right hand) counters all other cuts which are cut at you, whether from below, slantwise, or across. In contrast, the Wrath Cut or the Crossing Middle Cut counters or takes the High Cut.

Correspondingly, also take note that two identical cuts which are cut against one another (with their steps) always move one another aside and counteract. However, the one whose cut directs the other's farthest outward is the one who comes above the other with their weapon during the cuts. Therefore, *as often as an opponent cuts at you from below [or] across (regardless of whether from the left or right),* oppose them with |II.15| High Cuts. *If they cut at you from above,* take away their High Cut with crossing or slanted Wrath Cuts. This rule is truly to be observed in all fencing and especially applied with respect to the Before and the After, the Strong and the Weak.

How one should use the four openings
[Chapter 5]

You have learned previously in the Sword [Section] as much as was necessary about what the division of the opponent[183] is, why it is useful, and what belongs to it. However, as nothing less than the cuts are based on this division, it appears necessary to me to provide additional information about the application and use of the same, specifically about what can be of service for this weapon, because the opportunity in how and when you can usefully place the cuts must be viewed in part as based on this division.

Therefore, you must first learn here how one can recognize the openings quickly. Secondly, how one should position themselves against the same. However, because these [openings] can occur in many ways and therefore it would take too long here to explain all of the errors, I will

---

[182] These are not the Master Cuts (of which there are five) but rather the primary cuts that Meyer references in the rest of the text—that is, High, Wrath, Middle, and Low.

[183] See I.3^V.

explain this briefly, from which I hope that you will be able to perceive and learn sufficiently from the stated teaching about the use of the division (if you consider it seriously in addition to diligent practice of the same).

And first, take note of this teaching about perceiving the opening in your counterpart's cuts: that is, in the Onset, pay diligent attention to the side *from which they want to cut at you, and, in that moment in which they cut,* look to see how you can yield from their strike or how you catch it and rebuff it away from you without injury.

Then, cut quickly and precisely to the part *from which or through which they cut at you.* Then be diligent to note this: the part from which they guided their strike is always most exposed. This is a very significant rule, which you should consider diligently and give your attention to, so that with cunning and quick work, you fence precisely at the opening *from which or through which they came with their weapon.*

Secondly, the opening is easily and well perceived in their counterposture or in their stance. Thus, *if they hold their weapon too high or low, or also too far out to the sides,* you should guide your strikes carefully, powerfully, and long at this part which you perceive to be the most open. Before it correctly hits this, cut quickly over to the opposite (as long as they have not offered you this opening with intention).

|II.15ᵛ| *Regarding the [opponent] who offers you an opening with intention,* you shouldn't cut in at that opponent without caution, but instead consider your opportunity well. Because if you cut incautiously at the opening (as you will see afterward in the sequences), *they can quickly withdraw from you by stepping outward and simultaneously reach past you in their cut.*

For example, *if they hold their weapon in Wrath on the right* and you cut at them straight at their head: *they can step outward with their right foot to the side on their right toward your left and simultaneously cut from above in toward your head,* such that you not only cut in vain at their opening *which they withdrew from you,* but you exposed yourself more with this extension *so that they can reach past you even better with their cut.* Thus, you cut in vain with your cut in the Before and they hit with their cut in response.

Therefore, the second [rule] now follows: how you attack the openings through deceit.

Namely, *if they hold their dusack on the right side in a guard, in whatever guard they want, low or high:*

➤ Execute a powerful High Strike toward their left side (*which faces forward*) not because you want to hit them, but instead *because they must oppose your cut from their right to counteract it.*
➤ Don't let your cut touch on their counteraction but instead immediately pull your cut covertly back in a full sweeping motion and guide it quickly to the right of the other [fencer]—*from whence they had come to oppose your cut—so that they have counteracted to no avail.*

Perform this not only on both sides, but also toward all four openings so that, *if they hold their weapon in the quarter of one [opening],* you attack with cuts opposite to it—yet not to hit, but instead to lure them out of this quarter. *In that moment in which they move with their weapon from that part,* cut cunningly in at them at that point.

You should thus be careful and unobtrusive in all cuts in the Before.

Furthermore, you should set up your cuts in this way so that one out of two always hits (whether it is the first or the second), and therefore I will set you some examples:

If they stand before you in the Arch and hold their counterposture so low that you can view their face over their dusack:

➤ Undertake to cut two Middle Cuts from both sides (opposite one another) in front of yourself, like this:

➤ Cut the first from your right above their dusack, in front of their hilt (hard, across, with extended arm), through their face, and turn your body strongly toward your left following the cut.

➤ Cut the second through from your left under their dusack, across toward their right arm. *If they don't want to be hit in the face by the first cut, they must rise up.* This lifting provides you the space so that you hit them with the second strike (from your left [at] their |II.16| right arm).

➤ Cut a Wrath Cut quickly as the third, slantwise from your right toward their left and through their face.

If they hold their counteraction high:

➤ Cut the first through below their dusack from your left toward their right arm.

➤ *In that moment in which they drop downward in response to the cut:* cut the second quickly from your right, above their dusack at their face.

When you execute these two cuts quickly one after another, you hit either their arm in the first [cut] below their dusack, or their face in the second [cut] above their dusack. *They will be hard pressed to counteract both of your cuts.*

Likewise, if they hold their weapon too far to their right side:

➤ Cut the first powerfully through from above toward their left side.

➤ *In that moment in which they displace their weapon to their left side to oppose your strike,* step and cut the second back around outwardly (above their right arm).

If they hold their weapon too far to their left:

➤ Cut the first through toward their right from above, and the second at their left while retreating.

These cuts should always move quickly, both following one another (with their steps).

However, you will find sufficient instruction in the sequences as to how one should mislead from one opening to another.

About counteraction and how all cuts are divided into three parts, that is, the Provoker, Taker, and Hitter[184]
Chapter 6

The stances and cuts, and also the openings, at which the cuts are directed, have been explained at length. It is insufficient that you learned previously how you can cut the cuts nobly and long in front toward your enemy. Instead, it is also necessary that you can at least rebuff and counteract those cuts that are cut at you by your counterpart. Therefore, although I wrote in general

---

[184] *Reitzer, Nemer, Treffer.*

in the Sword [Section] about counteracting, necessity demands that I deal somewhat more clearly and technically about counterpostures in the Dusack [Section] than about the other primary techniques of fencing. Therefore, it is to be noted that the counterposture is preferably of two kinds: namely, one from above, the second from below.

From the first, which comes from the High Cut, the stance originates which is called the Slice or also the Simple |II.17| Brace. The second counterposture comes from the Low Cut, from which the Arch finds its origin. These two counterpostures are each completed in two ways: firstly, by catching or opposing the strike; secondly by cutting away. Catching is nothing other than that you oppose and stop your counterpart's strikes with the counterposture, whether using the Arch through the Low Cut or with the Simple Brace through or from the High Cut.

You should not understand this counterposture [to mean projecting forward] as some have the habit of doing—namely, that they simply hold their weapon and let it be struck.[185] Instead, if you want to receive their strike and counteract it, you should rise upward from below with extended arm, with your counterposture counter to their High Cut in the air—because the higher you catch their cut in the air, the more you weaken it, and you can apply your cut in response both more fruitfully and you can also complete it in greater safety.

Likewise, if you want to counteract the Low Cuts, you should move opposite the cuts from above and drop down on them with arms extended. These counterpostures both end in Long-point, namely like this:

If an opponent cuts at you from below or across:
➤ Drop down on them with a Simple Brace and take note:
➤ In that moment in which it touches or beats, turn your point with a step outward from their cut into their face.

If they cut from above:
➤ Catch them from below in Arch.
➤ In that moment in which the dusacks beat together, shove your front end forward into their chest.

This is sometimes called the Stork's Beak.[186]

The other type of counteracting occurs with simultaneous cuts, as one cut is then countered by the other. You should take note, however, that in this case the cut in the After always counters the lower cut, like this.

In that moment in which they cut in:
➤ Cut simultaneously with them.
➤ With this simultaneous cut, leap well out of their cut to the side so that their dusack arrives below and yours above when they touch together.

This was the purpose of the step, that they thus arrive below in the cut in the Before and you arrive above in the cut in the After.

---

[185] There is a potential pun here based on *verstehen/vorstehen*. *Verstehen* means "to understand", whereas *vorstehen* means "to place in front of". The prefix *ver-* held both meanings at the time, and Meyer often uses *ver-* to mean "in front of" in other situations. (Here it is negated: to not be understood, or something is not to be placed in front.)

[186] *Storken Schnabel.*

You can also counter their High Cut with your Low Cut. Namely, *in that moment in which their cut flies in:*

➤ Step to the side out of their strike.

➤ Cut powerfully through, upward and opposite their High Cut.

Even though the Low Cut is weak against the High Cut, the Low Cut is sufficiently reinforced by the step to ward off. Thus, one strike always counters the other (as was mentioned previously). The High Cut counters |II.18 [187]] all other cuts, as you will subsequently hear in the Watchtower.

So that you can understand these things better, I want to distinguish the cuts into three usages. Firstly, they are used to provoke, secondly to take or counteract, thirdly to hit.

Those cuts with which I infuriate and goad an opponent to move out of their advantage and to cut, these I call Provokers.

Those cuts with which I cut away and ward off their cuts (to which I have infuriated and moved them), these I call Takers.

Those cuts which I cut quickly to the closest opening (as the third [strike] after I have first provoked them to cut and secondly taken that strike to which I have moved them), and before they recover themselves back from their strike (which I have taken), these I call Hitters.

In the Onset, *when your counterpart rests in a stance, guard, or counterposture and doesn't want to hit* and you ought not cut at an opening while they are in their advantage, do this:

➤ Extend yourself long in front of them and cut through—with one cut, two, three—In front of them and toward their openings or through their counteraction, for example, with seriously angry body language, [acting] as if you had overcommitted in your cutting. [188]

Take diligent note during this:

➤ *As soon as they lift and cut in:* step to the side out from their cut and cut their cut away from you with a powerful cut. [189]

If they are not sufficiently weakened with one [cut], execute it with two or three cuts crosswise and away from you through their dusack for a long as it takes for you to feel that *they are sufficiently weakened.*

➤ Immediately and *before they regain their strength again or rise up or recover:* cut at the closest opening, and from the opening be quickly back on their dusack (whether with binds or with cuts).

If they stand before you in a guard so that you cannot cut through their counterposture, or they execute their counterposture so that you don't trust yourself to carry out your first provoking cut at the opening without injury, pay attention or take note that you cut through as quickly with the second strike to or toward their weapon (the closer to their hand the better) with one or two strikes opposite one another according to opportunity.

---

[187] Incorrectly labeled '19' in the first edition.

[188] *Verhauen* means things like "cut down, set off on a journey, cut to injure, lead into danger, cut to deceive". GRIMM, vol 25, col. 542.

[189] This repetition of cut as both noun and verb in this section accurately reflects Meyer's repeated use of *Hauw* and *hauwen*. This may be one of his least rhetorically polished paragraphs.

Furthermore, you should also know and take note that you should always alternate between the three cuts, so that sometimes the first, now the second, then the third is a Provoker, Taker, or a Hitter.

Therefore, if you can hit with the first, you should use the second for a counterposture. However, if you hit with the second, counteract with the third. Because if you want to fence safely with those weapons that are used in one hand, you should accustom yourself to always executing three cuts quickly, one after the other. It doesn't have to be one kind of cut, but instead it should be distributed among and change between the High, Middle, and Low so that of the three, one is always a Hitter (namely, the first, the second, or the third). Because the same examples will occur in the subsequent sequences, I will remind you of this then.

|II.18ᵛ| **The stances now follow, including the sequences**
Chapter 7

Since I have now taught all necessary material associated with this weapon, I will continue onward to explain the stances, including their sequences. You should note well that the stances ought not be understood as a guard[190] in which you tarry, awaiting your opponent's fencing, but instead as a beginning or end of the cuts and counterposture (as already mentioned previously).

Look at it like this: when you stand in Arch, if you now want to cut out of Arch then you will arrive into Watchtower or Bull (in a recovery for [executing] an upward cut), and in that moment in which you pull your dusack around your head to strike, you move through the Wrath Guard (during that pull around) and only from there does the first cut completely take place. When you have completed the High Cut, you will arrive at the end of the same in the Change [Guard] (if you don't move through with a cut to the side). If you want to counteract upward from there, you will arrive in Arch or Longpoint.

Therefore, take note how the body language is displayed in the pull around into a cut or recovery from a cut, which is thus named after the same configuration (like the Wrath Cut from its wrathful body language, the Bull from its thrust, and the Boar because it cuts from the side like a boar). This is now the most acknowledged reason for the invention of these stances, because when you pull a cut up into a stance, you can still change the same while in the air and turn or move to another opening. Thus, when you pull up to a cut, at the outermost point at which you arrive with your draw upward, you can tarry a bit to see *whether they want to cut at the opening in your cut*, so that you can observe [while] in the air whether you can arrive over their in-flying[191] cut by |II.19| simultaneously cutting over [their cut]. However, you should not tarry longer in any stance (only remaining as a recovery from a strike) and instead always change out of one stance into another until you observe an opportunity to cut. You should also well imagine the sequences that are assigned to each stance, practice them, and make them known to yourself, so that if you arrive in a stance in the middle of fencing, you are prepared and are ready with counter sequences.

---

[190] *Custos, wart, hüte.*

[191] Because German has directional prefixes *hin-* and *her-* to indicate movement away from or toward an object, I have used "in-flying" and "in-coming" as translations of the adjectives *herfliegend* and *herkommend* respectively, which Meyer often uses to describe weapon movements by both parties.

Furthermore, you should also know that even if I have assigned a specific sequence to one of the ſtances, this does not mean that one should not also fence these sequences from other ſtances, or that it cannot happen. The moſt common reason that I have assigned one sequence to one ſtance and another [sequence] to another [ſtance] is so that they would be dealt with in order. These sequences are not set in this way so that they cannot be changed in fencing, but inſtead are only an example from which one can learn to seek and take their opportunity (according to the sequences) and may prepare and change them to be useful to themselves. Because, as we cannot have one single nature, we also cannot have one single type of fencing; however, everything muſt flow together and be taken from one common foundation.

|II.20| **About the Watchtower**[192] **[Wacht] and the sequences assigned to the same**
Chapter 8

This High Guard is the beginning of the High Cut and is therefore named the Watchtower, because you observe with the prepared ſtrike up high and watch out for where they expose themselves in front of you with cuts so that you can reach paſt them with cuts from above. This is because your counterpart [can] cut at you as they want; you can reach paſt them from this guard or at leaſt suppress their cuts or convert them into nothing.

Position yourself in this guard like this: ſtand with your right foot forward, hold your dusack over your head and let the blade hang down behind you (as the portrait toward the right hand indicates in the Figure [B] printed previously). I will explain using some subsequently set examples so that you can also know what is to be fenced from this. Namely, for the firſt:

How and in what way you should reach paſt an opponent with simultaneous cuts

In the Onset, position yourself into the High Guard. *If your counterfencer cuts at your body from outside, whether at the arm, high, or low,* take note:
> ➤ *In that moment in which their arm extends for the cut:*
> ➤ Step toward their right and away from their ſtrike, and cut ſtrongly through in front of yourself and outwardly over their right arm and at their head, so that your dusack arrives in the Middle Guard on your left after the completed cut.

---

[192] This guard is called *Luginsland* in JML (like the other guard names, based on the 15th century messer teachings of Hans Lecküchner), which is historically a lookout in a crow's nest on-board a ship, and in the 15th and 16th centuries was an irreverent word for certain watchtowers, specifically the one located at the farthest edges of a fortress which looked out over flatlands and thus physically resembled a crow's nest in location and action. "Watchtower" is used in this translation as the more general term.

➢ From there, cut across toward their right arm through their face (this applies regardless of whether this is carried out with the flat or with the long edge), so that you arrive with your dusack at your right shoulder at the end of the cut.

➢ From there, quickly and immediately cut two Wrath Cuts from both sides (cross-wise[193] opposite one another) and through their face.

|II.20ᵛ| *If they cut in front of you at your face and toward your left:*

➢ *In that moment in which they cut in:* ſtep with your left foot behind your right around toward their left side out of their cut and follow with the right foot somewhat toward them.

➢ In that moment in which you ſtep, cut two long ſtraight cuts (with arm extended) slant-wise through from above at their face and toward their hand, both haſtily one after the other.

➢ You then arrive at the end of the second cut in Middle Guard on your left.

You thus expose your right side with these cuts. Therefore, take note during this.

If they were to cut at the openings you have provided:

➢ Cut their in-flying cut away, from your left toward your right with a powerful cut upward, one that is so ſtrong that your dusack shoots around above your head back into right Bull.

➢ Simultaneously with this cut outward, ſtep with your left foot toward their right and threaten to thruſt with the front end from outside over their right arm.

➢ However, pull your thruſt quickly back to yourself and cut in front through their face with a retreating ſtep.

If they don't cut at the opening you provided:

➢ Cut anyway with the long edge from your left upward through their right and complete the sequence as has now been taught.

➢ Then cut away from them through the cross.

Rule

In summary, when you ſtand in the High Guard and they cut at your body (from outside or inside, at the left or right), always ſtep out of their ſtrike and cut in equally long and above their ſtrike at their head. As often as you complete this High Cut from one side, also cut a Middle Cut from the other side and through their face (juſt as often as a counter). Subsequently, follow this with other additional sequences, or cut away from them using the Cross.

|II.21| **How you should fence from the Watchtower if your counterfencer doesn't want to cut firſt; the firſt sequence from a Misser**

Take note: *if your counterpart does not want to cut, but inſtead positions themselves in a counterpoſture in front of you:*

➢ Step with your right foot around toward their right and cut the firſt from above and outside at their right arm *so that they muſt defend againſt [this]*.

---

[193] The Cross Cut is formed from two Wrath Cuts. 'Cross-wise' is used to distinguish *Kreutzweiß* from the Crosswise Cut [*Zwerch*].

> As soon as you then perceive *that they are moving with their dusack to counteract opposite your cut*: don't let your cut hit or touch.
> Instead, *in that moment in which they lift upwards*, let [your cut] miss and drop down past them.
> Step quickly back around toward their left during this and cut again through their face from in front.

A sequence and example of how you should provoke your counterpart such that they lift upward so that you can damage their right arm

Another:
> Cut the first from above straight toward the center part in their hairline.
> In that moment in which your cut flies in from above, turn the short edge outwards in the air towards them, as if you wanted to hit with the same.
> Take note, *in that moment in which they rise to receive your cut:*
> Immediately pull back up, without contact, around your head and cut strongly through with the flat, outwardly from your left and across toward their right arm. This should occur *in that moment in which they drop down with their cut.*
> Then, follow with Cross Cuts.

This Misser sequence is quick, because by turning the short edge toward them, you recover yourself so that you can complete the Middle strike more quickly and strongly. Then, *in that moment in which they raise their arms*, your blow flies in across and hits.

|II.21ᵛ| Another, in which it is taught how you should provoke your counterpart to a cut, catch the same, and when they want to rise farther upward, how you should cut through their face or arm during that

When you observe that they quickly cut in response, you should carry out these sequences with which you provoke them and encourage them to cut.

Thus, position yourself with body language as if you wanted to cut powerfully. However, don't let it hit. Instead, withdraw that cut back to a counterposture so that you catch their cut (with which they intended to overtake you) and thus chase away their advantage.

To take one example of this:
> Guide a High Strike at them with a step forward.
> In that moment in which it should now hit, turn the short edge toward their left and thus recover yourself for a Low Cut toward their right.
> However, you should not complete this Low Cut.
> Instead, *as soon as they subsequently lift up and cut in:* cut the Low Cut (for which you have now recovered yourself) and catch their in-flying cut still in the air with this Low Cut from your left.
> Take note *as soon as they move away from the counterposture and lift up:*
> During their movement, cut a powerful Middle Cut from your right quickly through their face.
> Follow with a Wrath Cut long from your left.

How you should drive an opponent up and down with strength so that they must cede [you] space to cut with Middle Cuts to both sides, at their arms and through their face

➤ Turn your right side well toward them and cut two high, straight, and strong cuts through their face in one driving move forward (with two steps forward of your right foot).
➤ As soon as you have driven their dusack up high, cut two Middle Cuts quickly from both sides opposite one another. The first through from your right under their dusack toward their inside arm; the second |II.22| from your left toward their right, above their dusack through their face (*in that moment in which they lift up*).
➤ The third cut is again a High Strike from your right from above through their face.

Two techniques are specifically assigned to the Watchtower and should be carried out through the High Cut: namely, the first is reaching past, as taught in part previously; the second is to suppress[194] any cuts, as will follow after this in the Simple Brace.

About the Bull[195] [Stier] and its sequences
Chapter 9

This stance is not dissimilar from the Plummet [Cut] and is one of the best stances from which all kinds of techniques can be fenced using all of the cuts that are appropriate for attacking in the Before.

For this stance, position yourself like this: stand with the left foot forward, hold your dusack with your hilt to the right next to your head so that the front end extends toward your opponent's face (as the large portrait on the left shows in the Figure [B] printed previously, although they are in mid-step). And the stance is nothing more [or less] than a gathered thrust from above.

|II.23| **The first sequence, in which it is taught how you should reach past with simultaneous cuts from the Bull, with an appended Middle Cut including a thrust and a Cross Cut afterwards**

When you move into the guard of the Bull in the Onset, and *your counterpart cuts at you from their right, whether from above or below:*
➤ Leap well out from their strike toward their left side.

---

[194] "Reaching past" is *überlangen*. "Suppress a cut" is *dempfen/dämpfen*.
[195] *Stier* can mean its cognate, that is, a young, male (usually castrated) bovine. However, it is more commonly linked with both the bull—the mature, uncastrated male with powerful and wild actions. This is not a tame or gentle draft animal, but a feral and dangerous one. GRIMM, vol. 18, cols. 2846–51, particularly points 3–5.

➤ Cut through simultaneously with them, with an extended arm, at their face and toward the hand in which they hold their weapon (ensuring that you remain with your weapon above theirs).

➤ Cut strongly enough that your dusack shoots around above your head into the Plummet (that is into the left Bull).

➤ From there, cut a Middle Cut back through toward their face (with the outward flat facing toward their right), so that your dusack shoots around above your head into right Bull.

These two cuts should move quickly and strongly one after the other.

➤ Afterwards, threaten with a thrust from there.

➤ At the halfway point [of the thrust], pull the thrust back and cut two strikes through the cross.

The second sequence, in which is taught how you should cut through their arm (in that moment in which they lift), and how you should cut simultaneously and above their weapon at their head (in that moment in which they cut back)

When you move into this guard, pay attention. *As soon as they have cut:*
➤ *In that moment in which they lift for the strike,* cut through across or from below from your right toward their hand.

➤ Afterwards, *in that moment in which they cut back,* cut again quickly from outside and over their right arm at their head with powerful steps outward toward their right out from their strike.

|II.23ᵛ| ### The third sequence, how you should counteract and cut in response from the Bull if they cut outwardly at you (that is, at your right)

If they cut at your right side when you stand in Bull, whether from above or below:
➤ Step with your right foot toward their right and out to the side and toward them.

➤ Extend your dusack away from you toward their right. Let your front end hang toward the ground in this push forward and rebuff their strike with the dusack from your left toward your right.

This setting off and step outward should take place together with one another.

➤ *As soon as their[196] cut touches on the outward flat of your dusack in this counterposture:* withdraw your dusack back upward and away from theirs toward your left to strike.

➤ Cut outwardly above their right arm at their head, and step around toward their right simultaneously with this cut.

Or:
➤ If you have rebuffed their strike with the hanging dusack through the Arch, and *they pull up from your counterposture to cut:*

➤ Step with a doubled step well out on their right side.

---

[196] The text refers to *dein hauw an... deines dusackens... rüret*, which is impossible. I have corrected this to "their cut" and "your dusack".

➤ Simultaneously with the steps outward, pull your dusack upward around your head and cut slantwise upward from your right with the curved edge, under their dusack, at their inward arm bone or flat (as can be seen in the subsequent Figure [I]).

This Low Cut must take place *in that moment in which they pull up to cut*, so that your dusack arrives at your left shoulder.

➤ From there, cut two strikes long through the cross.

The fourth sequence is a rule in which it is taught how you should safely catch all cuts in all stances, whether they come from the right or the left

This is a good rule, therefore: whenever you have cut with diligence or have cut through into one of the stances *and they cut in response, swiftly and unexpectedly*, you shall move out of that same stance with the point toward their face (such that your long edge is turned towards their in-flying strike in this push forward |II.24| to catch [their strike]) in such a way that you stand in Longpoint at the end of the thrust. As an example of this, when you move in the Onset into the aforementioned Bull in front of your opponent and *your counterpart cuts from their right toward your left:*

➤ Step quickly with your right foot out toward their left (somewhat to the side).
➤ *In that moment in which they cut in:* thrust toward their face with arm extended. In this thrust, turn the long edge toward their incoming cut, so that you counteract their strike and thrust simultaneously.
➤ *If they defend against your thrust and guide it upward:* lift your hilt up toward your left.
➤ From there, cut[197] a Low Cut through their right.
➤ Then cut quickly back straight from above through their face, with another step forward of your right foot.

What you should fence from the Bull toward an opponent who does not want to cut first

In the first Onset, when you become aware that *they don't want to cut or attack first with their sequences*, take diligent note *how they want position themselves into a guard or stance*, so

---

[197] The text reads *Vauw* at this point, which must be a typo for *Hauw*.

that you can see whether *they hold their hand too high or low, or also too far outward toward one side in their fencing.*

Therefore, pay attention. As soon as you observe that *they hold their hand, including the weapon, too high*, you fence at them in this way, namely:

➤ From the aforementioned Bull, ſtep and cut through from your right, across from below and underneath their dusack (*while they hold it too high*), upward at their face and toward their right hand (*in which they hold their weapon*) so that that you arrive into the Wrath Guard on the left after the end of this Low Cut.

➤ From there, cut again with the long edge (as before), powerfully and ſtrong, upwardly toward their right through their face.

➤ In conjunction with this Low Cut, you should ſtep farther toward them with your right foot (which you have forward anyway) so that with this upward cut you arrive into the right High Guard.

➤ From there, cut a Cross Strike through their face quickly, long, and ſtrong.

|II.24ᵛ| **An example and sequence, how you bring someone down who holds**
their weapon high in fencing and how you should arrive above
the [weapon] at their head or face

In the Onset:

➤ Step toward them with your right foot and thruſt underneath their dusack toward their face or cheſt (as the upper and smaller scene in the Figure which is designated with the letter K teaches).

➤ *As soon as they drop down to counteract the thruſt (since they muſt counteract if they otherwise don't want to get hit in the face),* pull your dusack back around your head.

➤ *In that moment in which they drop down with their dusack:* cut ſtraight and across above the same through their face.

➤ Afterwards, cut the third ſtraight from above through the upright line with another ſtep forward of your right, so that you arrive in left Change [Guard] at the end of the cut.

➤ From there, move quickly to the counterpoſture outside of their right arm with the long edge upward and toward their weapon.

➤ Bind on their dusack from below and take diligent note:

➤ *As soon as they lift up from your weapon and pull upward to cut, in that moment in which they pull upward:* cut hard next to their hilt, from above, downward through their face.

For this cut, you should ſtep with the right foot farther around toward them, so that you bend your upper body far forward following the cut and you ſtand with your feet wide apart (so that you can arrive that much earlier back at the counterpoſture with your dusack).

How you should oppose an opponent who holds their weapon too low

If they hold their dusack low in their counterpoſture (namely, in the Arch) so that you can see and reach their face well above their hilt:

> From Bull, ſtep and cut from your right above their hilt, across through their face and thus close to their |II.25| hilt so that you touch and hit them with the abovementioned cut.

> In that moment in which your cut flies in through their face, simultaneously pull your hilt back upward.

> During this, ſtep with your left foot behind your right toward their left and follow with the right further toward the abovementioned side.

> During these ſteps, cut the second quickly ſtraight from above through their face.

These two cuts—namely, the Crossing [Cut] and then the High [Cut]—should always be completed with the ſteps juſt taught, quickly and faſt in series. Then the sequence comes off well.

An example and sequence about an opponent who cuts quickly in response, how you should cut through the Weak of their dusack (to provoke), then catch their cut, and then cut through their face (with the curved edge to the right and with the long to the left)

Furthermore, *if they hold or guide their dusack extended far in front of themselves and are also ready to cut quickly in response:*

> Step and cut the firſt from your right (from the abovementioned Bull) toward their left, this time through the Weak of their dusack.

They will want to quickly cut in response to this cut, in the opinion that they can overtake you.

> Therefore, pull your hilt quickly upward around your head after the firſt ſtrike and cut the second (also from your right), across toward their cut (to which your firſt cut has provoked them and shifted them upward), so that you catch them with the [second cut] *while they are ſtill flying in* (as the large scene toward the right side shows in the Figure which is designated with the letter G).

➢ *As soon as their cut touches or beats on the long edge of your dusack:* step around quickly with your left foot out toward their right side.

➢ Simultaneously with this step, pull your dusack back up from their weapon around your head and cut with the curved edge outside of their right arm at their head (as the smaller scene between the larger also teaches you in the abovementioned Figure [G]).

➢ *They must counteract that (if they otherwise don't want to be hit). In that moment in which they want to defend against this and counteract,* |II.25ᵛ| *they expose their face,* so you should quickly cut through strongly at it with a step backward of your left foot.

➢ After this cut, move quickly back into a good counterposture.

An example and sequence, how you should break through the counterposture of your counterfencer from below and above with force

If your counterpart opposes you in a stance (regardless of which stance it is), position yourself in the Onset into the guard of the Bull also on the right.

➢ From there, threaten them with a powerful thrust from above toward their left.

➢ *As soon as they rise to oppose your thrust,* pull the same uncompleted [thrust] quickly back towards yourself and upward toward your right.

➢ Catapult strongly and powerfully from there, with the curved edge upward, toward their left through their face and counterposture.

In association with all of this, you should always step in this way: namely, in that moment in which you threaten the thrust, lift your right foot to step, and pull the threatened thrust so quickly that you set the lifted foot back down with the completion of the Low Cut, so that the step forward and the strike are completed together.

➢ Following this Low Cut, cut quickly from above through their face, powerfully and long in front of yourself, with additional steps outward toward their left.

When you execute the outward rip from below, with the High Cut, swiftly one after the other, with their associated steps, the sequence moves well.

If another opposes you with this sequence, counter it like this, namely:

➢ In that moment in which you are moving upward toward their thrust to catch it, and you perceive that *they will withdraw and want to cut from below:*

➢ Step with your right foot well toward their right side and drop down with the long edge strongly from above onto their in-flying Low Cut, so that, in this drop, the front end of your dusack extends toward your opponent's right side.

Block their Low Cut like this, so that they cannot move through.

- ➤ From there, pull quickly back up toward your left shoulder.
- ➤ From there, cut again at their head, from outside and over their right arm.

If they were to break through with their upward rip:
- ➤ |II.26| Catch their High Cut above [yourself] on the long edge (assuming that you have deflected the previous Low Cut from above without injury) and take diligent note:
- ➤ *In that moment in which their cut beats or touches on your counterposture:*
- ➤ Your dusack should immediately be pulled toward your left shoulder (as before) and cut at their closest opening.

Or in that moment in which you have received their High Cut on your counterposture:
- ➤ Thrust at them quickly (*before they recover*) underneath their dusack at their face.
- ➤ *They must defend against this* and thus give you space at an upper opening.

An example and sequence, how you should goad an opponent upward with body language (including cuts) so that you can better injure their arm bone with the curved edge

Take note: when you are in the Onset, observe diligently *whether your counterfencer wants to position themselves into the Arch.* As soon as you have recognized this:
- ➤ Step and guide a powerful High Cut from the guard of the Bull (with serious body language) toward their left.
- ➤ However, don't allow the cut to hit or touch, but take diligent note.
- ➤ *As soon as they lift to counteract your cut:* pull it quickly toward your left, back around your head.
- ➤ During the withdrawal, rotate your grip even more than when you want to cut farther, with the curved edge forward in the hit.
- ➤ *In that moment in which they lift to capture your cut:* cut underneath their dusack from your right (from inside) at their arm and upward toward their left (as you can learn and take note of in the portrait on the right in the next Figure [I]).

However, in this cut, you should not come [any] closer to them than that you can just reach their arm between their elbow and their hand with the outermost [point] of your dusack.
- ➤ After this cut, let your dusack move up around your head and cut a Middle Cut, including a High Cut, through their face, etc.

|II.27| **A good sequence from the Dangerous Cut**
                which can also be properly fenced from the Bull

If you find your counterpart in Arch:
> ➤ Step and cut a powerful cut from above, strongly at their hilt. *They will cut quickly and soon in response.*
> ➤ Therefore, take note that you don't move through with your cut, but instead, as soon as your cut beats on theirs, let your blade drop downward next to their right arm and pull your hilt simultaneously back upward.
> ➤ *In that moment in which they pull up to strike,* cut the second quickly, next to their hilt and inwardly through their face (*while they are still rising into the air for their strike*).

Observe at the same time that you don't move closer with your foot than that point from which you can easily reach them with the outermost point of your [dusack].

Another from the Awakener

In the Onset, take note *when you find an opponent in Arch:*
> ➤ In order to counteract, step and cut a powerful cut from the right Bull.
> ➤ As soon as the cut beats or touches on their counterposture, turn the front end over their dusack and inward at their face.
> ➤ Thrust in on top of their dusack, *so that they must counteract and defend upward.*
> ➤ Therefore, take note: *in that moment in which they lift up,* pull your dusack around your head and cut from your right across toward their left at their face (*while they are still high with their dusack*).
> ➤ *If they counteract the second from above:* remain with your dusack on theirs in the bind and turn your point on their dusack and inward toward their face again.
> ➤ *If they also defend against this:* move your point through under their right arm and thrust outwardly over their right arm at their face.
> ➤ *If they defend against the thrust:* let your dusack move around your head and cut a Middle Cut across through toward the closest opening.

|II.27ᵛ| **A good sequence from the Bull with which you can break through powerfully**

In the Onset, take note as soon as you can now reach your counterpart:
> ➤ Step and cut a strong cut powerfully through their counterposture with arm extended, so that at the end of the cut, your dusack sweeps around next to your left arm with the front end behind you (when seen from the outside).
> ➤ From there, again cut strongly and powerfully upwards from below through their right.
> ➤ Cut the third from your right toward their left (whether from above or below) through their dusack, through their face.

A general rule to counter all kinds of sequences

Although I had initially planned to order and set a counter to be assigned to each sequence, it appears to me that it would be good to put this off, especially as in my other book (which I will possibly allow to be published in time and for the good of the art), I have set and described many nice sequences, including their counters. Therefore, I will merely present a general rule here, from which you can take and learn all kinds of counters. In addition, the art of fencing is such that you cannot easily know or hurriedly take note what kind of sequences they want to fence against you, to say nothing of the fact that you would have to be able to quickly know how it is countered.

Therefore, I always hold with the fencer, who knows many sequences and few counters, and also knows to carefully fence using the same in the Before and After, and allows the other to concern themselves with the counters. This fencer is so prepared with sequences that, if one of theirs is countered, they have two others at that point which are already in the works.

The general counter which I will set here is this: namely, when you have been overpowered with cuts by your counterfencer so that you must counteract them:

> Move with a strong counterposture under their cut and hold them back *so that they can't get through with the same [cut] and so that they must pull their dusack back around from yours.*
> *While they are thus* |II.28| *lifting from your dusack for another strike,* thrust straight in front of you at them and toward their face.
> While doing this, turn the long edge toward that point which you see that *they will cut in from again.*
> This will counteract them.

However, if you cannot hold up their cut because, perhaps, they are too strong for you and will break through with strength, take note:

> *In that moment in which their strike drops through toward the earth from your dusack, or moves out to the side:*
> You should thrust straight at their face (*while their dusack is still moving through*), and you should complete this thrust before their weapon has completely dropped down to the ground.

If they oppose you with untrustworthy cuts, in that moment in which they guide their dusack around from one point to another:

> Once again thrust at them straight forward from Longpoint toward their face or their chest.
> While doing this, pay diligent attention *to the point where they want to attack from with their cut.* Turn the long edge toward that same point (with the hilt upward). During this, remain with your front end equally in front of their face or chest.
> As soon and as often as you then observe your opportunity, let a cut fly at their closest opening.

And that is the correct summary and final opinion on all counters. Namely, as often as two cuts touch or bind together, *in that moment in which they beat in the bind,* you thrust forward on their dusack (regardless of whether their dusack moves away from yours).

You should consider and take note of this rule in all hits, thus you will easily counter whatever they fence at you and you will also be able to drive them and their sequences away from you.

Correspondingly, *if your counterfencer does not want to cut and positions themselves in the same stances before you* so that you absolutely can't cut in at an opening without [careful] consideration, the following will also be of interest. You should therefore practice and have good knowledge of countering all of the stances and driving [your opponents] out of [the stances] (namely, *if your counterfencer positions themselves in a stance*, regardless of the guard, thrust straight at their face from Longpoint).

From whichever side they strike out at your thrust, cut in at that same side. There will subsequently be more instruction about this.

|II.29|
About the Wrath Guard [Zorn Hut]
Chapter 10

The stance from which the strongest cut (called the Father Strike) is brought forth is also used on both sides. However, there is no other difference between this stance and between the Bull—

other than that the Bull presents the thrust and the Wrath presents the cut with wrathful body language. However, as relates to the sequences that are fenced from there, you can fence from one as from the other.

Although this Wrath Guard presents one side as completely open, you can still bring forth and fence many and various swift and strong sequences from there, about which I will explain and set down a few here.

The portrait on the right side of the Figure [M] printed previously shows you how you should position yourself in this guard, from which stance you should then rebuff all strikes that are cut at you and cut in response.

The first sequence teaches you how you should take their cuts (from above and below) and cut in response to the same with Twisting Cuts

Take note: in the Onset, when you move into the right Wrath Guard (with which you expose the entire left side), your counterfencer is thereby caused to fence at the same with cuts.

Therefore, take diligent note *as soon as they cut at you from above:*

➤ During the same, step with your right foot well to their left side and toward them, close to their side and out away from their strike.

➤ Cut simultaneously with them at their head and the hand in which they hold the weapon (however, so that your dusack arrives above their dusack in this simultaneous cutting), so that your dusack arrives next to your left in the Change [Guard] at the end of the cut.

➤ From there (*if they cut at your right opening*), remove their in-flying strike with the long edge strongly and upwardly toward your right.

➤ Simultaneously with this removal, step with your right foot well toward their right side and let your dusack move completely around your head (in the removal mentioned previously).

➤ Cut Twisting Cuts outwardly over their right arm.

|II.29ᵛ| **The second sequence also teaches how you should take their cut again, and additionally how you should cut in response from the same side**

If you ſtand in the right Wrath to wait for your opponent's attack (as before), *as soon as they cut at your presented opening (whether from above or below):*
- Step out toward their left and cut their in-coming ſtrike away from you with a Wrath Cut toward their hand.
- *Before they recover back from the removal of that ſtrike:* cut the second quickly *before they rise* (as ſtated) and also from your left at the opening, with another leap forward of your right foot.
- After these two cuts, you should soon arrive back into the counterpoſture by means of a Low Cut upward toward their right arm.

An example and sequence, how you should turn the point into their face in all hits

Furthermore, if they cut at your body from outside of your right arm (whether from above or below):
- During this, ſtep with your right foot well toward their right out of their ſtrike.
- *In that moment in which their cut flies in:* cut across above the [cut] from outside over their right arm at their head.
- Or cut a Suppressing Cut[198] from above at their hand.
  If they turn their counterpoſture toward your cut:
- Take note in that moment in which the dusacks touch and hit together, and push your front end in front of you in at their face (on their dusack or on top of their right arm).
- *They muſt defend againſt this thruſt and guide or beat your dusack outward to the sides.*
- *From whichever side they have now removed your thruſt:* cut quickly in at that same side, whether to the left or right (as you were also taught previously with the counters).
  If they guide your dusack upwards in the removal:
- Allow it to move above your head and cut across from below through their face with the curved edge facing upward (*while they are ſtill lifting their arm up in the removal*).

|II.30| **Another: how you should turn your point in front of you, away, and into their face**

In the Onset, position yourself into the Wrath Guard and take diligent note *as soon as they cut at you:*
- Step and cut simultaneously with them with an extended arm towards their left at their head.
- As soon as the cuts touch or beat together, push your dusack ſtraight forward with your point in front of you into their face.
- In this push forward, turn the long edge (including your hilt) upward toward their incoming weapon.

---

[198] The '*Dempfhauw* = Suppressing Cut' is a rapier technique and is described starting on II.56. *Dämpfen/dempfen* means both "to bring something to an end using violence or force", and "to suffocate or smother". Both of these meanings are supported in the text.

➢ *As soon as their second strike beats on your long edge:* lift your hilt up high and remain with the aforementioned hilt over your head.

➢ Indesly strike at their right arm from outside, with the inward flat and hanging dusack.

➢ *As soon as they swipe in response to your hit:* step backward and cut through their face from the front.

How you should catch your counterpart's cuts from the Wrath Guard using a push forward and immediately cutting in response

Position yourself in the Onset into the Wrath Guard again, and take note *as soon as your counterfencer cuts at you:*

➢ Lift your dusack over your head under their incoming strike and catch their cut with the long edge of your dusack, so that your dusack hangs down with the point downward at the ground and toward your left when you catch their cut.

➢ Afterwards, *as soon as their cut glances on your dusack,* cut two fast and strong strikes upward from both sides with the curved edge, through their face from below.

➢ After these Low Cuts, cut quickly back around across with a Middle Cut, together with a straight High Cut from your right (also through their face or wherever they are open) so that you arrive on your left in Change [Guard] at the end of the last cut.

➢ From there, raise the long edge back upward into the counterposture.

|II.30ᵛ| **Another**

When you move into the aforementioned Wrath Guard in the Onset and you perceive that *your counterpart does not want to cut first:*

➢ Guide a powerful High Strike toward their head with a step forward of your right foot.

➢ During this, take diligent note *as soon as they lift their dusack into Arch to counteract:* don't let your cut hit their counterposture, but instead pull your dusack back around and upward toward you.

➢ Thrust away from you with your front end, beneath their dusack at their chest (as the upper small scene shows in the Figure designated with the letter K).

➢ If you want, you can also hit the first strongly from above and afterward let your dusack snap around in the air and similarly thrust in front of the chest.

➢ *As soon as they drop down on your thrust,* cut quickly upward, across to their hilt, through their face.

A good attack from the Wrath Cut with four cuts

In the Onset, take note when you move into the Wrath Guard, and as soon as you can reach them:

> ➤ Lift your hand from the right shoulder and **Indesly** turn the point of your dusack toward their face. Threaten to thrust at them; however, pull your dusack quickly back around your head and cut the first Low Cut of your dusack upward through their face.
> ➤ Cut the second from your right upward through their face.
> ➤ Cut the third likewise through their face, slantwise from above back toward their left once again.
> ➤ Cut the fourth, however, from your left slantwise through toward their right arm.

Also take note: as many cuts as you cut from one side, you should step as many times with the right foot toward the same side. Step toward them according to your cuts, because you should always keep your right foot forward for stepping in these sequences.

|II.31| **Bull and Wrath from the left**

Because you have now briefly heard about these two stances on your right and what you should fence from there, it is easy to understand from this what you can safely fence from

these reversed stances (without special instruction)—namely, by just reversing the sequences. However, in order that I provide a bit of instruction for this reversal, I do not want to omit setting some sequences here as examples. And, therefore, I will take these two stances together because with few exceptions, one can fence from the one as from the other.

The portrait on the right side in the previously-printed Figure (designated with L) teaches you how you should position yourself in the left Wrath. Regarding left Bull, you should remember how you aligned the work to the right previously and then set that up on the left.

The first sequence from the left Wrath

In the Onset, take note when you arrive in left Wrath and carry out one cut, three, or four in close series—from above from your left, and from below your right over your forward right thigh through their face, according to the slanted hanging lines which are designated with the two letters H and B—until you perceive an opening. Cut in at the same. Afterwards, cut quickly long again and through the cross so that you protect yourself further from their cuts.

A good sequence, how you should compel them down with their counter-poſture so that you can get to their face

Item:

➤ Execute the slanted cuts over your right thigh as previously; however, guide all cuts underneath their dusack and at their arm or at their fingers.

This will provoke and compel them *so that they muſt lower their counterpoſture.*

➤ *As soon as they drop down with their cut,* cut quickly and unexpectedly |II.31ᵛ| above their dusack, from your left toward their right and slantwise through their face.

➤ Or cut a Middle Cut from your left (above their hilt) and across toward their right and again through their face.

➤ After these, cut the second quickly through toward their left.

A quick and ſtrong sequence to fence from the left Bull, which is completed with five cuts

In the Onset, position yourself into the guard of the left Bull and take note *as soon as they cut at you from the outside and toward your right side:*

➤ Step with your left foot well out of their cut toward their right.

➤ Cut slantwise toward their right and through their face yet also simultaneously with them (so that your dusack arrives above theirs in the cut), so that you arrive with your dusack next to your right thigh and with the point on the ground at the end of this cut.

➤ Following this, turn the long edge back upward and pull a ſtrong Low Cut from your right toward your left (with the long edge upward) so that your dusack arrives at your left shoulder at the end of this cut.

➤ From there, cut a Low Cut ſtrongly upward through their right side so that your dusack shoots into the guard of the right Bull at the end of this cut.

➤ From here, cut a swift cut across through their face.

➤ For the laſt, cut a long, powerful High Cut (direćtly from above) through their face with another ſtep forward, etc.

|II.32| **A good sequence which is fenced from both the left and right Bull**

Item, powerfully execute one or three ſtrong cuts (from both below and above) obliquely over your right thigh through your opponent's face, and take note:

➤ When you have arrived the third time at your left shoulder, cut two Low Cuts from there (one from your left and the second from your right), both ſtrongly upwards through their face, so that you arrive back into the left Wrath Guard after the end of the second Low Cut.

➤ From there, cut with the short edge or beat with the flat across through their face (with extended arm), so ſtrongly that your dusack shoots into the guard of right Bull after the end of the cut.

➤ From there, ſtep again with your right foot farther toward them and thruſt toward their face from above with extended arm.

➤ Afterwards, cut the laſt quickly after the thruſt (also from above) through their face. And in all ſteps, remain with your right foot forward.

Rule

Item, *if an opponent cuts at you toward your right* (when you stand in one of the left High Guards), step outward toward their right and cut a powerful Twisting Cut, from outside and over their right arm at their head.

Because as often as you arrive into these two stances, you should endeavor that you always step out of their in-coming strike and simultaneously cut at their head and above their dusack with a long extension.

|II.32ᵛ| ### Counter to the Bull and Wrath [Guards] on the left

When you see them in the Onset and perceive that *your counterfencer opposes you in the aforementioned left High Guards*, position yourself quickly into the guard of the right Bull.

➢ From there, thrust from your right toward their face with a broad step forward. *They must defend against that and beat your thrust outward.*

➢ Let the [thrust] move around your head toward the side to which they have rebuffed you with their outward beat.

➢ Cut toward the other side at the opening.

➢ If you perceive that *they also want to remove the same [cut] and counteract*, let your dusack drop down in front on that same side as a miss.

➢ Cut at their other side.

The Simple Brace[199] [gerade Versatzung] or the Slice [Schnit]
Chapter 11

Position yourself into this Simple Brace like this: stand with your right foot forward and hold your dusack in front with an extended arm so that the long edge is toward your opponent and the front end is upward (as the large portrait on the right hand indicates in the next printed Figure [F]). I praise this stance for the best of them all because you can wait for your enemy in this stance more safely than in any other.

---

[199] I have translated *gerade Versatzung* as "Simple Brace" instead of "Straight Counterposture". While Meyer uses *Versatzung* to refer to a stance in general, especially a previously-specified stance, the *gerade Versatzung* is one of the named guards and is not merely an aspect of a generic stance. The term "Simple Brace" for *gerade Versatzung* comes from the diagonal braces used in half-timbered building construction to connect a vertical post to a horizontal beam, which are called *Versatz/Versatzung* in German and "brace" in English. A *gerader Versatz/Versatzung* has no additional offsetting wedge to affect its angulation, thus it is 'simple' as opposed to complex. See: Grimm, vol. 25, col.1047. It is linked to the Arch in both its vertical movement and its architectural imagery.

|II.33ᵛ| **An example and a sequence, how you should work from below at the opening**

When you arrive into the Simple Brace in front of your opponent, take note. *If they want to cut from in front at your face:*
> ➢ Turn the long edge toward their cut and catch them in the air toward their right.
> ➢ In addition, step with your left foot behind your right and toward them during this and follow quickly with your right foot afterwards and toward their left.
> ➢ *As soon as their strike beats on your dusack:* turn your long edge away from their dusack and draw the long edge toward your right through their muzzle.[200]
> ➢ Pull quickly and Indesly back around and cut a Middle Cut with a good counterposture toward their left at their face, so that (*if they were to cut a second [cut]*) you would hold off or rebuff their cut with this Middle Cut.
> ➢ After this Middle Cut, pull quickly back around your head and flick a cut from outside at their right arm.
> ➢ Immediately cut Cross Strikes long in front of yourself.

You should complete the Slice and the step backward simultaneously with one another. The sequence proceeds well [when this happens].

Furthermore, *if they cut from in front at your face:*
> ➢ Turn the long edge toward their cut as before.
> ➢ As soon as the dusacks beat together, pull your dusack back up again (around in front of your face) and cut outwardly at their right arm.
> ➢ Afterwards, cut the second quickly from in front and through their face.

Another

Item, counteract their High Cut with the long edge (as before).
> ➢ In that moment in which the dusacks beat together, yank your hilt upward toward your left so that you arrive completely in the left Bull.
> ➢ From there, cut a Low Cut or Middle Cut (or even cut slantwise) away from you, powerfully and long, over their right arm and toward their right, through their face.
> ➢ Immediately cut additional Cross Strikes long and through their face.

|II.34| Item:
> ➢ Counteract their strike with the long edge (as before) and in that moment in which it touches, pull your dusack back toward your right around your head.
> ➢ Step and thrust from outside over their right arm at their face.
> ➢ Pull back quickly and cut at their face from in front again.

Item, if you stand in the oft-mentioned Simple Brace and *your counterfencer cuts outwardly at you to the right:*
> ➢ Counteract their cut with the long edge.
> ➢ Quickly cut a Low or Middle Cut from your right toward their left through their face.

Perform this cut so strongly that your dusack moves above your head in this Twisting [Cut], once on each side. You will confuse them with this moving around so that they are misled.
> ➢ Immediately cut a Cross Strike.

---

[200] *Maul* is the mouth that animals have. When used for people, it functions to dehumanize them.

How you should set off their cuts, hang inward, and pull upward through their face

Item, if you stand in front of your opponent in the Simple Brace and *they cut outwardly toward your right:*
> - *In that moment in which they cut in,* step with your left foot quickly toward their right and out away from their strike and turn the long edge toward their cut (while doing this).
> - Simultaneously (and while in the Simple Brace) raise your hilt and shove your point on their dusack over their right arm from outside into their face.
> - Hastily hang your dusack from outside over their right arm and inwardly in front of their chest.
> - Step farther to their right side simultaneously with this inward hang.
> - Draw your long edge back upward toward your left through their face.
> - Immediately cut the Cross [Cut] long in front of yourself.

A good sequence, how you should set off from the Simple Brace, step out, and cut through their face

Or, if you stand in the abovementioned counterposture and *they cut or flick at you outwardly (at your right arm):*
> - Step well toward their right *while they strike.*
> - Turn their cut well away from you and toward your right with your long edge as you step backward.
> - From there, draw the |II.34ᵛ| long edge above their arm and back toward your left, through their face and up into the Middle Guard.
> - From there, continue to fence (as you will subsequently be taught in the Middle Guard).

If they cut at you so quickly from both sides that you are completely unable to fence with them as quickly in the first cuts:
> - Turn away some cuts with the long edge toward both sides and observe:
> - Whenever *they offer an opening with their cuts,* you should hastily cut in toward the same.

Item, *if your opponent cuts a Wrath Cut from their right side toward your left, and if they guide the cut high:*
> - Don't respond to the cut with a counteraction.
> - Instead let your front end sink downward and move through under their right arm with a wide step outward to your opponent's right side (as you were taught previously in the Overwhelming Cut).[201]
> - Cut simultaneously with them from outside and over their right arm, straight to the side or opening from which they had cut.

You should use this against those *who cut too high and cut more at the dusack than at the body.*

If an opponent doesn't want to cut at you (because you stand in this counterposture), take note of *how they guard themselves against you. If they position themselves into a stance on one side (whether above or below):*

[201] See II.10.

➤ Thrust at their face with an extended arm with your Longpoint straight in front of you.

➤ *They must defend against that.* Therefore, take note *from which side they strike your thrust outward or ward off your thrust.*

➤ Let [your dusack] move around and cut in at precisely that same side *from which they warded off or counteracted your thrust.*

You can perform this cut in response from above, across, or from below.

➤ Then cut quickly afterwards with Cross Cuts.

Or, if you take note that *they want to counteract your thrust (including the cut):*

➤ Thrust straight at their face (as taught above).

➤ *In that moment in which they move the thrust aside:* act as if you wanted to cut at the side *which they counteracted you from.*

➤ Don't let the cut hit or touch, but instead pull it back up around your head and cut at the other side.

As often as you perceive that *they want to counteract your cut,* don't let your cut hit but instead pull it back up to the other side.

➤ *If they move to a counterposture in response:* pull up again and do this until you trust that you can hit an opening.

|II.35| **Rule**

If an opponent cuts at you to the right or left, counteract them with the long edge, and as soon as it glances off, pull back upward and cut straight from above and back at the closest opening while stepping outward.

Item, *if they bind on you or if they stand in a counterposture in front of you,* take diligent note of their rising upward. *As soon as they rise up to a strike:* cut through across at their forearm bone.

A swift, misleading sequence for fencing in this Simple Brace

Take note: *when you find your counterfencer in the Simple Brace,* position yourself into the abovementioned Wrath Guard or Bull.

➤ Use serious body language to guide a High Strike from your right shoulder toward their face.

➤ Don't let it hit, but instead, in that moment in which the cut should touch above, move your dusack toward their left, around and below their dusack through the Rose and completely around in a circle, so that you arrive with your dusack upward in the air toward your left (having moved under their right arm).

➤ Let it quickly drop down twice in the air—once toward their left, the second toward their right.

➤ Cut a Middle Cut through from your right and in front toward their face.

You should step forward twice for all of these moves. The one at the first Wrath Cut, with which you shouldn't hit but instead miss and move around. The other during the two drops downward (and including the Middle Cut). In that moment in which you bring your dusack into the air to drop down, you should simultaneously lift your right foot, and thus execute the two drops downward so quickly that you only bring your foot down when you hit with the Middle Cut.

This is a free-flying and fast sequence and moves well if you have previously learned to execute it well.

➤ As soon as the Crossing Cut has hit, cut again across from below at their right arm with a good counteraction.

In that moment in which you hit with this cut, take note as to *whether they are high or low with their arms.*

➤ *If they are high:* turn your point under their arm with a thrust at the chest or hip.

➤ *If they are low:* turn the point |II.35ᵛ| over their arm at their face with your palm upward (so that your long edge faces upward in this thrust).

Whichever of these two thrusts appears most proper to you (palm up or down), follow the same thrust quickly and powerfully with these cuts in response: namely, [follow] the High thrust with a Low Cut through their right. [Follow] the Low thrust with a High Cut through their left.

Item, if you find *an opponent in a Simple Brace:*

➤ Cut quickly from your right and across toward their hilt with a step forward.

➤ In that moment in which the weapons touch together, step and twist your hilt through below and use this to move outward over their right hand.

➤ Rip downward toward yourself, drawing the dusacks through their face.

You will find more information about how you should counter this Simple Brace in the Middle Guard.

Finally, when you stand in this Simple Brace, take note (as also mentioned previously): *if an opponent cuts at you to the right or to the left,* turn the long edge toward their cut and push the point forward at their face (in addition to this counterposture). You thus force them *so that they must rise to counteract,* and this provides you with space at their openings. *If they don't cut:* attack with any other sequences that you feel are properly the best (as you will find sufficiently here).

|II.36ᵛ| **How you should fence from the Arch**[202] [Bogen]
Chapter 12

Position yourself into this guard, which is the counterposture from below (as the portrait on the left indicates in the Figure [N] printed previously).

---

[202] When viewed from the side, this stance looks like an arc or arch, which matches the 16th century meaning of *Bogen* more closely than "bow". The only 'bows' of this type in English are rainbows and oxbows.

How you should entice their ſtrike onto your Arch and cut in response

Take note: when you thus arrive in Arch in front of your opponent, pay attention: *as soon as they cut from their right toward your left at your head:*

> ➤ Turn your hanging dusack toward their ſtrike with the long edge upward.
> ➤ During this and simultaneously with this counteraction, ſtep toward them with your left foot behind your right and out away from their ſtrike.
> ➤ Let their cut sweep downward on your long edge and off next to your left.
> ➤ Step with your right foot farther to their left and cut through their face from in front.

Or, if you have counteracted their cut upward toward your left using your Arch (as taught above):

> ➤ In that moment in which it glances off or beats, pull your dusack toward your left shoulder.
> ➤ Immediately cut from the left shoulder toward their right slantwise through their face or arm.
> ➤ Furthermore, cut with Middle and Cross Cuts away from you.

(Regarding this counterposture, view the small scene between the large portraits in Figure B.)

Another, with which it is taught how you should cut at their face above or below their dusack, depending on whether they have moved too high or low in the counteraction

Take note: when you thus ſtand in front of your opponent in Arch *and they cut powerfully through your counterpoſture,* pay attention to *how they rise back into their counterpoſture— whether they move too high or low.*

If they are too low[203] in this rising upward (so that you can see their face above their dusack):

> ➤ Cut |II.37| a Middle Cut quickly across at their face and above their dusack, *before they have correctly lifted through into their counterpoſture.*
> ➤ Immediately pull back around your head and cut a Cross Strike in front of you.

If they lift too high after they have cut:

> ➤ Cut immediately below their dusack across at their face *in that moment in which they are ſtill lifting.*

Rule: how you should let them miss[204] and cut in response

In the Onset, position yourself into Arch and observe that you are not too close to them.

---

[203] Corrected from *zu hoch* or "too high", as everything to follow occurs *over* their dusack.
[204] I.e. not contact.

- ➤ As soon as you take note that *they want to cut*, don't catch their ſtrike but inſtead *let them cut and miss.*
- ➤ That is, *in that moment in which they cut in*, pull your dusack upward toward you, and yield with your front foot toward the back so that they don't hit.
- ➤ *In that moment in which their dusack drops toward the ground*, cut quickly in response with a ſtep forward.

After you have observed that they can be hit, you can execute this responding cut from above, slantwise, across, or from below.

- ➤ You should also quickly follow this responding cut with Cross Cuts, unless you observe an opportunity for other sequences.

How you should cut at the forearm bone, in that moment in which *they want to rise upward to cut*

If you find your counterpart in Arch, position yourself also [in Arch] and pay attention: *as soon as they rise upward to hit:*

- ➤ *In that moment in which they lift*, cut a Middle Cut through and inwardly toward their forearm bone (as you can see the same in the Figure printed previously, designated with P, in the small scene to the left side).

You can also fence from outside and over their arm with Twiſting Cuts, *in that moment in which they rise upward.*

- ➤ |II.37ᵛ| Or take note *in that moment in which they rise upward:* ſtep out toward their left and let your dusack move well around in your hand.
- ➤ Cut next to or under their hilt inwardly at their face *in that moment in which they rise upward.*

Counter to the Arch

If they don't want to cut from the Arch, perform the sequence from the High Cut at them like this:

- ➤ Cut a swift High Cut toward their face through the Weak of their dusack.

You provoke them to hit with this cut.

- ➤ *As soon as they rise upward to cut:* cut their in-coming ſtrike away from you using a Middle Cut toward their hand.

➢ Then quickly [cut] the third in response.

Take note: if you stand in Arch along with your counterpart, pull your dusack toward your left shoulder.

➢ From there, move the outward flat next to their hilt and upward toward their face.

You frighten them with this *so that they move upward.*

➢ In this sweep upward, let it move around your head, and cut the second through from your left and across toward their right arm.

Item, when you arrive in Arch in the Onset, pull your dusack out of Arch into the Middle Guard toward your left.

➢ From there, move the outward flat from outside of their right arm and upward toward their face and toward your right shoulder.

Complete this in a strong fluid movement, so that your dusack additionally snaps upward around your head and threaten them with the inward flat to flick out at their right ear.

➢ *As soon as they swipe at this Flicking Strike,* immediately pull back upward and around your head and cut across through from your right with the curved edge, toward their face or arms.

This is a very good sequence. When you do it correctly, it will not fail you.

The Flicking Cuts

Take note: when you stand before an opponent in Arch, *if they don't want to cut:*

➢ Pull upward into Watchtower.

➢ Use body language to pretend that you want to cut from above; however, don't do that.

➢ Instead reverse (while in the air) and flick with the long edge from below at their right arm.

➢ Twist the dusack |II.38| back toward your left shoulder.

➢ From there, cut back and around through their right shoulder (either below or above their arm) and through their face.

➢ Cut Cross Cuts or a straight [diagonal] Driving Cut[205] long in front of yourself.

Item:

➢ *If they cut from above:* counteract upward toward your left.

➢ *In that moment in which they pull their dusack up from their executed cut:* cut quickly through, during this, from your left toward their right, either above or below their dusack, depending on whether they lift quickly or slowly. Thus, you arrive with your dusack next to your right side at the end of the cut.

➢ From there, cut through across quickly and strongly, from below with the long edge, so that your dusack arrives back at the left shoulder.

➢ From there, cut in straight and long from above.

When an opponent cuts at you from above:

➢ Take note Indesly *as they rise* and pull your weapon to your left shoulder again.

➢ Cut across with the long edge from your left simultaneously with them [their cut].

---

[205] *Treiben,* as a specific action and not the more general synonym for executing an action, appears in the Dusack Section in relation to diagonal strikes executed from Left Wrath over the right thigh and back, and in the Halberd Section in relation to diagonal strikes executed from the wielder's high left to their low right and back. The *Treibhauw* is not explicitly defined in either section, so I have interpreted it as describing a similar action. See the chapter on the Wrath Guard on the left.

➤ Step well out toward their right with this cut.
You thus hit them and take away their cut using the Cross.

A good sequence using three Middle Cuts

➤ Cut your first Middle Cut from your right (above their hilt and through their face).
➤ The second: from your left, also strongly through (under their right arm).
➤ The third: from your right again toward their left (at their face).
When you cut these correctly, one among the three will hit.
Take note when you arrive in front of an opponent in the Onset *and they don't immediately strike, but instead wait for your strike:* observe until you can provoke them long and greatly with your body language *(until they lift and hit).*
➤ During this, take diligent note: *as soon as they lift and hit in,* pull your dusack around your head to strike as well.
➤ Strike simultaneously in with them, such that you receive their strike on the strong of your dusack.
➤ Simultaneously and in that moment in which your dusack beats on theirs, you also hit them on their head with the outward half edge (as you can see in the two smaller portraits between the larger in the Figure with the H).

➤ From there, let [it] snap quickly back around and thrust with inverted hand under their dusack in front of their chest (as, indeed, the smaller scene on the [upper] right |II.38ᵛ| teaches you in the Figure designated with the letter K).

➤ Afterwards, cut through the cross, long and away from yourself.
In the second part, you will find that which can subsequently be fenced from these two counterpostures.

How you should change through, pursue, slice, and fence from the binds

Namely like this, if you (including your counterpart) stand in Arch:

➢ Bind onto them in the middle of their dusack—this applies regardless of whether that occurs using the Arch or Simple Brace.

➢ Remain with the bind on their dusack and provoke them with your front end above or below their dusack, depending on how you have bound on, for as long [as it takes] *until they rise upward to cut.*

➢ *Immediately after they pull away or around from your dusack:* cut through next to their hilt, at their face or toward their arm *while they are guiding their hand to cut in (and are thus still up high)*, with a step backward.

It should be observed that in all binds (regardless of which cut was used to cause it), you can quickly rush to an opening by twisting on their dusack, as you can gather from the following rhymes which I have set here from my didactic poem.[206, 207] I will explain it somewhat to you (before I then describe the other stances) and it follows like this:

> Then, as often as your cut touches in binds /
> The point is guided to the opening by turning.
> And you feel correctly to remain in all,
> Likewise, to slice off and counterslice.
> You pull the cuts away straight and smoothly /
> If you move through quickly, you will correctly find them.
> If you change through quickly, moving with steps /
> You will injure their chest and face greatly.

|II.39| The first technique to be understood here is this: namely, as often as you bind with your dusack onto theirs (whether from above or below), you should turn your point inward toward their body in that moment in which your weapon touches theirs. Likewise, as often as two cuts hit one another, always turn the point quickly inward on their dusack (as mentioned), and likewise in that moment in which the dusacks touch together or bind. That should or can be completed in all cuts. You should apprehend that you cut in upward and bind on against a High Cut, and bind from above against the Crossing or Low Cut. Now, as often as two cuts touch one another in this way, turn your point inward (while remaining on their dusack) and thrust in at their body. *If they rush at your opening during this,* twist back toward their dusack and turn them away from yourself with a counter slice.

---

[206] Historically, a *Zedel* was a written instrument, in MHG it was specifically a legal instrument, a record or complaint/pleading. ENHG both reduces this original meaning, in the judicial sense, to information presented as a list that was less informative than a legal brief. However, the concept is expanded to an abbreviated or abridged way of communicating that is contextually comprehensible (the implication being that it is incomprehensible without this context) and written using terms that are limited to specific locations or to specific technical language.

The fencing *Zedel* (recitations, didactic poems) meet this definition in that they often contain lists, they are highly abbreviated or abridged, they are contextually comprehensible, and use highly specified terminology.

[207] Meyer didn't record a longer Dusack *Zedel* anywhere, so all we have are the twenty verses here.

To understand the second technique in the aforementioned rhyme: it teaches you how you should correctly pull the cuts away from the bind to the body, and from the body to the bind (that is, to the dusack). Namely, as often as the weapons hit together in the bind or you have received their cut with a counterposture, that you slice from the same bind with your dusack toward their body with a pulled slice, and then from the body back swiftly on their dusack with a counterslice, so that you pull the weapon to the body [and] from their body back to the weapon using the slice (as was taught previously in the Simple Brace).

The third technique, about which is taught here, is the change through. This change through, even though it is also otherwise used for fencing outside of the binds, is very serviceable and clever to also to fence from the binds. This is because, as often as two cuts hit one another, you can, in that moment in which the weapons touch together or bind, move through properly under their [weapon] and step out to the other side, and fence at their weapon and their body with all kinds of sequences.

The fourth is how you should suddenly withdraw [your] cuts and counterposture. Namely: *when your counterfencer cuts in at you,* you should let it drop in front [of you] as a miss *in that moment just prior to when their cut should hit,* and then cut quickly in response. I have previously spoken about this before. Or, if you guide a powerful cut in toward their opening, and *as soon as they move to counteract the same,* you pull back and guide the same to another opening.

Finally, you should also diligently learn to correctly step and to feel, which is the best among these techniques for fencing with speed, because the word fúlen is understood here [to mean] that you learn to recognize the correct and proper time for each technique.

|II.39ᵛ| Since the aforementioned techniques are treated now and then in the stances,
it is unnecessary to discuss them comprehensively here, and I will then only
briefly discuss the most necessary

And firstly, you have heard previously in the two guards (namely the Bull and Wrath) how you should turn the point inwardly to the body; therefore, in relation to the cuts that are pulled away, take note of this example.

If you find *an opponent in Arch:*
➢ Bind on them with your Arch at the front-most part of their dusack.
➢ In that moment in which it touches, guide your front end outward around theirs and in to their body.
➢ Draw the long edge upward between their body and dusack and through their face (Figure I).

Even if you move too far forward with your draw upward into the air and expose yourself, you can ſtill hold them back and recover with downward slices or counterslices.

Item, bind your Arch on theirs at the Strong.
➢ In that moment in which the weapons touch together, ſtep forward and turn your hilt over their dusack in toward their cheſt.
➢ Shove your dusack haſtily downward.
➢ Draw your long edge behind their dusack through their face.

Item, if you bind them close to their hilt:
➢ As soon as your bind touches (*if they otherwise maintain their counterpoſture*), cut from outside and above their right arm, inwardly through their face.

If they hold their counterpoſture high and you have bound them with an equal Arch at the middle of their dusack:
➢ Turn your short edge under their dusack, inward toward their left (and remain with the bind on their dusack during this).
➢ Turn the short edge quickly back toward their right so that the short edge arrives on their head or through their face.
➢ However, pull the hilt quickly back upward toward you and cut long in response (as the smaller scene in the Figure designated with K teach you).

This sequence appears to be impossible, as presented; however, when you perform it at its correⒸt timing—which is quickly, in the firſt meeting of the bind—you will truly have completed it before they perceive it.

Item, *if an opponent binds you with their*[208] *Arch on yours and is high in their counterpoſture:*
➢ Turn your short edge under their dusack again, inwardly toward their left so that you offer a front opening.

---

[208] The text has *deinem Bogen*; however, the opponent's movements are indicated, therefore the first Arch is the opponent's.

➤ *When they hurry from above to this opening*, slice from your right at their arm with a step outward, *in that moment in which they move with their strike* (as the larger tenants in the Figure designated with K teach you).

➤ From this slice, move your point quickly toward their face.

Item, bind on their Arch with Simple Brace (that is, |II.40ᵛ| with the High Cut).
➤ In that moment in which that cut touches, turn the long edge either downward or upward (remaining on their dusack in the bind) and either toward or through their face (as you can see in the small tenants on the right side in the Figure designated with P).

➤ This will drive them upwards; therefore, cut across through their arms *while they rise.*

You can learn how you should slice off and counterslice at several places in this text. Namely:
➤ *When an opponent cuts at you from above:* cut across counter to it.
➤ In that moment in which your cut touches theirs, step quickly out to the side and toward their left during this.
➤ Draw the long edge away from their dusack toward your right through their face.

➤ *If they rush to your opening during this:* slice quickly back around toward their weapon.

➤ *If they chase your dusack:* move quickly through from below, as occurs in the change through.

Regarding the change through, take note of this rule:

➤ *If an opponent cuts from their right at you:* cut also from your right toward their [cut].

➤ In that moment in which the cuts should juſt hit together, move through[209] under their dusack toward the other side with a wide outward ſtep.

➤ Throw your blade from outside over their right arm at their head, etc.

Or, *if they don't want to cut:*

➤ Cut again toward an opening with serious body language and take diligent note.

➤ *As soon as they lift their dusack to receive your cut:* don't let it hit, but inſtead, in that moment in which it should juſt touch, move through under their dusack and drop your point from outside over their right arm at their face.

➤ *If they defend againſt this and rebuff the thruſt away from them:* draw the long edge inside their right arm and upward through their face.

➤ Cut quickly from your right back opposite to that.

If it were to become necessary for you to drop down to counteraƈt during this, you should not let anything prevent you from doing so.

Item, *if your counterpart ſtands before you in the Slice:*

➤ Cut from your right toward their left.

➤ In that moment in which it should juſt hit, ſtep well to their right side and simultaneously move through below with your dusack.

➤ Thruſt from outside of their right arm at their face again.

➤ *If they defend againſt that and rise upward:* move outside and around their arm with the point, to move back underneath the same and at their right cheſt.

➤ *If they defend againſt that again and move downward:* move the point around from outside of their right arm again and thruſt again from above and from outside over the same arm at their face.

You should always move your point around their arm in this way (so that you are now below, and soon above their dusack with your point facing their body).

|II.41| Learn pursuit[210] with every fencing partner[211] / Whether they are soft or hard in the binds /

---

[209] Although this is supposed to address the change through [*Durchwechsel*], the descriptions use a more ambiguous verb, '*durchfahren*, = move through'. In this first sequence in particular, this motion is directed at the entire body with the wide step, whereas in the second sequence this movement is directed at the blade and the third includes both a step and a blade movement.

[210] *Nachreisen.*

[211] *Gefert* or *Gefährt* has a number of meanings, but in this verse, it is the antecedent to *Er* (your fencing opponent in the bind). This accords with the meaning of '*fahrtgenosse* = companion on your journey', which is a common meaning of the word. The pun with *gefahr/gefähr* for 'danger' is probably intentional. *Gefährte* is also applied to strenuous activities in general and military/combative actions in particular. Grimm, vol. 4, cols. 2087–2095.

**If you pursue and follow with slices /
Observe their arm, be quick with ſteps.** [212]

You have heard enough previously as to what pursuit is and that it is a special skill; therefore, I will only give you a short inſtruction into how you should use the Strong and the Weak in pursuit, likewise how you should fence hard and soft againſt it.

And take note, *when they are hard on your dusack in their resiſtance in the bind:*
➢ Move or change through quickly underneath, or let [your dusack] move back and snap around. (You have been taught this at length previously.)
If they are not hard againſt you, but inſtead soft on your dusack in the bind:
➢ Shove them away from you with one movement. You shouldn't fall too far forward with this movement, so that you can be quick with a slice or with cuts at the opening (*before they escape them again*).
If you have bound them on the Strong of their dusack (however that were to happen) *and they ſtrike around from there:*
➢ Follow them with a slice toward their arm and the opening. Observe that you don't anticipate and move in front of their arm (*if they were to move through*).

You should underſtand all of this from both sides. You will find here and there in the sequences [inſtruction on] how you should pursue from the Weak with slices or pulls. These [offer] sufficient examples for underſtanding.

About the Boar [Œber]
Chapter 13

You have heard somewhat about the high ſtances, including their sequences. Now the low ſtances follow, from which the High Guards are countered—because *when your counterpart fences at you from above*, you should fence against them from below. The guard of the Boar is used only on the right, in the way the portrait on the left indicates in the Figure designated with the letter M.

---

[212] This assumes that Meyer is following his A-A, B-B rhyming scheme, which is the pattern from all other verses in this text. In his Liechtenauer citations in the Sword Section, the quotes have covered only one couplet, which appears in this two-column format. Otherwise, when Meyer has included verse, it has appeared in a centered column; the exception being the extended verse section that introduces the third part of the Sword Section on I.44^V.

If this follows by columns (A-B, A-B), then the verses read: "Learn pursuit with every fencing partner / If you pursue and follow with slices / Whether they are soft or hard in the binds / Observe their arm, be quick with steps."

|II.42| **The first sequence states how you should let their strike slide off your**[213]
dusack and cut long in response

When an opponent opposes you in the High Guards on the right:
➤ Position yourself in Boar.
➤ As soon as they cut at you from above, step with your right foot well toward their left and out of their strike.
➤ Simultaneously with this step, lift upward with a hanging dusack so that you let their strike slide off of your Arch.
➤ Cut two strikes, one after the other, quickly and long through their left in response.

Another, how you should step through and fence from outside at their head and arms using Twisting Cuts

If they cut from above:
➤ Step with the right foot toward their right and push your dusack with the point straight toward their face. Simultaneously, catch their cut on your long edge using this push forward.
➤ *In that moment in which the cut beats or touches:* twist the hilt through under their right arm and upwards toward your left (as the smaller portrait on the left side between the large tenants shows in the Figure designated with the letter F).

➤ Step quickly toward their right and cut powerful Twisting Cuts from outside over their right arm.
➤ During this, as soon as you observe that *they have moved their counteraction too far away from their face*, immediately fence in front at their face.

How you should run under their cuts and thrust in front of the chest at the face so that they have to expose themselves

In the Onset, when you arrive in Boar and *an opponent cuts at you from above:*
➤ Rise up with the Arch to catch their strike in the air with a step forward of your right foot.

---

[213] The text has *seinem Dusacken*; however, it is impossible for their strike to slide off of their dusack.

➢ Quickly lower your body downward and thrust away from you with the front end and under their dusack at their chest.

➢ |II.42ᵛ| Step back quickly and cut through their face with Cross Cuts.

Or, after you have received their cut with the Arch from the Boar:

➢ Step quickly with your left foot as well, far around to their right.

➢ Thrust from outside over their right arm at their face.

They must defend against that or be hit.

➢ *If they defend against [it], they expose their face in front;* therefore, step back quickly with the left [foot] and execute High Cuts powerfully through the [face].

Counter

➢ Counteract the thrust with a strike.

➢ Cut a Middle Cut through their face.

➢ Cut through the cross in response.

How you should thrust underneath their dusack and at their face from a completed counterposture

➢ Catch their High Cut on your Arch high in the air.

➢ Pull your dusack back up high and out from under their strike.

➢ Thrust up from below (and next to your right side) and underneath their dusack at their face. During this, keep your left hand above your head, until you turn your hilt back upward into Arch.

➢ From there, you should immediately pull around your head to cut from above.

The thrust from below must be carried out quickly *before they have recovered back from their cut.*

How you should thrust from the Boar in the Before

If an opponent does not want to cut at you:

➢ Draw your dusack back out of the Boar and next to your right side, and step and thrust at them from above.

➢ In the same step, pull the uncompleted thrust back quickly and thrust next to your right and up underneath their dusack from below (as before).

➢ Pull back upward and complete it as [instructed] previously.

|II.43| A good sequence, how you should flick over from out of the counter-posture and seek their right arm with Twisting Cuts and flicks

Item, position yourself in the guard of the Boar. *If your counterfencer cuts at you from above:*

➢ Step with the right foot toward them and move powerfully upward with the Arch.

➢ As soon as their cut beats on your dusack, flick the short edge at their left ear (over their hand and above their dusack).

➤ Step quickly with your left foot across (out toward their right) and cut a powerful Twisting Cut from outside over their right arm.

Or, immediately after the flick has occurred:

➤ Let your dusack snap back around and flick them from outside at their right arm with the hanging dusack and inward flat.

➤ Afterwards, cut through their face from in front with a step backward.

The Roaring Cut and the Awakening Cut both counter the Boar.

About the Middle Guard [Mittelhut], how and what one should fence from there
Chapter 14

I call this the Middle Guard because it arises from the Middle Cut. However, you can arrive in this guard at the end of three cuts:

When you cut a Crooked Cut from your right through the wrath line and allow it to flow back next to your left into the Middle Guard; then, using the Middle Cut itself; thirdly, when you cut a Crooked Cut from below through the upwardly slanted line from your right toward your left.

These three cuts always flow to the Middle Guard as the closest [stance], unless you wrench with effort into another.

Regarding it, position yourself as the portrait indicates on the right side in this Figure [C]. From this guard, you can also fence all sequences that were taught for left Wrath and Bull. Therefore, I only want to discuss a few sequences so that you are sufficiently taught the use of this guard.

|II.44| **How you should rebuff your counterpart's cuts from the Middle Guard and cut in response**

And firstly, when you arrive in the Middle Guard in the Onset in front of your opponent *and they cut first at an opening:*

➤ Cut their in-coming strike away with the long edge (from your left toward their right, and from above through the slanted hanging line) so that you arrive on the right side with your dusack.

➤ Step with the left [foot] well out to their right side with this cut.

➤ As soon as this has occurred (*before they have recovered again from their first committed strike*), step with your right farther toward them and cut quickly from your right and above their dusack through their face (or over their right arm).

➤ If *they are so quick after the first strike* that you can't arrive with your second strike over their arm, cut likewise through from your right at their arm or hand (*while they are pulling upward*).

➤ You then arrive back in the Middle Guard as before.

How you, in the Middle Guard, should cut their ſtrike away
and upward from below and follow with Twiſting Cuts

If they cut at you from above when you ſtand in the Middle Guard:
> ➤ Ward off their in-coming ſtrike upward from below with the long edge so ſtrongly that your weapon flies around above your head to ſtrike.
> ➤ Step quickly with double ſteps well to their right.
> ➤ Cut a powerful Twiſting Cut from outside over their right arm or at the side (wherever you can reach them in haſte).

These two cuts should be performed swiftly so that you hit with your second ſtrike *before they recover from their firſt committed ſtrike.*

Or if they were so swift and were to rise upward before you had completed the Twiſting Cut, ensure that you arrive at leaſt simultaneously with them with the Twiſting Cut (over their right arm from outside).

|II.44ᵛ| **How you should throw your dusack over their right arm from outside,**
and draw the long edge upward through their face

Or take note when you ſtand thus in the Middle Guard. *In that moment in which they cut at you:*
> ➤ Step out of their cut and to their right. During this ſtep, throw your curved edge toward their right from outside and over their right arm and in at their face.
> ➤ During this throw, duck your head well behind your dusack away from their ſtrike.
> ➤ Take diligent note *if they don't hold hard in the firſt hit:* haſtily push with the Strong of your dusack back downward and away from you.
> ➤ Draw your long edge upward through their face upward into the air (as the upper small tenants show in the Figure designated with L).

> ➤ In the air, pull your dusack forward into a Middle Cut toward their right (back on their dusack).
> ➤ When you have bound back on with a Middle Cut from your left toward their right, pay attention: *as soon as they lift from the bind,* cut in front at their face *while they lift* or cut inwardly at their arm with a ſtep back (which is safer).
> ➤ Defend yourself with the Cross.

If they had used their cut in a counteraction against your throw:
➤ Pull your hilt back upward toward your left and let it fly around above your head in the air.
➤ Cut strongly from your right and upward from below with the curved edge, so that your dusack arrives back at your left, whether in the Wrath or Middle Guard.
➤ From there, quickly cut a Cross in response.

How you should fence from the Middle Guard against your counterpart who does not want to cut

If your counterpart does not want to cut, fence from the Middle Guard against them as follows. Take note:
➤ As soon as you can reach them, cut a Cross through their face.
➤ *If they have extended their dusack in the counterposture:* fence a Cross at the hand in which they hold their weapon (in this case).
Using these Cross Cuts, you will drive them backward or provoke them *so that they will also cut.*
➤ *As soon as they do that,* get ready quickly |II.45| and cut two powerful Middle Cuts through from both sides and toward their in-flying strikes (opposite one another).
By this means, you weaken not only their strikes but also tire their arm to a degree that you can arrive at the opening with other subsequent cuts.

Another, how you should attack from the Middle Guard against an opponent who doesn't want to cut

Take note. When you find *an opponent in Arch or else in Simple Brace* and you have your weapon in the Middle Guard:
➤ Step with the left foot well out to the side (toward their right) and cut across from outside at their arm simultaneously with this step.
➤ Pay attention Indesly.
If they want to defend against this or counteract:
➤ Don't let your cut touch, but instead pull back around your head and cut inwardly through their face while you step back toward their left side.
Or, if they counter you with a Simple Brace:
➤ Throw the curved edge out over their right arm at their face (as the upper small scene toward the left indicates in the Figure printed previously, designated with C).

They must defend against that, which means that they will remove their face so that you can cut at that well while stepping backward.

How you should flick from outside at their right arm from this guard

Another. *If an opponent encounters you in Arch or in the Simple Brace:*
- ➤ Position yourself into the Middle Guard.
- ➤ From there, strike outwardly at their right arm in a flick (with the outward flat).
- ➤ Pull your hilt quickly upward (so that your blade hangs downward in this upward pull) simultaneously jerking your dusack around your head.
- ➤ With a step forward of your right foot, cut through once again with the curved edge (from below and from inside toward their right arm) so that after the end of the cut, you arrive with your dusack |II.45^V| in the left Wrath Guard.
- ➤ From there, cut powerfully from below through their face.
- ➤ Afterwards, cut quickly through the cross in response.

From this guard, you can also cut through the Roses, flick, and attack with other misleading attacks. The Rose Cuts move particularly well through misleading from this Middle Guard.

If you find *an opponent in Arch who is smaller in stature than you:*
- ➤ Move the outward flat from your right and toward their right, over their hilt and outside of their right arm, and (in a single motion) below their dusack, back around, and upward toward your right, letting it fly over your head in the air while you lift your foot.
- ➤ Strike immediately in a flick (with the hanging dusack and inward flat) from outside at their right arm.
- ➤ Following that, pull upward and cut the Roaring Cut across through the middle line (as you have learned in the previous section on cuts).

And this must be carried out in one motion, so that you raise your right foot in the first motion and set it down as a step forward when the Roaring Cut hits.

About the Change [Wechſel] [Guard] and its sequences
Chapter 15

Position yourself in this guard like this: stand with your right foot forward, hold your dusack next to you and at your side with extended arm, with the tip on the ground so that the half edge faces toward your opponent (as the large portrait on the right teaches in the Figure printed previously with the N).

It is called the Change [Guard] because you arrive in this guard through the Change Cut, and it works on both sides.

How you should rip upward at an opponent in the Arch
and cut at their face before they recover

If you are physically strong:
> ➤ Position yourself in the Change [Guard] against the Arch.
> ➤ Rip strongly upward at their forward-hanging arch with the half edge.

You thus force them to rise up.
> ➤ |II.46| *In that moment in which they rise upward and are still holding their dusack up high:* during this, cut quickly down from above at their face or chest.

Only one step belongs to this sequence, which you should complete with the step outward as a leap with your right foot.

This sequence is also a counter against the Arch

Take note *when an opponent in Arch encounters you:*
> ➤ Position yourself in Change [Guard] on your left.
> ➤ Step and thrust long away from you (upward from below and under their counterposture) and toward their face or the chest.
> ➤ As soon as you sense that your point has hit or has been set on them, **Indesly** lift your hilt quickly in front of your head and remain with the point at their body during this.
> ➤ *They will defend against that or beat it outward.*
> ➤ So pay attention and *immediately when they lift to strike:* step to their left side and strike in at their face next to their counterposture.

Counter

Take note *if an opponent thrusts in under your counterposture at your face* (as taught above):
> ➤ Rebuff the thrust, which exposes your face.
> ➤ *As soon as they cut at it [your face]:* cut off their strike close to their hand (and between your two hands) and thrust the hilt into their face (as the small scene toward the left indicates in the subsequently printed Figure designated with the letter B).

|II.47| **Another from the Change [Guard]**

When you approach an opponent, cut through in front of them from your right into the left Change [Guard]. Use body language to imply that you cut in vain.

> - *As soon as they rush at your opening from above:* lift quickly upward to counteract with the long edge facing toward their right arm.
> - As soon as the dusacks beat on one another, turn the front end quickly upward and thrust (with your palm upward) outwardly over their arm at their face.
> - Pull quickly back upward and let the blade snap around.
> - Strike them from outside on their elbow (with the inside flat of a hanging dusack and your palm downward).
> - Before that is correctly completed: step back and cut in front through their face.

This is a fine misleading sequence with which you expose your opponent (when you carry it out quickly).

Counter against the Change [Guard] on the left

If your counterpart encounters you and they are also in the left Change [Guard]:
> - Move up into the right Bull from the [Change Guard].
> - From there, cut a Low Cut through their left.
> - Step with the second [cut] and cut from your right (from above and also through their left).

Step out well toward their left with doubled steps with these Low and High Cuts.
> - Immediately after, cut a Cross Cut long.

Counter against the Change [Guard] on the left

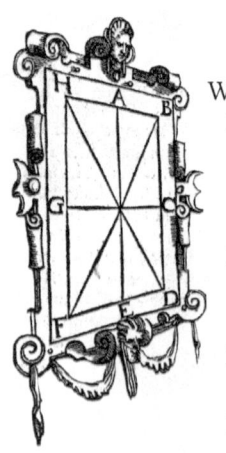

When you find *an opponent in the Change [Guard]* in the Onset:
> - As soon as you can reach them, cut from your left shoulder slantwise through their face (designated for your information as line H and D), so that you arrive at the end of the cut next to your right with the point on the ground.
> - Turn your dusack and *(in that moment in which they lift)* cut strongly through from below through their arm and under their dusack in one pull (so that your dusack arrives back at your left shoulder).
> - From there, cut a Low Cut across through their face. At the end of this cut, your weapon should arrive at the right shoulder to strike.
> - Cut a long Cross from the same shoulder in response.

|II.47ᵛ| Position yourself into the guard of the Boar.
> - From there, step and thrust forward toward their face with an extended arm.
> - *They must defend against that, and thus expose their face.*

Counter against the right Change [Guard]

➤ Move with Longpoint into their face.
➤ *As soon as they lift,* cut a Low Cut with the long edge from your left through their right. Follow with Middle Cuts.

Take note, when you cut in front into right Change [Guard] *and your counterpart rushes after you:* cut through strongly and upward toward their cut with the long edge. Let it move above and around your head and cut back from your right (through their left from above) while stepping around.

If you don't arrive through in the upward cut, turn your dusack on theirs to thrust.

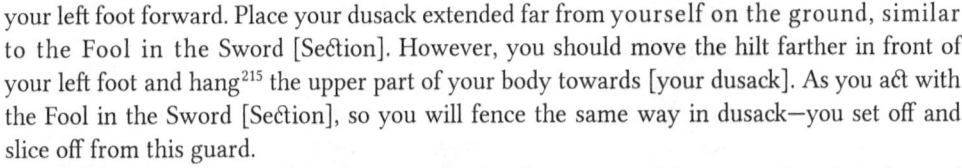

Bulwark [Bastey][214]
[Chapter 16]

I consider the Bulwark to have been named by the Ancients because the lower part of the body is shielded as well as the upper part, just as a bulwark protects and shields the lower part of a city wall.

Position yourself into it like this: stand with your left foot forward. Place your dusack extended far from yourself on the ground, similar to the Fool in the Sword [Section]. However, you should move the hilt farther in front of your left foot and hang[215] the upper part of your body towards [your dusack]. As you act with the Fool in the Sword [Section], so you will fence the same way in dusack—you set off and slice off from this guard.

The Bulwark is executed like this: namely, stand as previously, set your dusack in front of your foot with the point on the ground and your hilt vertically above. *Regardless of what your opponent wants to cut at you:* step out of their cut and cut simultaneously with them from above, or catch their cut on the long edge and work at the closest opening. There are many sequences to be fenced from here; however, you will find them [located] both previously and subsequently.

A sequence for Running In

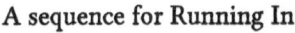

In the Onset, cut a High Strike from above at their head.
➤ *If they counteract the cut on their Arch and rise upward:* lift your hilt and drop the front end over their counterposture in at their face (as is taught above about the Awakener). By this means, you drive them higher |II.48| to counteract it.

---

[214] The '*bastey* = bulwark' is the lower part of a wall that is angled outward from the top to support and defend the wall. A bastion (the cognate) is an area projecting outward from the wall where the defenders are located to protect the wall. Its defensive purpose lies in providing a location for human defenders—unlike the bulwark, whose intrinsic purpose is to provide additional support and defense without human intervention.

[215] In the meaning of 'hold a body part diagonally away from the vertical center axis'.

- ➤ Lower your body Indesly a bit with a leap forward and move with the butt end (that is, with the hilt) in under their dusack at their face.
- ➤ *If they drop with their dusack in response:* thrust over their right arm from outside again with the front end (as you can see in the small tenants on the left in the Figure printed previously, designated with the letter O).

- ➤ Afterwards, cut away from them using the Cross.

Or, if an opponent wants to overrun you with High Strikes:

- ➤ Catch them (still high in the air) from below on your Arch with another leap forward under their weapon.
- ➤ In that moment in which it glances off or touches: thrust with the hilt under their dusack at their face.
- ➤ Then complete the sequence as previously.

There are some who (after they have reinforced their arms with all kinds of fabric rags) are accustomed to placing their head between their arms and running at [their opponent] close under their opponent's weapon. With regard to these: you ought to concern yourself with their attack from above and not expose yourself with any sequences; therefore, you ought to use these three techniques:

Firstly, *when they run under your weapon:* lift your arm simultaneously with this and remain high in your counterposture. While you have both of your arms high, hit them with a flick in their face with the curved edge.

Secondly, *if they defend against this:* move your point outside and around their arm and strike with the curved edge outside of their right arm on their head (as the small scene between the larger portraits show you in the Figure G printed previously).

Item, remain with both hands high as well and keep your left hand above your pommel and close to their hilt. Hit them quickly under their left arm and behind at their neck *before they observe it.*

(Regarding this, view the smaller scene on the left side in the aforementioned Figure [G].)

Accordingly, when you again approach your opponent (as now taught), you should cut short or cut them off in front, but absolutely don't cut through unless you can then withdraw from their High Cut with a step outward.

If you want to safely cut short away from them, position yourself with body language like you wanted to cut seriously through their opening in front of them.

> Don't do that, but instead turn your cut in full motion into a counterposture (into Longpoint), so that the long edge is turned toward their in-flying strike (*which they rush to cut at you in response*).
> *As soon as they take note that you want to cut through in front of them at their opening, they will quickly cut from above* |II.48ᵛ| *in response.*
> Catch the same cut with the extended counterposture.
> As soon as it touches or beats, cut completely through in response.
> Withdraw using the Cross.

Or *when they approach you like this, such that you both stand with arms high* (as just mentioned): thrust with the front end away from you, in front at their chest (as the tenant indicates in the Figure [O] printed previously).[216] This is called the Stork's Beak[217] because you extend long away from you with the thrust.

If they defend against the thrust, cut long in response.

If you find another [who is] stronger than you are, don't approach them too closely and don't let them run in at you. In addition, pay attention to where they cut from so that you catch their cut high in the air and move quickly through under their weapon. You thus change their cut with a counteraction and can hold it away and rebuff it. However, it is better if you can yield from their cuts and let their cuts miss. You should do this in the way that was taught above in the cut in response.[218]

There now follows more about running in, so view those rhymes from my didactic poem which were included previously and which are also placed here.

---

[216] This fencer has been reversed for clarity, since they are left-handed in the original.

[217] *Strocken schnadel* (heavily misspelled).

[218] See II.37.

Also, when you approach your opponent closely,
Attack their right with your left.
Learn both grips, ſtraight and reversed,
Use quick ſteps as a defense.
Quickly grab their unspecified [anatomical bits];
If they reverse their hand on your cheſt, turn [it].
If you want more revenge on them,
Then you can thus break their arm.

With regard to all running in, [you should] primarily pay attention to grappling, wreſtling, counters, and throws, which, although there are many kinds, are briefly summed up in the above rhyming lines.

Therefore, take note about the firſt: as soon as you approach your opponent or have moved under their weapon: you should immediately reach with your left hand at their right arm (in which they hold their weapon) [to grab] their joint[219] close behind |II.49| their hand. Immediately rotate[220] it [or] jerk it toward you (depending on where you consider your advantage).

You should also know that grappling is completed in two ways (as mentioned): namely, ſtraight and reversed. Straight grips require no explanation. There are also two types of reversed grips: firſtly, when you grip with the thumb inward; secondly, when you turn [it] outward in the grip. How you should use all of them will be described more fully in the Dagger [Section]. Therefore, I will only explain those sequences here that belong to running in or to throws.

The firſt

When an opponent wants to reach paſt you with High Strikes:
  ➤ Counteraᴄt the ſtrike with a high counterpoſture.
  ➤ Quickly and simultaneously [reach] your left reversed hand under your counterpoſture and ſtrongly grab onto their right hand.
  ➤ Rotate it away from yourself and ſtep forward with your left foot behind their right while doing this.
  ➤ Thruſt your hilt away from you and in front at their cheſt.
  ➤ *Thus, they fall on their back.*

Or, *if an opponent runs over you from above:*
  ➤ Counteraᴄt high as before.
  ➤ During the counteraᴄtion, ſtep with your right foot between their two legs, bend your body forward and reach with your right arm under their right [arm] and around their back from outside.
  ➤ Grab under their right knee with your left hand.

---

[219] The German term *glied* means, with respect to human bodies, "a joint or movable body part". In this case, the wrist. At other points, the knuckles, elbow, or knee are indicated by *glied*.
[220] Under the assumption that '*umtreiben* = move it around, back and forth, unpredictably', is actually a typo for '*umreiben* = rotate'.

➤ Lift[221] simultaneously upward and throw them.

Because I don't exactly praise running in with the dusack, I will refrain from any more.

Take note now in conclusion, when you want to fence with an opponent, pay attention. *If they attack quickly with their sequences and guide their cuts wide around:* direct all of your sequences [to that point] so that you will still cut them if your opponent were to go astray, and if you were to rush at the openings *in that moment in which they cut falsely.* Yet don't be so eager during this that you lose your advantage.

Secondly, *if your counterfencer does not want to cut first but instead is diligent to counteract and cut in response:* use misleading and guide your cut up to their counterposture, but pull it back uncompleted and cut at another opening.

You should also pay attention to their stance and so do not cut in at their opening to hit, but instead to bring them out of their advantage so that you can then more hit certainly with the second cut (after they have moved high or low).

Take this as a small example |II.49ᵛ| of this. *If they hold their weapon in the Arch too far to their left:*

➤ Cut powerfully from your left across from below toward their right arm.

If they counteract with a hanging dusack (so that they expose their face):

➤ Pull your hilt quickly back upward toward your left and cut from the same [location] back over their dusack at their face (as the smaller scene in Figure A shows you).

You can also provoke them with body language from one point to another and cut them quickly, cunningly, and with advantage at the opening while they are thus swiping around.

If an opponent comes before you *who pays attention to your lifting, and cuts at your opening* (while you draw up to strike), you should deceive them about your cuts.

➤ Pull as if you already wanted to cut.

➤ Take diligent note *as soon as they want to cut:* turn your pulled cut into a counterposture and catch them on it.

➤ As soon as their cut touches your counterposture, you should cut in response (as you then find all types of sequences sufficiently gathered here in this work).

---

[221] *Heben,* while the common word for "lift" in German, is uncommon in this text.

I have dealt with this weapon extensively because the youth are generally instructed too quickly in it. This leads to the fact that it is difficult to understand if a thing is not presented accurately, especially in this art. Also, some techniques cannot be taught with an expectation of comprehension without repetition or engagement with some other techniques. Therefore, may the generous reader find favor with my services in this.

END

Contents of fencing with the Rapier,
and in what order it is presented and described
Chapter 1

Concerning rapier fencing, which is currently a very necessary and useful exercise, there is no doubt that so much of it is a newly discovered practice for the Germans (since it was brought to us by other peoples). This is due to the fact that although our ancestors permitted thrusting in serious matters against a common enemy, they did not permit the same in trivial exercises nor did they permit any kind of use of this by their mutually sworn men-at-arms or others who had come into conflict outside of the mutual enemy, and the same should be maintained even today by honorable men-at-arms and other German citizens.[222]

Therefore, rapier fencing would be superfluous were it not for interactions with foreign people, such that thrusting—and many other habits that were unknown to the ancient Germans—have now taken root among us. However, because such foreign customs increase daily in many locations, it has also become more necessary for us that such foreign and alien customs of those people become known and familiar to us—indeed, that we practice and equip ourselves not less than they (as much is proper for necessary defense) so that we can shield ourselves from them (when it becomes necessary), to properly oppose them and be victorious.

Therefore, I will present and describe rapier fencing in an orderly fashion—as I have learned it from well-considered people and |II.51| through daily practice of the same—and about how one should comport oneself with this or similar weapons.

In order that this can be completed more usefully for the student, I wanted to first explain one sequence after another, each in isolation, in a specific order.

Namely, to present at the beginning in an orderly and understandable way how the opponent[223] is further divided (which is different from previously) and the use and usefulness of the same, including the division of the weapon. Then, how one should direct the stances, cuts, and thrusts (including their circumstances) in the work [of fencing]. Following this, how one converts the cuts into thrusts and the thrusts into cuts; item, misleading; correct steps; also how one should use all kinds of counteractions. And this should all be dealt with and taught in the first part.

Then, in the second part, I want to undertake and deal with the exercises in and of themselves: how one should fence against the counterpart using the sequences learned previously.

I will initially use as an introduction (with a necessary and useful teaching): how one should obliquely slice off, set

---

[222] A *Burger* is a citizen of a specific urban area and indicates an economic class, and may also denote a length of residency in a location. The poor were not citizens, nor were those in rural areas. It is less a civilian/military distinction than an upper (non-noble) class, as distinguished from the less well-off.

[223] As previously, the German word *Man/Mann* will be translated as "opponent", and references to pronouns will be the gender-inclusive "they/them" unless the pronoun refers to a specifically named Master. In addition, Meyer uses several other terms to refer to the opponent, which will be reflected in the translation.

off, and provoke from one stance into another, in order to mislead and also to change from one stance into another.

Afterwards, [I will] describe fencing using the common and straight counterposture (or Simple Brace).[224] And because one must move, arrive, or drop into the abovementioned stances with each cut, strike, or counteraction, I want to show and teach (even before I have finished with the abovementioned counterposture) how you can recover quickly from one stance into another—as you would arrive in one of the guards in the full course of your fencing—and can oppose them, so that you are not overpowered. Afterwards, I conclude the Simple Brace with an introduction of many swift and advantageous instructions and sequences. And finally, I will append a short teaching about how one should use a secondary weapon (like daggers, capes, and the like) as necessary. Therefore, I hope that when you take this weapon in your hand and read with attention, |II.51ᵛ| that you will understand it as described in the aforementioned order and can make use of it.

About the division of the opponent and the weapon, and about their use
Chapter 2

Although the division of the opponent has been dealt with previously so that anyone could easily direct themselves using this weapon as well, it appeared necessary to me to treat this

division more fully according to the circumstance of the rapier (as it differs from other German weapons in use) so that you can learn to guide the subsequent cuts more surely against the opponent's body, high or low, with greater understanding.

And the division is like this: the opponent is divided into left and right by an upright line (just as previously). However, two similar lines are added to that initial upright line, with which both the right and left shoulder are cut through vertically on each side.

Secondly, the opponent is also divided by three obliquely hanging slanted lines into four parts, so that the first line begins at their left shoulder close to the neck, crosses the upper part of their chest, and ends below their right arm. The second begins above their left hip, obliquely crosses their stomach, and ends at the top of their right thigh. The third, however, begins at the thickest part of their left thigh and ends on the other side at their [right][225] knee. When you now draw three obliquely slanted

---

[224] *Versatzung* is a term that is used in different contexts with highly different meanings. In the context of carpentry, it is a connective piece that diagonally links a vertical post to a horizontal crosspiece as a brace against shear forces. The straight or simple *Versatzung* requires no angled offsets in the vertical/horizontal pieces to fit it into place. It is architecturally apparent in half-timbered houses, which makes it a common image, in particular in reference to the view of the weapon and arm from the side in the stance Meyer calls *gerade Versatzung*. *Versatzung* by itself, as a reference to a specified or unspecified stance, becomes a defended position in response to a particular action by the opponent, thus a counterposture.

[225] The text says "left", but this is obviously wrong.

lines opposite to these (from the other side) through the opponent  in the way just mentioned, three intersections occur in such a way as |II.52| you will see painted later.

Three other lines are likewise drawn across through the opponent (so that they are again divided into four parts).

This previously-taught division functions firstly and primarily so that you know to position yourself differently (as necessary) with the movements of the body following your cuts, which you then either guide at their upper or lower body. Because if you guide a cut toward their upper part (whether it arises from above, slantwise, across, or from below), you must also remain upright and high with your own body, so that your shoulder stands equal to that upper part toward which you cut or thrust (as much as your height allows).

This should not happen with the other cuts which you guide toward their lower body, but instead, the lower you cut, the more you should sink down with your upper body—which then must be achieved with steps, as you will subsequently find reported in more detail with the cuts.

If you were to cut those cuts which are directed toward their upper body with your body sunk down low, you would shorten your strike. The same applies if you were to cut low and remain with your body upright and high. Your cut would not only be shortened, but you would also completely expose your upper body.

The weapon, however, is not divided any differently than the way it occurred previously in the Sword Section—namely, in four equal parts. From these divisions you can learn how you remain close or far from the opponent and which sequences you should use to fence in a specific part.

Thus, if you are so close to them when fencing that you can reach the outermost part of their blade with your outermost, you can then fence well against them with sweeping |II.52ᵛ| cuts and thrusts (whether with misleading or otherwise pulled cuts). Because *if they want to reach in towards your openings* (while you were moving around with your weapon), *they cannot rush you* because you can likewise be as ready to follow with your in-flying[226] strike as they are with theirs.

However, if you have more closely approached one another so that both blades touch together in the middle in the bind, you should not ever cut around in this case, nor rise from their blade without specific advantage because, as soon as you were to rise from their blade, they could rush you with pursuit.[227] Instead, be aware of the sequences which can be fenced on their blade, and take diligent note: *if they were to cut through or otherwise expose themselves,* that you force them backwards in response.

If, however, you move even closer so that you have bound with the middle of your blade on the middle of theirs, be quick with grappling, wrestling, and throwing, because you have no other means ([unless] you step back from them).

---

[226] Because German has directional prefixes *hin-* and *her-* to indicate movement away from or toward an object, I have used "in-flying" and "in-coming" as translations of the adjectives *herfliegend* and *herkommend* respectively, which Meyer often uses to describe weapon movements by both parties.
[227] "Pursuit" in the translation is only used for the technical term *nachreissen*. Other verbs that indicate a movement that follows an attack will be translated according to their specific meanings.

About the guards and stances with the rapier
Chapter 3

The stances in rapier fencing are primarily counted as five, of which each one can be executed and brought through the lines on both sides (and straight in front of you). I want to explain how they are named and how they should be completed, and set them in order, as follows.

|II.53ᵛ| **High Guard [Oberhut] (including the Oxen [Ochssen])**

The High Guard is considered and executed at the sides in two ways—namely, once to thrust, the second to cut. Position yourself like this: stand with your right foot forward. Hold your hilt next to your right, high in front of you, extended upwardly at your side. However, as the larger portrait on the right side in Figure B indicates, your front end (which is the tip)[228] extends towards your opponent's face. This is named the Ox because you threaten a thrust from above with your weapon in this guard, because the Ox is, in itself, nothing more than a thrust from above.

If you now hold the weapon with the hilt as taught (with the arm extended upward to the side), but the blade is not extended toward your opponent and instead is extended behind away from your opponent, this is the High Guard to strike (as the other one is the High Guard to thrust).

High Guard on the left

Stand with the right foot forward (as at first). Hold your weapon with the hilt high next to your left with the arm extended upward, so that the point extends again at your opponent (toward their right, at their face). You then stand correctly in the High Guard of Ox on the left. If, however, you hold your hilt upward next to your left and forward of your previously-extended arm and you turn or reverse your blade with the point upwards behind your left shoulder, you stand again the High Guard to cut (next to your left, just as previously on your right).

The High Guard is also completed directly in front of your face with your arm extended upwards and forwards so that the point faces out in front of you, yet not to thrust but instead to cut (even though the [cut] can also be converted into a thrust).

Thus, you have the High Guard on the right and on the left, to thrust and cut, and likewise also straight in front of you.

---

[228] The terms *vorderen ort* and *hinteren ort* come from the polearm section, where they refer to the two ends of the staff, but occasionally Meyer also uses them as synonyms for the tip and the pommel of other weapons.

|II.54| **Low Guard [Unterhut]**

The Low Guard also extends from below in three ways, namely, straight in front of you and on both sides.

However, the straight is nothing other than the end of a straight High Cut, while those on the sides are the ends of the slanted Wrath Cut, since you arrive with your weapon at the end of the aforementioned High Cut so that your blade extends long in front of you (with the front end on the ground toward your opponent and your hilt held well in front of your bent knee), with arm extended and body inclined toward it (that is, sunk toward the ground). Otherwise, it is seldom used for a guard or ward.

Low Guard on the right

Position yourself into this guard like this: stand with your right foot forward. Hold your weapon with an obliquely hanging arm next to you outside of your right thigh. Let the tip (that is, the front end) lie on the ground out in front of you (as this can be seen in the larger portrait in Figure D).

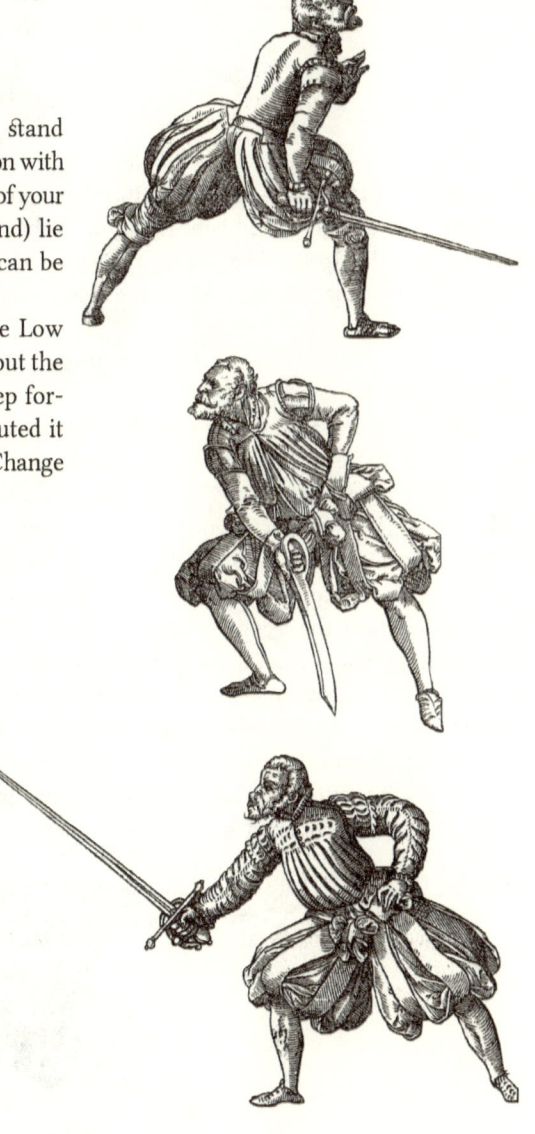

As you have now been taught about the Low Guard on the right, you also understand about the Low Guard on the left: that you always step forward the right foot, so that you have executed it correctly if you stand as described about the Change [Guard] in the Dusack Section.[229]

Iron Gate [Eisenport]

Position yourself in this [guard] like this: stand with your right foot forward (as always). Hold your weapon with an extended arm hanging in front of your right knee, so that the tip extends forward toward your opponent's face (as the portrait shows in the subsequently-printed Figure designated with the letter C).

It is called Iron Gate because you are well protected in this stance from your counterpart's thrusts and cuts—just like behind an iron gate—but also because you can safely harass your counterpart from here with all kinds of techniques if you correctly guide the weapon in this stance,

[229] See Figure N and Chapter 15 in the Dusack, and Figure G in the Rapier.

including the sequences (each according to its opportunity). You can also keep your weapon in this form and shift toward the right and left sides, or hold it on the sides just as well as straight in front of you. Thus, you have the Iron Gates in front and on both sides.

|II.54ᵛ|

Plow [Pflug]

The Plow is, in itself, nothing more than a low thrust; however, regarding a stance, use it like this: stand with the right foot forwards (as previously). Hold your weapon with the horizontal crossguard below in front of your right knee so that, in holding the weapon, your thumb lies over the crossguard out onto the flat of the blade. This flat should then face upward towards you, the other [flat] downward away from you turned toward the ground. In this stance, you should stand with the feet wide apart and with the forward knee well bent forward, so that the body hangs slightly forward following the weapon. The tip should also be extended well toward your opponent's stomach. During the work, this stance is also directed at both sides, namely to the right and left (like the Iron Gate).

Longpoint [Lang Ort]

The Longpoint in rapier is the end of all thrusts (understood as Flying Thrusts) which occur long in front of you. All thrusts which do not end in the Longpoint (with contact) are too short. However, concerning reversed (or otherwise shortened) thrusts: these can be completed outside of Longpoint; however, you immediately turn back from them into Longpoint.

This stance is also established at three points—not to the sides along the transverse line, but instead according to the height of the upright opponent: namely, the first Longpoint extends toward their face, the second toward their belt, the third toward their stomach or genitals.

Position yourself generally into Longpoint like this: stand with the right foot forward (as always). Hold your weapon with your arm extended at its longest toward your opponent's face, so that you shoulder is at the same height as that location toward which you would thrust. If you now guide your thrust toward their face, you are not permitted to step too far; instead, it is sufficient if you merely step so far that your upper body leans forward after the thrust, so that you do not drop down with your shoulder.

However, if you want to thrust at the belt or even lower, you must step so wide with the feet that your shoulder is equally low with the point at which you have thrust. You will find more information about this later.

|II.55| **About the classification of the four cuts; also, how they should be directed
in the work (including their circumstances) and how they should be
fenced against the opponent**
Chapter 4

While there are not more than four primary cuts *per se* (as has been mentioned often), yet in
this weapon many other additional cuts are used and fenced (as in those discussed previously),
so I will classify for you the aforementioned four cuts on both sides and high and low, according to the indication of the lines shown previously. And so that you may have a fundamental
instruction into all cuts, you will be instructed and taught how such should be completed and
set up differently (high or low, and toward the opponent's body).

And firstly, three cuts are carried out and derived from the
High Cut, among which the first is cut straight down the center
upright line (at the head from above) and is called the Hairline
Cut or Brain Strike. The second, in which you reverse the hands
while cutting downward so that you hit with the half edge (or
rear edge), is called the Squinter Cut. Subsequently, the third,
which is guided in the subsequently-described way: by cutting
downward along the two [vertical] side lines, is called the Suppressing Cut.

Furthermore, the second cut is also cut three different ways
(high and low, according to the indication of the slanted hanging lines), of which the first and highest is named the Shoulder
Cut and Defensive Strike, the second is named the Hip Cut, and
the third is named the Thigh Cut.

Likewise, the Middle or Crossing Cut is also carried out at
three locations on the opponent's body, which are then named with three different names:
namely, the Neck, Belt, and Foot Cut.

However, the Low Cuts are likewise cut upward and through the same lines |II.55ᵛ| through
which the High Cuts were guided (straight from above or slanted) without acquiring specific
names (except as is stated about Winging and Dividing).²³⁰

You now understand that these four cuts remain singular in their type regardless of how
they are cut by you, yet their names change according to the limb [targeted] or their effect
when you cut in (depending on whether you guide them high or low toward the body). I will
now explain to you all of the cuts that are derived from these, set in an orderly way one after
another.

---

²³⁰ Neither of these cuts are described in this book. According to Grimm (vol. 3, col. 1842), *fliegelhauw*
is a fencing term that appears in Sachs (1558), in which it might be a diagonal rising cut. *Flügel* appears
as the name of a sequence that includes a rising strike from above in Paurenfeindt (1516) and reprinted
in Egenolff (1530s): 5ᴿ.

Scheide means "sheath" or "vagina", *Scheidel/Scheitel* means "skull", and *scheiden/scheidelich*
means "to separate or divide"; this term is used in JMM: 5ᵛ and JMR: 121ᴿ, where it describes a rising
strike from the Ox with the half edge.

The first Hairline Cut [Schedelhauw] or High Cuts [Oberhauwe]

You have been sufficiently instructed in the Sword and Dusack [Sections][231] about what a High Cut intrinsically is; therefore, it remains merely to teach you how it is used and in which circumstances it is useful and of service.

Thus, *if an opponent stands before you in the Iron Gate or Simple Brace,* also position yourself in the same in the Onset.

➤ From there, lift your weapon straight upward with arm extended (so that your weapon remains in front of your face during this lift) and while doing this, take note on which side they expose themselves to you.

➤ Cut a slice at that same side, straight from above yet close to their blade so that it appears as if you wanted to cut at them in front at the point.

➤ Leap quickly to the other side and pull your weapon back upward (toward that same side to which you have leaped) to strike and quickly cut a straight High Cut down at the same, close to their blade, like a slice.

In this cut, you should arrive with the feet well apart due to the steps and you should have the front knee bent well forwards, so that your upper body sinks down well forward with the cut, so that your weapon drops to the ground with the blade as flat as possible.

➤ Lift your weapon quickly into Longpoint as the counterposture. During this, draw the front foot back to you and move your |II.56| body back upright. However, in that moment in which you right yourself, drop your weapon into Iron Gate (with the hilt below). Then you stand as in the beginning.

That was about the cut in the Before. Now take note of this opportunity in the After.

This if you two had both moved into the previously-mentioned guard or counterposture in the Onset and you were aware that your counterfencer was ready to cut first. Observe:

➤ *In that moment in which they cut in:* pull your front right foot back to the left [foot] and (simultaneously with the withdrawal of your front foot) lift your extended weapon upward high in front of your face and *let them miss with their cut and drop toward the ground in front of you (or, if they have already contacted, they cannot reach farther than your hilt).*

➤ As soon as their cut has moved past in front of your hilt, cut a slice at their head (down from above) with a leap forward with your right foot. And this should occur so quickly that your cut hits before their cut has completely dropped to the ground.

➤ Afterward, move back into the counterposture (as recently taught).

Suppressing Cut [Dempfhauw][232]

As it happens in fencing that you cut sometimes in the Before, sometimes in the After, it thus often occurs that both fencers cut simultaneously; therefore, because the Suppressing Cut is derived from the High Cut (due to its point of origin), I will provide you instruction in the same, namely like this: when you observe that *your opponent wants to cut at you from their*

---

[231] See I.11 and II.3.

[232] *Dämpfen/dempfen* means both "to bring something to an end using violence or force", and "to suffocate or smother". Both of these meanings are supported in the text.

right *(whether slanted, across, or from below)*, take note. *In that moment in which they lift their weapon to cut:*

➢ Pull your weapon simultaneously upward.
➢ During this upward pull, leap quickly |II.56^V| out toward their left side,
➢ *In that moment in which their cut flies in:* guide your High Cut toward their right shoulder (so that your hilt moves somewhat prior to their blade in this downward movement).

You also arrive with your feet wide apart due to the steps so that your upper body is dropped following the cut (as stated above). Thus, you will hit either on their right arm or on the Strong of their blade. With this cut, you should suppress their blade to the ground, and also weaken them so that you can give them a cut or thrust *before they recover again.*

If, however, they were able to work immediately forward under your blade and urgently follow you with rapid cuts so that you cannot fence at their openings without risk:

➢ Quickly step out twice toward their right side and pull back into a High Cut next to your left side.
➢ *In that moment in which they cut in:* execute the [High Cut] outwardly over their right arm and toward their left shoulder. You again hit either the right arm or their blade down from above (as previously, [but] from the other side).

The Suppressing Cut is also completed in this way: position yourself into the Iron Gate in front. *In that moment in which they lift to strike (whether from the right or left side):*

➢ Quickly lift your weapon and cut simultaneously with them from above at the Strong of their blade—but so that your blade points upward as it drops and your hilt hangs downward towards yourself.

Cut this way—with an extended arm and lowered body, which can then take place [with your feet] standing far apart from one another—*at all of their cuts that they execute toward you (both from the right or left)* until you feel that they are sufficiently weakened, so that, *before they recover again and rise up,* you can fence at their openings.

In all of this, take note: *the lower they guide their cuts at you,* the lower you should move your upper body through wide steps, so that your pommel is lowered well toward the ground in the cuts and that you meet all of their cuts correctly with your Iron Gate |II.57| (high or low, depending on how they guide their cuts).

The third follows, namely the Squinter, which is a High Cut with reversed hands.

Squinter Cut [Sd̨ielḩauw]

[This] is correctly applicable against those who remain with an extended arm stiffly in front of their faces to counteract, because you drive them upward out of their counterposture with this.

Execute the cut like this when you have pulled your weapon upward with arm extended in front of you into the High Guard to strike and *they guide a cut toward your body during this (regardless of the side):*

➢ Step to the other side (away from their cut), cutting downward from above at the Strong of their blade such that your hand rotates in the downward movement so that you hit their blade with the short edge or flat instead of with the long edge.
➢ Immediately and in that moment in which the weapons touch together (or if you have not hit them with the outermost part of the short edge of your blade) thrust on their blade in

front of you at their face. In this thrust, turn the long edge downwards so that you stand in Longpoint at the end of the thrust.

Take note of these rules when turning into the cut.

If they cut from their right toward your left and you want to attack them with a Squinter: turn your cut outward in the cut [so that] the half edge faces downward and away from you.

If, however, they guide their cut from their left toward your right: turn your hand inward toward your body while cutting down (with the half edge downward).

And this applies whether you contact their blade with the flat or half edge. However, the closer the two Strongs |II.57^V| arrive together, the earlier your Weak or outermost part of your blade contacts behind their [blade].[233]

Slanted High Cut [Oberhauw schlims]

This cut has two names in rapier fencing, namely the Wrath Cut and Defensive Strike.

It is called a Wrath Cut when you cut this at their body in the Before without obstruction. You will be able to gather from the following examples how you should usefully fence with this cut at their opening, and it will depend on how they hold their weapon.

Thus, *if they hold their weapon low:* cut quickly, unpredictably, and slantwise above their weapon and through their face, then follow back quickly from the other side.

However, *if they hold their weapon high:* cut slantwise underneath their weapon and through their body (as quickly as before), and then from the other side. And that is about the Wrath Cut.

The Defensive Strike is what it is called when you use this cut to rebuff their cut and thrust away from you. *When they cut or thrust from above (whichever they desire):* this slanted High Cut (which you guide through their face and toward their hand) will deflect it away.

However, *if they cut lower at you, namely at the middle of your body:* guide your cut lower as well, slantwise toward their hand so that you hit their blade.

If, however, they cut at you even lower, namely at your feet: cut toward their feet as well, with a lowered body and hanging blade so that the blades meet together below in a cross (as is observed in the Figure designated with the letter B).

---

[233] The final clause in this sentence is problematic. The text reads: *ehe er dein schwech oder eusserste theil deiner klingen hinder der seinen antrifft* or "before they contact your weak or outermost part of your blade behind their own". This doesn't make sense. My translation assumes that *ehe er* is a typo for *je eher*, which the initial *jhe neher* would set up as the following clause.

The Cross Cut originates from these two Wrath Cuts: namely, when you guide the two from both sides toward one another, high or low (as the three crosses on the portrait indicate in the subsequently-printed image).[234]

Lastly, you should also learn [to execute] these two slanted Wrath Cuts, high or low, from both sides through |II.58| the three crosses (as mentioned previously concerning one side) with their steps, skillfully and long in front of you (and not impetuously)—but cut and completed as slices with drawn cuts.

Regarding the first and highest, position yourself like this: stand with your right foot forward and with your body upright (as mentioned above). Cut with an extended arm from both sides slantwise through their shoulder. In these cuts, turn the right side well toward them following the cut.

Regarding the second, middle cross, position yourself in this way: move with the right foot forward, but so that in stepping, you stand with the feet one shoe-length or more farther apart from one another, and bend the knee well forwards so that you drop your right shoulder down somewhat (from which you bring forth your cuts) and you stand at the same height as the intersection of the middle cross. Cut slantwise from both sides together through the middle of your opponent's body (just as you cut through their upper body previously).

Regarding the lowest cross,[235] you must stand with the feet even wider apart and you should also bend your front knee farther forward than previously, so that you stand with your upper body farther forward than before (sunk downward) and with your right shoulder |II.58ᵛ| the same as in the others—that is, that you arrive at the same height in this. If this is impossible for your body, you should leave off with the lower cross (as it is not for everyone, etc.) because if you cut at their feet and yet want to remain upright in body, they can immediately rush to your face with the straight thrust even though one can unpredictably complete a Foot Cut to the side before they become aware of it.

You should learn to cut these three crosses in their particulars, including their steps (forwards and backward) so that you will be well practiced in them before you should need them. Yet also such that you always remain with the right foot forward in the stepping.

You can also practice these three crosses to be of further use in this way:

Cut the first against the high cross (through the same oblique left hanging line).

[Cut] the second, with another step forward, from your left toward their right (obliquely through the middle cross).

[Cut] the third again from your right toward their left ([through the] lower cross, slantwise through their foot) with another step forward of your front foot.

These three cuts should be completed quickly with three steps forward of the front right foot. As you have now alternated from the high to the low, you can also alternate again from the lower up to the upper line, from one to the other.

---

[234] The human target on II.58.

[235] Note that the lowest cross on the image on II.58 is actually centered on the fencer's forward knee. Meyer later defines the 'foot cut' as the entire lower leg, from the knee downward, whichever point is farthest forward and can be reached.

Hip Cuts [Hüft hauwe]

Execute them like this:

➤ Guide a powerful High Cut toward their head; however, do not let it hit or touch but instead, turn the half edge outward and away from you and toward their left ear during the descent of your cut, to thus drop the front end deeply at their face. You both force them to counteract and also recover yourself for the following cut.

➤ |II.59| *In that moment in which they rise to counteract:* quickly pull your hilt around in front of your face and upward toward your left.

➤ Cut from your left outside of their right arm at their right hip (slantwise down from above), *in that moment in which they have lifted up*ward.

This should take place in a single step and movement, so that you lift your right foot with the first pulled High Cut and set the [foot] down again (in a step forward) simultaneously with the completion of the Hip Cut. You have thus completed it correctly. This works from both sides.

You can also execute the same cut with a thrust:

➤ Thrust from the Iron Gate straight upward at their face.

➤ *As soon as they raise their hilt:* cut at their hip as before.

Round Strike [Rundstreich]

Execute this [cut] like this:

➤ If you stand in the Iron Gate, pull your hilt upward toward your left and into the guard of the left Ox.

➤ From there, draw a Middle Cut across toward their right and through at their face.

➤ In the same movement, [draw] the second cut likewise from your right and toward their left, also completely through their thigh or knee.

These two cuts should be completed in one movement: the first above through their right, the second low through their left, quickly with a single step, in a circle (which strikes high through the opponent's face and low through the thigh). It is then called a Round Strike because of the circle that it makes in moving around.

|II.59ᵛ| ## Double Round Strike [Doppel Rundstreich]

Position yourself in this [cut] like this:

➤ Cut the first across from your right toward their face; however, do not let this hit.

➤ Instead, when cutting in, pull your hilt between you and them toward your left into the same guard of the Ox and turn your right side well toward your left following the hilt.

➤ You should not remain in this position even for a moment but instead, in that same pull upward, drop your head downward and cut the second around your head, toward their right side, and across at the middle or shoulder.

➤ However, this cut doesn't move through, but only up to their counteraction (*if they have turned [the counteraction] forward*).

➤ [Afterwards,] lift your hilt upwards in this cut and toward your right again.

➤ Let your blade move back around and cut the third completely from your right across through their foot.

Regarding the steps, hold yourself like this: for the first two Middle Cuts, set your right foot only a little forward (yet not firmly on the ground), but recover or take the weight with this step so that you can step more properly and farther with the same foot in the third cut. Thus, as soon as the foot touches the ground in the first step, it should be lifted again, and set forward with the third cut.

You should carry out these three cuts in one motion like an upright capital S[236] with two quick steps. In addition, the last cut should be the strongest and be cut completely through. The other two should be smaller, only up to the opening, and be pulled back again from the same (and, as said, everything in one swooping motion). When you learn to make these two Round Strikes, namely the single and the double, correctly and well, (since |II.60| advantage and not power is part of all pulled cuts), you will be able to fence very many nice and skillful sequences from them.

Neck Cuts [Hals hauwe]

Fence this [cut] this way: hold your weapon on the right in the Low Guard or the Iron Gate and wait [to see] *if they want to fence at you from their right.*
  > *If they thrust or cut from the same toward your left:* leap well toward their left and away from their thrust or cut and (simultaneously with this leap) hit across with the inward flat high on the Strong of their blade.
  > In that moment in which your flat blade touches on theirs, leap quickly farther around toward their left side (during this).
  > Indesly draw your sharp edge toward their right through their neck (after you have pressed their blade downward with your flat).

As you have now completed this from this side, you can also direct it in the work from the other side. You can also complete the cut freely, without any preparation, *after they have dropped down or cut through and missed,* using a Middle Cut across toward their neck (as you will have sufficient examples in the following sequences).

Foot Cut [Fußhauw], and about the Middle Cut [Mittelhauw]

The foot is understood here to be the entire leg from the knee down to the foot, which can be cut across and slantwise. The across is nothing more than the Middle Cut, executed high or low. The slantwise is the Wrath Cut. However, you should not cut at the foot unless you have weakened them with Suppressing [Cuts] (as taught previously) or |II.60ᵛ| have taken their blade using another sequence. Exceptions are that they lose out [to you] due to long waiting or have otherwise moved too far upward.

Hand Cuts [Hand hauwe]

The Hand Cut can be completed in many ways, as you will gather in the sequences. However, take note: *as often as they cut at your feet, they must extend their hand far from themselves.*

---

[236] Note that this was printed in Fraktur, so the letter S is formed like a wide 6. The pages in JMR attributed to Meyer's own hand also form the S this way (see folio 123ᴿ for a prominent example).

Therefore, you can evade them well with the feet and simultaneously cut at their hand (as the two portraits teach in the Figure designated with the letter B).

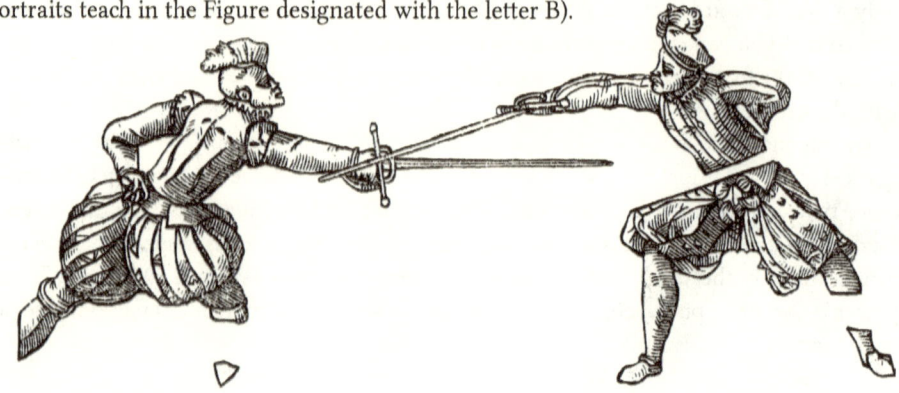

In addition, you can also cut at the hand as often as your opponent sweeps around too high or too wide. This Hand Cut in rapier is the most prestigious one because, if used against an opponent, *they must defend themselves* and they have halfway lost (if not completely lost). You will also hear about the misleading and reversed cuts enough in later sequences in the second part.

Double Cut [Doppel hauw]

Execute this [cut] like this:
- ➤ *If an opponent cuts at you from their right toward your left:* cut to oppose their cut with a Horizontal or Low Cut and catch them (the higher in the air the better).
- ➤ Take note, *as soon as their cut glances on your blade:* turn your half edge inward on their blade and draw your long edge rapidly up from their blade and through toward their face.
- ➤ In this cut, pull your hilt upward and let the blade move through and under their right arm toward your left. Leap simultaneously well out toward their right side with a bent body.
- ➤ With the long edge, cut outwardly over their right arm |II.61| at their head.

These three cuts, when you execute them correctly, are completed swiftly in one motion. It is easily gathered from this Double Cut how one should double all other cuts. Now, in further consideration, the Foot Cuts (slanted and across, including all strikes) then are also four like the cuts—namely High, slanted, Crossing, and Low Strikes—and can also be completed with the inner and outer flat.

Since these can all be understood well in the subsequent sequences, it isn't necessary to deal with them here specifically. Therefore, I will proceed to describe and list the thrusts, how many there are and how you should fence with them.

About thrusts
Chapter 5

There are primarily three thrusts from which all others arise and originate: namely, the first from above and the second from below (of which each is executed from both sides); the third moves from your center straight in front of you into Longpoint. As mentioned, I will set something down here about these three primary thrusts from these starting points, from which you will be able to sufficiently learn and understand all other thrusts.

The High Thrust should be directed from the right Ox toward your opponent's face or chest. Execute it like this:

➤ Position yourself into the High Guard of the right Ox (as you were taught above). Raise your right foot for a step forward and (simultaneously with this lifting of your foot) pull your hilt back behind you, over your right shoulder, to gather for a strong thrust.

➤ From there, thrust toward their chest with a wide step forward of your lifted foot.

➤ In that moment in which your thrust should just hit, turn the long edge downward toward your left (as in a slice) so that after the end of the thrust, you are lowered toward the ground with your front knee bent well forward and your upper body bent forward following the thrust such that, after completing the thrust, your blade arrives down on the ground with arm extended far in front of your foot in a slice off.

➤ From there, move back upward into the right Ox with the long edge and recovered foot— indeed, just as you stood at the beginning. That works on both sides.

|II.62| **Face Thrust** [Geſicht ſtich]

Initially learn to thrust this from your left with your palm upward, like this:

➤ Position yourself into the guard of left Ox.

➤ *If they subsequently thrust toward your right:* leap well out toward their right and away from their thrust.

➤ *In that moment in which they thrust in,* thrust in with an extended arm from your left and above their right arm at their face.

Item, position yourself into the guard of the right Ox, and take note. *As soon as they thrust in at you:*

➤ Step with your left foot to your left side and out away from their thrust. Quickly follow with your right foot out toward their right side.

➤ Simultaneously with these steps: let your blade drop down toward your left [side] and snap around into the guard of the left Ox next to the same [side].

➤ From there, *in that moment in which they thrust in,* thrust over their right arm and at their face (as before).

You will hit (as can be seen in the smaller scene in the Figure A printed previously) and you will [arrive to] stand in Longpoint. This must occur quickly *in that moment in which they thrust in.*

Throat Thrust [Gurgelstich]

This thrust is carried out in many ways, of which I will present one to you.

If you find your counterpart in the Onset in the Iron Gate:

> Threaten them with a thrust from the left High Guard of the Ox while stepping out, over their right arm and toward their face—but such that you remain high with your hilt.
> Take note while doing this. *If they lift their hilt toward their right intending to divert or counteract your thrust:* let your front end drop downward next to their right shoulder and move or change through with the same point under their right arm.
> Only after you are completely inside of their right arm, thrust up at their throat from below such that, when guiding your thrust in, the long edge is downward, the short edge upward, and your weapon ends high in Longpoint after completing the thrust.

This [thrust] is correctly completed in this way. This must be carried out completely and unexpectedly.

|II.62ᵛ| ### Heart Thrust [Hertz stich]

The Heart Thrust can be correctly used in the work from above, from the middle, and from below; however, from all of these take note of this point. *If an opponent cuts from their right:*

> Cut from your right across toward their weapon.
> With this cut, step with your right foot through and well under their blade toward their right, so that you catch their strike in the Strong of your blade.

(The closer to their hilt and the higher in the air that this occurs, the better.)

> In that moment in which the blades glance together, turn the point inward toward their left breast (so that your blade remains on theirs and so that the half edge is turned onto their long edge) and thrust in [while] remaining on their blade (as the larger tenants show in the Figure which is designated with the letter G).

In all of this, take diligent note about *whether they want to move away from your blade.*

> As soon as you feel this, turn your long edge back toward their blade, moving with a slice on their blade.
> Keep their blade in front and toward their body and observe where you could set on during this to your advantage.

Genital Thrust [Gemecht stich]

Execute this like this:

If an opponent cuts at you from outside at your right thigh, knee, or foot:
➢ Catch their blade with a countercut from your left. During this, step with your left foot well out toward their right.
➢ As soon as the blades touch together, step farther forward with your right foot toward them. During this step, turn the point upward and inward under their blade and thrust at their genitals.

Or, *if they initially thrust at you from below:*
➢ Step with your left foot toward their right out to the side (as before) and guide their in-flying thrust out with a hanging blade (from your left toward your right).
➢ As soon as your blade touches theirs, step quickly with your right foot farther toward them and thrust under their blade at their genitals as before.

|II.63ᵛ| ## Reversed Thrust [Verkehrter stich]

Although this may be initiated in many different ways, there is only one way to end it, so position yourself like this:

If your opponent stands in front of you in Simple Brace or Iron Gate:
➢ Thrust at them straight upward from the right Low Guard, inside of their weapon and hard on their blade at their face. During the thrust in, turn the long edge toward their blade and upward to your left.
➢ *If they press or guide your blade out away from them toward their right (upward or to the side):* let your blade snap back around above and toward yourself, so strongly that your blade sweeps downward next to your left (with the point down) and back upward, in under their blade.
➢ Thrust in with the reversed hand, under their blade at their right arm (as you can see in the smaller tenants on the left in the Figure designated with C). The point must be aligned using the snap around in the first fluid motion, otherwise the thrust is too weak.

➢ Immediately withdraw your weapon toward your left and cut a Defensive Strike through their right shoulder, face, or side.

> Or, let it snap around over your hand again, and thrust from outside of their right arm back at their face so that in this additional thrust, the palm is up, the half edge is turned toward their weapon.

In all of this, bend your head downward and well toward your left out from their weapon.

> *They must defend against and divert this thrust. As soon as they execute this:* guide the point around outside of their arm again, so that in this move around, your hand reverses as before, and thrust with reversed hand as before under their right arm at the body.

> Follow with a Defensive Strike as taught before.

Double Thrust [Doppel stich]

If you encounter an opponent in the Iron Gate (straight in front of them):

> Thrust from the Low Guard (from your right and on the inside), hard onto their weapon and close to their hilt, and upwards at their face.

With this, you force them *so that they must lift their hilt upward.*

> As soon as you perceive *that they are lifting their hilt upward to counteract:* during the thrust upward, raise your hilt to transform your low thrust into a high [thrust] behind or next to their hilt and across at their body.

|II.64| **Another**

Or thrust inwardly at their body.

> *As soon as they oppose the thrust with a counteraction:* turn the half edge inwards toward their body and let the blade run through under their right arm and toward your left side under their blade, letting it snap around up high next to your left side into the guard of the left Ox.

> From there, step out toward their right and thrust outside of their right arm and at their face.

These two thrusts should move quickly one after the other.

Or thrust from outside and over their right arm at their face.

> *If they turn your thrust out toward their right out to the side:* immediately let your blade snap around toward your left back into the guard of the left Ox. During this, step with your left foot well behind your right foot[237] toward their left side.

> Thrust from the left High Guard of the Ox, with a step forward of your right foot, inside of their weapon and toward their face.

These thrusts are doubled in many ways elsewhere, about which there will be more in the second part.

Misleading Thrust [Verfierte stich]

In the Onset:

> Guide a powerful thrust from the right High Guard of the Ox towards their face.

> During the thrust, turn your thrust up from below and thrust under their hilt and upward at their stomach (with a large step forward of your foot).

---

[237] The text reads *rechten Arm*; however, that doesn't make sense. Ergo, right foot.

When you correctly turn this high thrust into a low thrust using the Rose, at the beginning it will appear no differently than if you were thrusting from above, *yet before they perceive it,* you will have hit below.

Item, *if your counterpart stands before you in Iron Gate:*

➤ Thrust inside of their weapon and up-ward toward their face and miss so that your blade snaps back toward your right in right Ox.

➤ Act as if you wanted to thrust from the outside and over their right arm, but misleadingly change your thrust in the air and thrust from above, underneath their weapon, inwardly at their face.

Flying Thrust [Fliegender stich]

This Flying Thrust is the most distinguished and it is necessary for every fencer to know it; therefore, execute it like this:

➤ Position yourself in the Onset in the Low Guard on your right (so that you stand with your feet not too far apart |II.64ᵛ| so that you can step forward with each thrust); you should also hold your right arm stiffly at the elbow and as straight[238] as possible.

➤ With body language and movement, pretend as if you wanted to thrust in here and there; and, while you appear to thrust the point somewhat toward them, cunningly retain your weapon close to you and under your complete control.

➤ Immediately when you perceive your opportunity (after you have misled them somewhat with your serious body language), whenever and wherever they least expect it, fly toward them, thrusting unexpectedly, and execute this quickly with an additional step forward (as if it had been shot from a crossbow).

➤ As quickly as you have thrust in, you should just as quickly pull your weapon back into the abovementioned Side Guard.[239]

➤ From there, *if they continue to thrust in,* you should slice off from both sides.

So that you can be better practiced in this thrust, select a certain point at which you can thrust in front of you. Thrust, with a step forward, from the right Low Guard in front of you so that, when the thrust hits, you stand as the scene indicates in the Figure designated with A. Then pull your weapon back into the abovementioned stance, during which you also pull your foot back to its previous location. From there, thrust swiftly with a step

---

[238] The German is actually *ungebogen,* or "unbent".

[239] As Meyer mentions on II.91ᵛ, Side Guard is another name for the Low Guard on the right, which is why he uses "aforementioned" to describe it here.

forward toward it. Pull your weapon back quickly from the thrust, including the movement of your foot back to its previous position. Execute one thrust or six.

Then pull your weapon back into the abovementioned stance, during which you also pull your foot back to its previous location. From there, thrust swiftly with a step forward toward it. Pull your weapon back quickly from the thrust, including the movement of your foot back to its previous position. Execute one thrust or six.

However, because this cannot be described (or shown with a living body), you should practice this seriously with the help of the fencing master and study it even more diligently.

A good teaching and rule as to how one should convert the cuts into thrusts, the thrusts into cuts [240]
Chapter 6

Since you have been taught and briefly presented with the cuts, including the thrusts, I will subsequently also demonstrate briefly how one should transform the cuts into thrusts and the thrusts into cuts, as this is a notable master technique to correctly carry out this conversion and to be able to use it according to [your] opportunity. Since this can be fenced in many and diverse |II.65| ways, for which reason it would be pointless to describe them all here at length, I will present them solely using some examples from the four cuts and teach them this way.

Namely, in the Onset, when you can reach their outermost [point] with your outermost [point]:

➤ Cut a powerful High Cut slantwise toward their left and, in that moment in which your cut flies in, turn your hand so that the half edge is rotated inward toward them.
➤ During this cut, slightly slow the movement of your hilt in the air (quite clandestinely) until your blade has shot forward into a thrust.
➤ Thrust the remaining part of the thrust toward their chest, sinking down with a broad step forward in the way you were taught previously concerning the High Thrust.

Item, cut a Middle or Low Cut from your right toward your opponent's left, and take diligent note:

➤ *As soon as they move in opposition to counter act*, and just as your cut is about to hit, quickly turn the same cut into a thrust (before it hits).
If you hit their blade with your cut, whether from a High, slantwise, Crossing, or Low Cut:
➤ In that moment in which the blades touch together or glance off, immediately turn the point inward toward their body.
➤ Thrust in on their blade as you were taught in the Dusack Section (for example, about the Awakening Cut). [241]

---

[240] Also known as *mutieren*, which means "to transform, convert, or change".
[241] See II.11.

To convert thrusts into cuts, carry them out like this:

➤ Guide a powerful high thrust toward your opponent's face and, if you perceive halfway along the path that *they want to lift and counteract:*
➤ In that moment in which your thrust should just hit, swiftly pull your hilt upward a little bit.
➤ Cut through from the side next to or under their hilt.

Item, guide a straight thrust toward their face.

➤ In that moment in which it should just hit, turn your hilt upward toward your left and let your blade move around your head.
➤ Cut at them slantwise from outside and through their right (whether from below or above).
➤ If, however, a cut overpowers you so that you have to counteract, then as soon as *their cut touches on your blade:*
➤ |II.65^V| Turn your blade on theirs with the point inward and toward their body so that you recover to cut at your pleasure.

From these sequences explained previously, you can understand enough about how the cuts are converted into thrusts and the thrusts into cuts, if you want to diligently ruminate on it.

The misleading flows from this as follows.

About the misleading
Chapter 7

The cuts, including their circumstances, were sufficiently explained previously. However, because misleading will be discussed often in subsequent sequences, it has become necessary to mention something about it so that I am not pulled backward in the sequences or held up in my writing.

There are two types of misleading. The first is carried out with the weapon; the second through body language. It is not my intention to deal with misleading with the weapon for long here, because this has already been discussed often in the two preceding weapon sections. Namely, misleading is nothing more than [this]: as I guide my strike toward an opening, I perceive that *they are moving to counteract the same [strike]* (so that my cut will be worthless). Therefore, I allow [my strike] to pass as a miss and quickly pull it somewhere else to the closest opening in the same fluid motion.

It is imperative for this type of misleading that you are well instructed and practiced in the four openings (including other divisions) so that as you cut in, you masterfully, skillfully, and imperceptibly delay a cut which you directed at a high opening and can then complete it below on the same side in its initial path (or at another opening). Take this as an example:

➤ |II.66| Guide a powerful Wrath Cut toward their left.
➤ In that moment in which and before this cut has traveled half of the distance (while still in the air), turn the half edge inward toward them while the blade is moving in so that it appears as if you wanted to thrust at their face.
➤ Yet, with this conversion in the air you have recovered for another cut, which you should also complete through their left (whether above or below).

In summary, if you want to touch your opponent above, threaten them before that or cast a meaningful look below; or if you want to hit them on the left, first threaten towards their right so that *they must swipe after with their weapon and give you space on the other side* (as such will be further taught in the sequences).

The second is now included in this misleading, the one that is completed with body language, and many and various wonderful sequences are fenced from both types [of misleading]. However, in order that you can actually have an understanding of this misleading with body language, I want to explain this to you with much more extensive examples and sundries. While misleading through body language, as with the weapon, is nothing more than showing one cut or sequence and completing it in a different way, in this case you must consider well and take note of the Provoking, Taking, and Hitting (about which you were taught previously in the Dusack [Section]).[242]

The initial focus of all provocation (with body language or otherwise) is that you bring them out of their advantage to cut or thrust. Secondly, *as soon as they cut or thrust in*, you can forcibly hold off or ward off their cut (to which you have incited them through your provocation) and can weaken it to such an extent that, thirdly, you can easily reach them and touch them without injury *before they recover themselves. If they recover back thereafter,* you would be ready to oppose them before they could approach (even with wild, useless strikes).[243]

However, |II.66ᵛ| in order that you can understand it better, I will set some examples for you—not so that it must follow like this, but as an introduction into better sequences, namely:

If you now want to mislead another with body language, use the following meaningless movements[244] without risk. Thus:

➤ If you find an opponent in the Low Guard on the right:
➤ Position yourself in the Iron Gate and express with body language as if you seriously want to thrust at their face.

With regards to this body language: lift your right foot; keep your eyes fixed on their face; with a moving arm and clenched fists, with your nose turned up and foot lifted, guide the point toward their face as if you seriously wanted to thrust.

➤ In the thrust inward, turn the long edge up toward your left. Frighten them with your thrust *so that they suddenly lift to counteract.*
➤ *In that moment in which they lift:* let your thrust move around your head and cut outwardly at their right leg with another step forward and with your body bent forward.
➤ Guard yourself quickly with Defensive Strikes to protect yourself.

---

[242] See II.16.

[243] Although formed from the compounding of 'protection' and 'strike', the meaning is more akin to the air guitar of fencing: *Luftstreichen und Possen* = wild strikes, flailing in the air, and antics (GRIMM, vol. 15, column 224). All of the examples provided in GRIMM are of this type of flailing, or fighting against a non-existent opponent, and not actual defensive work. This would dovetail with the emphasis on body language in misleading.

[244] *Ceremonie*, GRIMM, vol. 31, column 670: "Since the Reformation, considered to be empty and external, therefore superfluous customs, and simultaneously directed at the body language and gestures, employed to deceive the unlettered people." *Seit der reformation als leere und äuszerliche, daherüberflüssige gebräuche angesehen und zugleich das augenmerk mehr auf die gebärden und gesten gerichtet: das unverständig volck zuo betriegen.*

Item, position yourself in the Iron Gate (as before) and position yourself with skillful body language as if you wanted to cut at their foot, and do that in this way:

➤ In the Onset, look seriously with wide-open eyes fixed at their forward foot.

➤ Indesly[245] lift your weapon and bend your body with the foot raised as if you wanted to cut low sometime during a step forward.

➤ However, in that moment in which you set your foot down in the step forward, thrust at their face from that point at which you have lifted your weapon.

During this thrust, keep your face and your body language focused hard on their foot. *They won't take note of the thrust* until it has occurred, because raising your weapon to cut is your recovery to thrust.

➤ The step and the thrust end simultaneously with one another.

And you should learn all of this and use it against those who stand stiffly in their counter-posture without any work, waiting for your cuts or thrusts. You must incite these types and provoke [them] from their |II.67| advantage because you really shouldn't attack them without an advantage, since you must be concerned *that they will reach past you or catch you in their own sequences.* Therefore, you must observe how to bring them out of their advantage. In order that you can understand this better, I want to briefly repeat the Provoking, Taking, and Hitting presented above.

If you find your *opponent in a guard or quarter in which they wait for your thrust or cut,* cut through at the closest opening with no intention of hitting them. Observe as well that you are not too close to them, and pay attention that you do not give way or allow your weapon to move too far from you out of your control; instead, hold your weapon close to you (in good strength and full control), but conceal this and mislead with your body language as if you had cut awry. *As soon as they rush to your exposed opening with cuts or thrusts,* immediately revitalize yourself back to the strike and cut their in-coming cut or thrust strongly away from you; or cut a Suppressing Cut from above (as this may become possible for you). And that is called the Taker, because you take away the weapon with equal force (*which they had not predicted*). As soon as you have thus taken their strike or thrust, rush to the closest opening with cuts or thrusts. These following cuts are then called the Hitters.

Thus, you have Provoking, Taking, and Hitting, which are nothing more than: providing [them with] a reason to cut, counteracting or removing the same cut that you caused, and immediately cutting back at the opening. And this does not mean that you shouldn't hit them with the Provoking Cut or with the Taker if you can. Instead, these are named solely because the most proper meaning is either to provoke them out of their advantage or to take away and

---

[245] On I.25 of the Sword Section, Meyer explains his interpretation of *Indes* to be a small word fraught with meaning: in that moment in which you strike, you should simultaneously observe the other openings that you could also hit, and also all of the techniques your opponent can use to strike or counteract your action. It is therefore a point of hyperfocus on what you are doing and what you and your opponent could be doing at the same time and subsequently. Meyer states that he will use *Indes* as a shortened form to call all of this to mind.

Within his text, Meyer tends to use *Indes* to indicate a disruptive action, that is, because he is describing actions sequentially that occur simultaneously or disruptively. Thus, *while* you are performing the above action, execute the following action *at the same time* and *in the same space*. Because *Indes* is an adverb of time in German, it will frequently appear adverbially in the translation, thus *Indesly*.

remove their cut. Whether you can hit immediately during this is unimportant, because you can also complete the aforementioned three techniques in one single cut. For example:

|II.67ᵛ| Take the Wrath Cut (which is the Defensive Strike in front of you) and observe *whether they move their weapon above or below the belt. If they hold their weapon above the belt:* cut the firſt underneath their weapon through the lower line from your right; cut the second through the upper line (also from your right); cut the third through the upper or lower line, depending on where you find them open.

Also, this process is not always maintained because you can provoke well, and hit, and take as the laſt. However, because experience (which can only be learned through daily pracſtice) muſt inform the majority of such techniques, I will let it reſt with this example.

Namely, if you find an opponent in one of the Low Guards in the Onset: cut a Middle Cut unexpecſtedly through their face from your right. With this cut, you force *them to move quickly upward, and thus they become exposed below.*

Therefore, quickly cut the second through their foot from your left before they become aware of this. That is the Hitter. *They will quickly rush at that [ſtrike] and at you.*

Therefore, cut the third as a Defensive Strike from your right, so that you take out their in-flying blade. That is then the Taker.

These two Middle Cuts should follow quickly one after the other, so that your second cut hits before they have correcſtly lifted to defend the firſt, and thus the third becomes the Taker.

About counteracſtions and how many of these are used in rapier, as this differs from other weapons

Seven types of counteracſtions[246] are found here, with names like set off, slice off, move through, suppress, hang in front, block, [and] guide outward or away using an upright and/or a hanging |II.68| blade.

[Setting off]

You set off when you turn the long edge againſt their weapon from one of the four ſtances and convert into Longpoint.

If you hold your weapon in the Low Guard to the right *and your counterpart cuts or thruſts at you:*
➢ Step to the side and out from their weapon.
➢ Lift your extended weapon upward and in front of you into Longpoint to catch their incoming thruſt or cut on the long edge.
➢ In that moment in which you catch their cut, simultaneously thruſt forward away from you with the Longpoint.
Pracſtice this from all four ſtances.

Slicing off [Abſchneiden]

Execute like this. Position yourself in the Low Guard on the right, and take note:

---

[246] *Absetzen, abschneiden, durchgehen, dempfen, verhengen/vorhängen, sperren, aus/abfahren.*

➤ As soon as your counterpart pulls up their hand to cut or thrust toward you, simultaneously raise your weapon as well and extend your hand (including the weapon) from your right toward their left.

➤ Simultaneously with this extension, drop your hilt down toward your knee (or lower [than your knee] if it is possible for you), so that your blade points upward with the point and somewhat in front of you.

➤ Receive their blade on the long edge and guide it downward in front of you toward your left using a slice.

This works on both sides.

Suppressing [Dempfen]

Suppressing is derived from the High Cut, as there is nothing intrinsically different to it. All other cuts can be suppressed and countered using this High Cut in the subsequently described way:

Position yourself in the Low Guard on the right. *When they* |II.68ᵛ| *cut from their right toward your left, [whether] from below or above,* take note:

➤ *In that moment in which they extend their arm to strike,* raise your weapon simultaneously with them so that in this lifting, your blade points upward away from you and your hilt points downward toward you.

➤ During all of this, step out and away from their cut twice (toward their left).

➤ Cut from above (according to the indication of the upright line) at their right with the long edge and lowered hilt, including another step forward of your right foot.

➤ You thus hit on the Strong of their blade.

➤ Therefore, in this Suppressing, lower your upper body well forward over your bent knee (following the cut) so that your hilt, in moving downward, comes toward the ground with your extended arm somewhat before your blade.

➤ *However, if they withdraw their blade out from under yours and guide another [strike] toward your right:* hastily leap twice, well toward their right with both feet, and cut again from above and from outside over their right arm (in the same way as before); however, [do it] so that the cut is directed toward their left upright line (as previously toward their right line).

➤ You will then hit on their Strong.

Thus, you can use the High Cut to suppress from all stances until you weaken and tire their arm with this so that you can then easily fence at their body.

Moving through [Durch gehen] [247]

This is nothing other than: *in that moment in which they guide their blade in,* you move through under their blade toward the other side. You then guide their strike, flying in from that same [initial] side, away toward that side on which you originally held your weapon.

---

[247] Meyer uses '*durchgehen/durchfahren* = move through' instead of '*durchwechseln* = change through' from the Sword Section. The translation reflects the terminology of the original.

Thus, if you hold your weapon in the right Low Guard, take note:

> *In that moment in which they guide their weapon to thrust or cut,* |II.69| hastily move your blade through under their blade or shift[248] your blade across toward your left to hang a little toward the ground.

> From there, slice off their in-coming cut or thrust toward your right (before they have fully completed it), so that you arrive back in the right Low Guard using this slice off.

And you can also execute this from all stances on both sides.

Hanging in front[249] [Vorhängen]

Hanging in front is a type of counteraction in which you guide your hilt with arm extended in front of you and above your head, so that your blade hangs downward in the direction of the ground, and you forcefully push their thrust or their strike away on both sides using your flat. It is therefore called "hanging in front" because your blade hangs in front of your face in this counteraction (in order to protect the same) and, although it can be brought into play from all of the stances, it is used most commonly and comfortably from the stances on the right side.

Position yourself in the Low Guard on the right this way.

> *If your counterpart thrusts or cuts at you:* step with your left foot behind your right, out toward their left, [and] quickly follow with your right [foot] farther toward them.

> In that moment [in which you step], lift your hilt upward and guide their blade away from your right toward your left on the flat of your hanging blade (as you can see in the two upper tenants on the right side in the subsequently-printed Figure designated with the letter E).

---

[248] The original text has *thransforiere*, which is either a typo for '*transformieren* = to transform, change' or '*transferieren* = to shift across'. Due to the directional contexts, I have chosen the latter interpretation.

[249] Meyer distinguishes *verhengen* in the Rapier Section from *hengen* in the Sword Section. In the Sword Section, hanging is related to Ox, while in rapier, hanging in front is achieved from any stance by moving the blade so that it 'hangs in front of your face'.

|II.70| **Blocking** [Sperren]

Take note, *if an opponent cuts outwardly at your right foot* when you stand in Iron Gate:
> ➤ Lower your blade with the point extended downward to the ground directly in front of your foot and yield slightly to the side by stepping out from their cut toward their right.
> ➤ Block their blade so that they can't move through.

Thus, you can block and stop all of the cuts which they want to cut through below.

Regarding more about blocks, you will find it previously in the Sword [Section].[250]

Beating outward [Ausschlagen] with a hanging blade

This outward beat follows from the block like this. *If an opponent cuts or thrusts straight at your lower body:*
> ➤ *In that moment in which they guide their weapon in,* pull your right foot back to your rearmost and simultaneously lift your weapon upward.
> ➤ Cut their blade downward from above and from your left outward toward your right, with your weapon extended so that the blade hangs a little downward.

Or, position yourself into a high Longpoint.
> ➤ *If your counterpart thrusts low at you:* let your blade drop down somewhat from your left toward your right.
> ➤ Cut their blade away from you and outward toward the side, with your lowered weapon between you and them.
> ➤ Quickly follow with your sequences.

|II.70ᵛ| **Taking out** [Ausnehmen] with the half edge

Position yourself in the Low Guard on the left, similar to the Change [Guard], so that the half edge faces your counterpart.
> ➤ *If your counterpart thrusts at you toward your face:* take it out with the half edge in one slash strongly upward (from your left and toward your right) so that your blade shoots around above your head back into right Ox.
> ➤ From there, thrust toward their chest, and in this thrust, turn the long edge downward and come back into left Low Guard.
> ➤ From there, take out as before.

Just as you have now learned to take out and upward with the half edge, you can also beat outward and upward with the long edge and with the flat. This can be done from both Low Guards.

You have previously heard at length in the Sword and the Dusack [Sections] about whatever else is necessary regarding counteracting.

---

[250] See I.22ᵛ.

This chapter will deal with changing, pursuit, remaining, feeling, pulling, and twisting[251]
Chapter 8

There are two ways to change:[252] one is through and under their weapon, which is therefore called the "change through"; the other is then completed above around the blade from one side to the other, which is called the "change around".

Execute the change through like this: firstly, when you guide a cut at your counterpart (whether high or low on the body), in that moment in which you cut in, take note of *whether they lift counter to your cut to counteract it.*

➢ If you perceive this, don't let |II.71| your cut touch their counteraction, but instead pull and guide your cut through under their blade.

➢ Thrust at the other side.

Thus, at their counteraction, always move or pull your cut through (right when it is about to touch), under their hilt or blade (regardless of which side you want to cut at), and work at them on the other side, whether using thrusts, cuts, Suppressing Cuts, or slices.

Secondly, execute the change through from your counteraction like this:

➢ *If they cut at you, regardless of which side,* take diligent note *whether they cut too high or toward your blade or otherwise do not sufficiently guide [their cut] at your body.*

➢ *When their cut flies in toward your counteraction,* immediately yield a little with your front foot back to your rearmost and pull your blade through and under theirs toward the other side (and thus let their cut miss).

➢ Thrust likewise at the side from which they guided their strike, and do that so quickly that your thrust hits before they have recovered back from the ruined cut.

This change through can be executed many different ways, as will be provided in the sequences.

The change around requires no further explanation, as it was dealt with quite often previously.

Execute pursuit like this:

➢ *If your counterpart holds their weapon low on their right,* wait.

➢ *As soon as they lift from there:* quickly thrust *in that moment in which they lift their weapon.*

➢ Likewise, take diligent note *if they hold their weapon on their left. In that moment in which they move it back from that same side:* quickly and cunningly thrust *while they move.*

➢ Likewise, *if they rise up from below:* follow them with a thrust from below (*in that moment in which they lift*).

➢ Also, *if they miss with their cut:* follow them quickly from above.

---

[251] *Wechseln, Nachreisen, Bleiben, Fühlen, Zucken, und Winden.* Most of these techniques are described in the Sword Section starting on I.17ᵛ.

[252] Literally "change through", but this seems likely to be an error since the chapter title just says '*wechseln* = changing' and the text goes on to list '*durchwechseln* = change through' and '*umwechseln* = change around' as subtypes of it.

However, take diligent note in this pursuit that in that moment in which you |II.71ᵛ| hit with your thrust, you always turn the long edge toward their in-flying weapon (if they would otherwise cut or thrust), because *as soon as they perceive that you are pursuing, they will immediately rush with their weapon back at yours and additionally and simultaneously rush and attack your exposed opening* (which you have exposed with this pursuit and extension [of your arm/rapier]).

➢ You can then deflect this attack with counter slices, setting off, slicing off, and by taking.

Item, if you bind them with your blade on theirs, take note and feel a*s soon as they lift their blade from.the bind in order to hit or to thrust at another side.*

➢ *In that moment in which they lift:* follow with a straight thrust directly at their body, and always turn the long edge toward their blade, both when thrusting in and when moving out.

Thus, to feel is nothing more than to test, to sense how and when they want to lift from your bind so that you know and can quickly and safely follow them (as was already mentioned in the Sword [Section]).

You were also previously taught to remain and pull in the Sword [Section]. Although you were also taught to twist previously, it is used differently here, like this: whenever you bind on your counterpart in the middle of their blade, you should not lift from there unless you have a special advantage. The reason is that *they would like to pursue and rush over you* (as taught previously). Therefore, remain hard on their blade with the bind, and turn the half edge or the front end inwards towards their body to set it on [their body]. *If they defend against this and push your blade out to the side*, quickly pull through downward and thrust at the other side with a retreat.

Alternatively, if they do not guide your [point] out to the side but instead push forcefully straight ahead at your body with a thrust as soon as they take note of your twist inward, then when you perceive this, remain similarly with your point on their body and use your hilt to turn the long edge down toward their blade and twist their point outward, and push forcefully farther away with your thrust over |II.72| your hand [palm upward] with another step outward.

I have written about this as a reminder to you so that you should diligently consider it, and so that when one of these is mentioned subsequently in the sequences, you can understand and take note of it more quickly and can therefore also grasp the sequence itself faster.

About steps

With regards to steps: it is unnecessary [here], because the associated steps will also actually be dealt with in each sequence in the subsequent fencing.

End of the first part in the Rapier [Section]

The second part of fencing with the rapier

Up to now, one technique after another has been presented and explained, such that anyone who has read, diligently attended to, contemplated, and seriously practiced the same can understand and learn it readily (assuming that they have previously had a master).

Therefore, I now intend to describe the practice in itself, and how you should direct the same against your counterpart in the work. I will take up the stances again (as previously), and because it is necessary and very useful, I will teach how one should obliquely slice off, set off, and alternate from one [stance] into another.

Thus, position yourself in the High Guard of the right Ox, and step so that your right foot is always forward towards them with your body presented upright. Provoke them to work with playful or threatening |II.72ᵛ| body language.

If they thrust toward your body during your provocation:

> ➢ *In that moment in which they guide their thrust in*, step with your right foot farther around toward their left, and simultaneously turn your long edge downward toward your left.
> ➢ In this downward turn, sink your body down following your weapon (well in front of you with bent knee), and slice or guide their blade downward and away from you with your long edge.
> ➢ After this slice off, you will arrive with your weapon and your body bent down into left Low Guard.
> ➢ From there (*and as soon as they thrust or cut back at your exposed right opening*), swiftly move your hilt and long edge back upward, and also move your body upright into the abovementioned High Guard.
> ➢ If you want, you can thrust toward their face as a response from this High Guard (with a broad step forward).
> ➢ If you do want to do this, powerfully turn the long edge (with a strong rotation) downward during this thrust and step forward, and bend your body down after it.
> ➢ After this, swiftly turn your weapon back next to your left for a strike, and cut slantwise from there toward their right and through their face (with arm extended), so that you arrive in the Low Guard on your right at the end of this cut.

In this guard, however, you stand before them with your face exposed. Therefore, pay diligent attention. *As soon as they thrust at you at this opening:*

> ➢ Step out with your left foot behind your right and toward their left.
> ➢ Lift your hilt and long edge upward from this Low Guard (with an extended arm) toward your left into the same High Guard of the Ox.
> ➢ From there (*if they thrust again at you*), swiftly turn the long edge back down and toward your right. During this turn, step well out from their thrust and toward their right side.
> ➢ Guide their blade downward with your extended long edge, from the left Ox back toward your right Low Guard.
> ➢ In that moment in which you |II.73| slice their thrust away from you (as mentioned), you should extend with your body lowered well over your knee (which is bent forward following the cut).

After this has occurred, you will be back in the right Low Guard, as in the beginning.

If they want to thrust at you again:

> ➢ Rise with the long edge up (in the way described previously) back into the left High Guard.

> From there, thrust quickly long in front of you toward their face, with a leap outward toward their right.

In this thrust, you should turn the long edge strongly downward and toward your right, and in this slice, you should turn your blade toward your right to strike.

> This allows you to subsequently make a swift cut slantwise through their left shoulder as a Defensive Strike.

> After you have completed that [cut], you will hold your weapon on the left in the Low Guard, from where you can counteract with the long edge back upward into the right High Guard and so on.

Thus, you have now been taught how you should counteract from the right High Guard and from the left Low Guard, upward and downward and obliquely opposite, using the indications of the slanted and hanging lines. Then, how you should use a Defensive Strike through their right to change around from the left Low Guard to the other side into the right Low Guard. And accordingly, to counteract obliquely upward and downward just as from the other [guard/side]; and, according to your opportunity, to change around again using a Wrath Cut or Defensive Strike. In this way, you can always counteract, slice off, and move upward, from one stance into another, cross-wise and back again, with the long edge upward and downward.

Furthermore, you can also set off into Longpoint from all four stances on the sides by stepping out, as follows: |II.73ᵛ| position yourself in the High Guard of the right Ox (in the manner taught above) and step toward them. *If they thrust toward your face:*

> Turn the long edge out of the High Guard toward their blade with your arm extended up into Longpoint.

> During this, and while you set off, step simultaneously with your left foot behind your right (out to the side and away from their blade) as you thrust forward on top of their blade and at their face or chest.

> As you have now set off downward from the High Guard, you should also set off upward from the right Low Guard, also up into Longpoint, and you should fence this from both sides.

Now follows how you should additionally alternate from one stance into the other, like this:

Alternation [Abwechseln]²⁵³

When you come before your opponent, position yourself in the Iron Gate or in the Simple Brace and then move upward with your weapon from the [guard] into right Ox.

> *If they do not thrust yet:* move back down again through the strike line, obliquely down into left Low Guard.

> *If they do not want to work yet:* move upward again from the [Low Guard] into the left High Guard.

---

²⁵³ '*Wechseln* = change', '*durchwechseln* = change through', and '*umwechseln* = change over'. '*Abwechseln*', however, means to alternate, usually from side to side, which is reflected in the movement between stances that are diagonally or vertically opposite.

➢ From the same High Guard, move obliquely downward again into the right Low Guard. Maintain the point in front of you at all times in this alternation.

If they were to thrust between [stances] whenever they want: slice it away from you with the long edge (according to the abovementioned form), through the crossed lines from one stance into another (as you have been taught previously).

Thus, you can step around in front of them with a proffered opening and alternate from one stance into the other until you perceive your opportunity.

Regarding what is additionally necessary to know about the use of the stances, you have heard about that in the Dusack [Section].[254]

|II.74| How you should fence from the Simple Brace and
how you should protect yourself

I will now explain the common [counterposture] or Simple Brace. First, I will teach you how you should counteract and thrust in response or cut in response toward all four points. Subsequently and secondly: how you should guard yourself from the misleading. And for the third: how you should comport yourself against them (*if they don't want to cut or thrust*) and how you can fence against them in the Before. And [I will explain] as clearly as possible as an introduction into the other sequences.

How you should catch their cut which they guide at you from their right
and how you should thrust quickly and straight toward their left

Position yourself in the Simple Brace in the Onset (as the center and single portrait teaches you in the Figure which is designated with F). Step toward them with an extended and stiff counterposture.

➢ *If they cut or thrust slantwise from their right toward your left:* turn the long edge (including the hilt) upward toward their in-flying cut or thrust.

➢ During this (or while you are counteracting), step with your rear (left) foot behind your right and out to the side toward your counterpart's left, away from their thrust or cut.

➢ In the step outward, catch their blade on the Strong of your [blade] close by your hilt (as taught).

---

[254] Chapter 7 in the Dusack Section covers the stances and begins on II.18ᵛ.

- ➤ In that moment and while the blades ſtill touch together in the bind, ſtep farther toward them with |II.75| your right foot (toward their left side).
- ➤ Simultaneously with this ſtep forward, thruſt on their blade (or ſtraight in front of you and away from [yourself]) toward their face and up into Longpoint.
- ➤ As soon as the thruſt hits or is completed, turn the long edge back toward their blade and jerk back into the previous counterpoſture, with which you shield yourself until you perceive your opportunity for another sequence.

The way in which you should catch the thruſt and cut from their left, and in which you should quickly thruſt in response toward their right before they recover back

If they cut or thruſt at you from the other side (namely from their left) toward your right, again slantwise from above:
- ➤ Turn your long edge (including your hilt), with your arm extended toward their in-flying blade to counteract it or to catch it.
- ➤ In that moment in which you extend your hilt toward their weapon to counteract, ſtep during this same action with your left foot out to the side toward their right and away from their blade.
- ➤ Subsequently, *as soon as their blade glances on yours in this counteraction,* pull your hilt out behind you back and above your right shoulder (into the recovery for a powerful thruſt).
- ➤ Thruſt ſtraight toward their face with a ſtep forward of your right foot (toward their right) so that you again stand with weapon extended in the high Longpoint at the end of the thruſt.
- ➤ After this thruſt, move assiduously from the abovementioned counterpoſture to turn away their cuts and thruſts directed at you, until an opening is opened for you.

|II.75ᵛ| **How you should hold yourself againſt an opponent who wants to confuse you with loud rebukes and ſtrikes**[255]

If an opponent were to cut and thruſt at you from both sides so quickly, that you were unable to introduce any thruſt in the midſt of these:
- ➤ Remain in the extended counterpoſture[256] with a ſtraight arm ſtrongly in front of you, and use this to turn all of their cuts and thruſts at both sides away from you.
- ➤ Take diligent note while doing this.
- ➤ *When they have soon exhauſted and tired themselves (as they cannot continue this advantageously for long):* turn your hilt upward toward one of their in-flying cuts or thruſts (whichever one appeals to you as opportune).
- ➤ Cut slantwise quickly and completely through the shoulder on the side *from which they have guided their cut or thruſt,* such that your weapon shoots back around to the other side for a High Thruſt.

---

[255] *Überboldern/überpoltern* means "to rebuke loudly/rigorously in order to confuse, to force someone to do something by rebuking them", GRIMM, vol. 23, col. 447. The point here is to be both aggressive with one's voice, body language, and weapon, but there's no simple way to express that in English.
[256] *Außgeſtreckter versatzung.* Longpoint?

Namely, if you cut through their right shoulder, let your blade (in the complete course of this cut) move back through next to your right and snap around into the right High Guard to thrust.

If you cut through their left shoulder in response to their cut, let your blade move back through next to your left side and snap around into the left High Guard to thrust.

As you have now allowed your blade to quickly snap around into the High Guards using this cut, you should also swiftly and quickly thrust powerfully toward their face or their chest from the same [guard], and you should lower your upper body well forward over your bent knee following the thrust (in the extension), because you will make also make space at another opening if this cut and thrust follow one another quickly and strongly.

|II.76| How you should oppose and thrust in response from the Simple Brace at an opponent who cuts at you from below

Furthermore, *if they thrust or cut at you from below or across, from whichever side they thrust or cut (whether from the right or left):*

➤ Step with your rearmost foot (that is, with your left) toward the other side and out of their incoming thrust or cut, rebuffing it with the (extended) long edge downward and away from you and out to the side.

➤ As soon as your weapon touches theirs in this counteraction, step with your right foot toward them, and thrust quickly, away from their blade, straight toward their face *before they recover upright.*

All of this—namely, the counteraction, step, and thrust—should move quickly together in the blink of an eye. This is understood [to be used] for those Low Cuts which they cut toward the center of your body or even higher.

How you should counteract from above and thrust in under their weapon

Item, *if they cut or thrust at you again from below, or beneath your weapon at your body:*

➤ Out of the counterposture taught above, drop down from above onto their blade with the long edge so that your blade hangs downward and out to the side in the counteraction (as this counteraction is illustrated in the smaller portrait on the left in the Figure with the C).

➤ Simultaneously, in that moment in which you counteract, step well toward their right side with two leaping steps, |II.76ᵛ| out and away from their blade.

➤ Thrust quickly under their right arm at their body (as the other portrait opposite this one shows [C]).

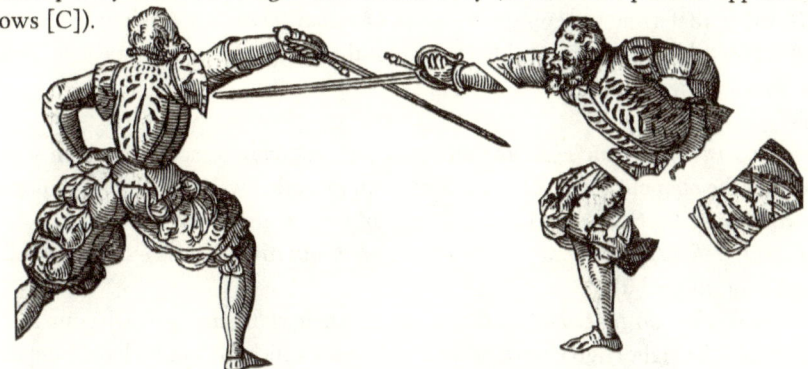

➤ From this thrust, pull your weapon quickly toward your left shoulder.
➤ From the [shoulder], cut a Defensive Strike obliquely through their right shoulder such that you arrive into the right Low Guard at the end of that same cut.
➤ From this guard, you are quickly back up in the Simple Brace to further protect yourself with it.

However, *if they guide their cut quite low toward your feet:*
➤ Don't counteract the same, but instead remove or yield backward out of their cut with the front foot up to the rearmost and *while they are still cutting,* thrust at them straight toward their face (as you can see in the central tenants in the Figure which is designated with F).

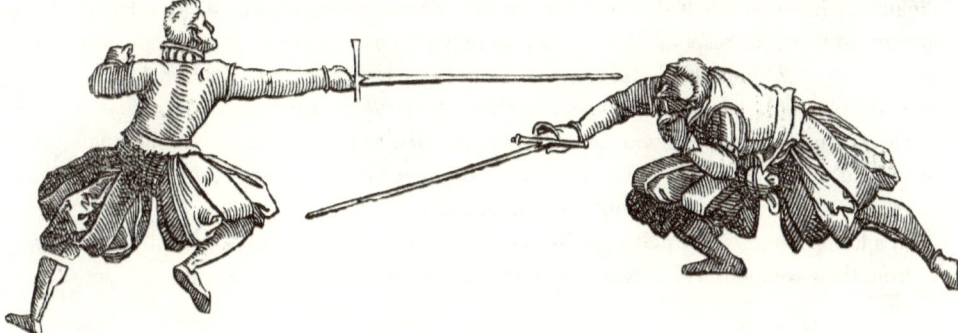

Because, as with all cuts which *they cut at you below your belt, they have to expose an opening above (in that moment in which they want to extend).* Therefore, according to the generally-stated rule, you can thrust or cut at their face quite safely (*in that moment in which they extend and reach out their hand with the weapon*).

How you should continue to hold yourself when they beat outward or move aside your thrust in response

You have learned previously how you should thrust quickly in response from the counterposture while stepping outward. *However, they can take this thrust out or move it aside.* Thus, take note of the second [part of] this teaching, namely:
➤ If you receive a thrust or cut on your counterposture and you then thrust a thrust in toward their face, *but they quickly beat yours outward,* then take diligent note *from which side they struck you[r thrust] out.*

- ➤ Step toward that same [side].
- ➤ |II.77| And in that same sweeping impetus (*which they gave your blade with their outward beat*), let your blade snap around to thrust and thrust equally at the same side *from which they struck outward.*

Namely,

- ➤ *If they thrust from their right toward your left:* you counteract them from your right, including the step out toward their left with an extended weapon (as taught previously), and you thrust directly toward their face out of the counterposture.
- ➤ *They must defend against that and beat your blade out toward their left* (as they otherwise do not want to get hit).
- ➤ *As soon as they do that,* step well out toward their right side and let your blade snap around into the right High Guard of the Ox (toward which they provided an impetus with their outward beat).
- ➤ Thrust quickly and powerfully in from outside of their right arm (from the abovementioned High Guard).

As you have now completed this toward their right, you can also direct it from your left in the work.

How you should change through and thrust at the other side of an opponent who wants to beat your first thrust outward

Take note: when you rebuff their cut and thrust it away from you with an extended blade and you want to thrust in response (as previously taught), but you perceive in the meantime that *they want to counteract,* take care that you do not step too closely toward them.

- ➤ Take diligent note, *in that moment in which they lift their hilt to counteract:* don't complete your thrust (which you should have indicated or threatened with body language).
- ➤ Instead, *while they are lifting,* move your weapon through under theirs and thrust powerfully in at the other side with an extended arm.
- ➤ In addition to |II.77ᵛ| this change through and thrust, you should have leapt out and away from their weapon, well toward the side toward which you have thrust.

How you should pull the thrust as if you were going to thrust in somewhere else, and then, in that moment in which they want to counteract, you thrust in again as you had indicated at the beginning

Again, when you observe or perceive that *they want to counteract the thrust* that you want to carry out from the first counterposture:

- ➤ Pull the thrust seriously toward yourself, as if you wanted to move through beneath and thrust at the other side.
- ➤ However, as soon as you observe that *they move their hilt away, intending to counteract,* thrust quickly back where you initially wanted to thrust (*while they are moving away*).

You will find more about this in its place in the sequences about misleading.

How you should thrust in response while they want to cut

Take note: when you are in the abovementioned counterposture in the Onset, and *your counterpart opposes you in the same Simple Brace:*

➤ Step vigorously and approach closer to them with a strong counterposture so that you can reach the middle of their blade with yours and can bind on (as the two top portraits in Figure C teach).

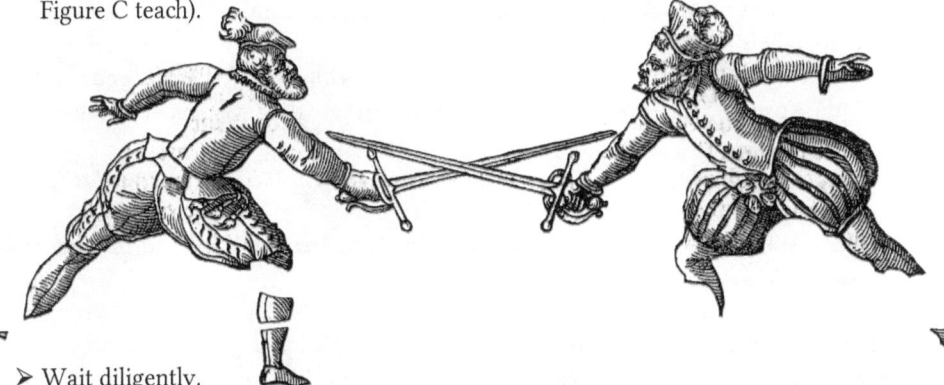

➤ Wait diligently.
➤ *When they want to lift their weapon from yours, either to cut or to thrust:* thrust |II.78| straight in front of you in at their face, chest, or stomach (*while they pull their weapon around*).

In order that they are provoked to rise and cut earlier, you can deceitfully offer an opening, yet retain your advantage. More about this later.

How you should catch your counterpart's cut and thrust, and cut in response

In the oft-mentioned Simple Brace, step toward your opponent and take note from which side they want to thrust or cut from (from below or from above).

If they guide their cut and thrust from their right toward your left from above:

➤ Turn your long edge (including the hilt) with weapon extended toward their in-flying blade.
➤ *In that moment in which their blade is still flying in,* step with your right foot out to the side and toward their right (during this same time) so that, *as soon as their blade drops down on or contacts yours,* you can quickly cut through their right side (with your weapon pulled back around) after that same step.

Whether [it is] high or low through their body depends on your opportunity.

➤ Your weapon then arrives in the right Low Guard at the end of this cut, from where [you] move back up to the counterposture.

Also execute the same *if they were to cut in from their left toward your right:*

➤ Counteract in the Strong of your blade with the weapon extended.
➤ Likewise, in that moment in which you extend your hilt for the counteraction, step with your right foot toward their left as soon as *their blade glances on yours.*
➤ With the abovementioned step, let [your blade] move back around your head and cut likewise toward their left side (*from which |II.79| they had brought their weapon*) and through their face.

➢ Then return quickly back from this Low Guard (in which you have arrived with this cut), back into the Simple Brace.

How you should conduct yourself using cuts in response to the Low Cuts

If an opponent cuts in at you from below (against whichever side that occurs):
➢ *In that moment in which they cut in,* you should step outward toward the other [side], away from their blade and toward them.
➢ In this outward leap, drop from above on their blade with your weapon extended downward (the closer to their Strong the better) so that you stop their weapon even below your belt.
➢ Afterwards, cut quickly across through toward their neck or face (*before they bring their blade any farther upward*).

However, if you cannot stop their cut but instead, *they break upward and through your counterposture with their Low Cut:*
➢ Remain with your weapon likewise extended downward and to the side and strongly in front of you in the counteraction so that they cannot injure you.
➢ *In that moment in which they are [still] rising with their weapon,* cut across from below [and] completely through their face or chest (*while their arm is still raised up high*).

This cut must quickly follow their break upward and through [your counterposture] so that you hit them from below *before they have completed their Low Cut.* And this cut [must be] so strong that at the end of it, you have swung your weapon toward the other side at the shoulder so strongly that your blade hangs down behind you after this swing.
➢ From there, quickly cut two slanted Wrath Cuts through the upper cross.

|II.79ᵛ| ### Another, how you should pursue with a thrust from below

You can also follow them with a thrust. Thus, if you want to counteract their Low Cut (as you have been taught) but *they break through with force,* take note.
➢ *In that moment in which they have broken through using their weapon against yours, and are still rising,* turn your point out of the counteraction toward their body and thrust at their chest from below up into a middle Longpoint (*before they complete their Low Cut*).
➢ Once there, observe *where they guide their blade from:* turn the long edge out of Longpoint toward their blade to receive the same with a counteraction.

Then, after you have been able to gain advantage, fence further with a sequence taught previously.

How you should attack from the counterposture using two Middle Cuts and a thrust in response

Or, if you have stepped toward your counterfencer in the Simple Brace, remain strong in the counterposture with an extended arm, always turning the long edge toward their in-coming cuts or thrusts. *If they cut from above, across, or from below,* move these outward and aside and away from you with your hilt, and take diligent note about your opportunity. *Whenever they slip up the smallest bit:*

- ➤ Pull around your head and cut the first Middle Cut across from your right through their face, so that you arrive in Middle Guard on your left at the end of this cut (which you [can] observe previously in the Dusack [Section] on the 5th and 43rd pages, |II.80| in the Figure designated with the letter C).

- ➤ Cut the second [Middle Cut] from the same [guard], also powerfully and across toward their right and through their face.
- ➤ [Cut] through so strongly that your blade sweeps out next to your right into the right High Guard of the Ox.
- ➤ During this, and while your blade moves around over your head, step with your left foot well to the side and out toward their right.
- ➤ Immediately follow with your right foot ([moving] closer toward them).
- ➤ During this step forward of your right foot, thrust powerfully from above toward their face.
- ➤ In this forward thrust, bend your front knee well forward while you set your foot down and bend your upper body well forwards and down (following the thrust).
- ➤ Now, when you turn the long edge downward with a strong rotation in this thrust, you will arrive at the end of the thrust in the Low Guard on the left side.
- ➤ From there, move quickly back into the Simple Brace.

How you should suppress and fence in response when counteracting

Take note with regards to your counterfencer: when you perceive that they want to drive you back powerfully with vehement cuts, counteract them (as taught previously)—with one cut, two, three, or four—into an extended counterposture until you have the opportunity to cut (as subsequently described).

As soon as you have observed [your opportunity] *and they want to pull their weapon back toward themselves or to themselves from a completed cut:*

- ➤ *In that moment in which they draw up their weapon for another cut:* simultaneously raise your weapon upwards and during this, quickly leap out and away from their cut toward their other side and somewhat toward them.
- ➤ *While they are cutting in,* cut straight down |II.80ᵛ| from above, between their head and their blade, as if you wanted to cut off (at the shoulder) the arm in which they hold their weapon.
- ➤ Let your hilt slightly precede the blade in the cut while dropping toward the ground.
- ➤ Also, let your feet step far apart from one another, so that you can better bend your upper body forward and downward over your forward bent knee following the cut.

If you leap out away from their in-flying cut in this way and, *in that moment in which they cut in,* you cut straight down from above at their right shoulder (between their head and weapon) so that you arrive with your upper body bent down forwards following this cut—not just because of the wide step, but also because your hilt preceded the blade when guiding the cut down toward the ground with an extended arm (as mentioned previously)—you will hit them close to the blade near their hilt in the Strong (if you otherwise miss their right arm) and weaken them to such an extent that you can thrust or cut at their body before they recover.

If you can't sufficiently weaken them, the cut [still] suppresses one or two [cuts] from both sides when you execute it correctly, so they can't complete more than three cuts without being weakened.

I have described this sequence in detail with some repetition, not merely because it is a very good sequence but also because it is difficult in and of itself: it is scarcely comprehensible without a living body to demonstrate, and it must be directed in the work with particular dexterity.

|II.81| How you should guard yourself from misleading, and also how you should act in response to the misleading cut and thrust

Since you have now heard directly and comprehensively how you should catch, move aside, and quickly fence in response to your counterpart's cut, as a second point, it is now necessary to know how you should act against *those who pull away their cuts* so that you are not misled in this counteraction.

If your counterfencer is somewhat experienced and practiced, *they will not always let the cuts hit (if they take note of your rigid counteraction), but instead will pull them away, change around, and cause you to completely miss.* Therefore, so that you cannot be deceived by these but instead can defend [against them], I want to explain this [defense] through the following rule.

Thus, take note of this as the first: namely, that in all counteractions, you don't move more than a hand's span out from the point at which you hold your counterposture. This applies to the sides (and includes upward and downward). Instead, you always remain with your hilt within a circle [with a radius] extending a good handspan (or the length of a work shoe) around the point of your counterposture.

However, because your body is not always sufficiently covered or defended by this counterposture, you should remove your upper body by stepping out from their cuts, or by ducking your head, or withdraw the opening from their blade, so that you always guide your hilt in front of you with an extended arm as a shield, and *wherever they cut from or thrust from,* you turn the long edge to counter the same and withdraw your head and face from their blade (behind yours). However, *if they were to thrust extremely low,* you should not merely drop down with the weapon |II.81ᵛ| but also lower your entire body with the extended weapon (unless you have observed an advantage in fencing a sequence), or *if they thrust quite low at you,* pull your lower body far back and evade and thrust straight in front of you at their face during this.

The second rule is part of the first, which is: namely, in all counteractions, guide the point straight ahead or toward their face, and wherever *they cut from,* turn the long edge likewise (as taught previously) to counter their cut, and while doing this, take note very diligently *whether they want to pull their incoming cut or thrust back.* Therefore, you should not swipe with your hilt in response, but instead, merely turn the long edge [to counter]. However, follow [this] with steps, or step quickly toward the side from which they have pulled their cut back, and while *they are still pulling back or are moving around,* thrust directly in front toward their face.

Thus, you don't counteract at all (namely, in [the way] that you would move your hilt to counter). Instead, as soon as you perceive that *they will pull back*, you step to the other side and merely turn the long edge [to counter], and simultaneously thrust straight in front of yourself; thus, you have counteracted. However, *because* your *opponent wants to mislead or deceive by pulling around or changing through, they expose themselves or shorten their own reach (unless they are very quick to complete the misleading with the correct body language)*. However, against unexperienced and unpracticed fencers, this is a certain hit. More about that later.

You have previously heard: firstly, how you should move each cut and thrust aside and fence in response; secondly, how you should comport yourself against their misleading and thrust in response or shove in (*while they change around or through*). Therefore, I will now take up |II.82| and teach [you] how you should position yourself in the Before and how to fence against those *who neither cut nor thrust, but instead merely counteract and wait to fence in the After*. However, it is necessary to place a small sequence here, which I will teach first and deal with briefly, like this:

How you should let your counterpart miss with their impetuous cuts and fence in response

When you have approached your counterpart in the Simple Brace, remain in Simple Brace with your weapon extended stiffly in front of you, and observe closely: *if your counterfencer crowds at you impetuously with cuts and thrusts from both sides,* then turn the same cut, thrust, or whatever away from you with extended and good counteraction, for as long and as many until you spot your opportunity during this. This opportunity arises *when they fence most seriously and surely with their cuts.*

Therefore, take diligent note. *When they guide a cut* that is appropriate for this:

➤ Yield with your front right foot back to the rearmost [foot], and *precisely in that moment in which their cut should now hit*, withdraw your counteraction upward towards yourself and to the side toward which *they guided their cut, so that they don't hit your counteraction with their cut but instead fall quickly (through) right in front of the same [your weapon].*

➤ Quickly cut toward their face, chest, or hand with another step forward of your right foot (*before their cut has fully arrived at the ground*).

This is a very good sequence with which you should familiarize yourself through practice and which you should carefully study with diligent attention.

|II.82ᵛ| How and in what way you should attack your counterfencer and fence against an opponent who doesn't want to cut or thrust (in the Before)

In this attack (in the Before), you must pay attention to the four openings and take diligent note on which part they guide their weapon in the Onset, as I will guide you with examples in the easiest and clearest way from one opening to the other.

Thus, in the first Onset, observe. *If they hold their weapon high in front on their right side:*

➤ Thrust below the [weapon] and inwardly at their stomach, executing that with a far-extended and reaching arm.

➤ *They must turn this thrust aside, counteract, or defend (if they don't want to be hit)*. Therefore, take diligent note.

> *In that moment in which they drop their weapon (to counteract the thrust):* pull the thrust back and move your blade through under theirs while they drop. In addition to this, move through and step with a leap (well toward their right and out from their weapon).
> *Before they reset their weapon,* thrust quickly at their face or chest from outside and over their right arm (as the smaller two tenants on the left hand in Figure F show you).

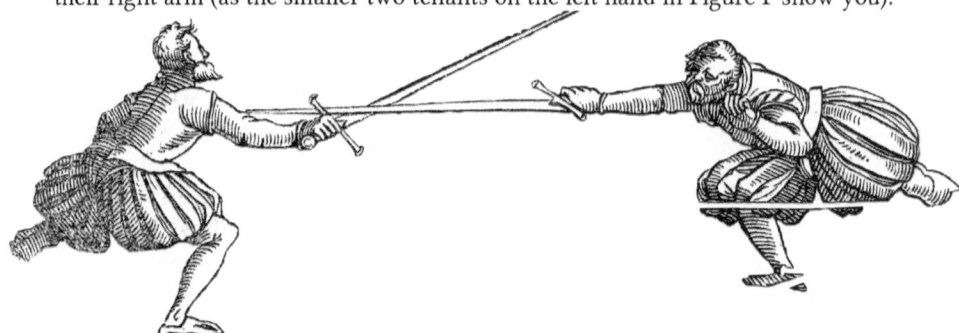

> As soon as you have hit with this thrust or have completed it, let your blade hang down toward your left from the same point, pull your weapon toward your left shoulder with the blade hanging, and cut a slantwise Defensive Strike from the same [location] downward through their face.

This pull, move through, and thrust (including the steps and the last Defensive Strike) should be executed quickly one after the other. This will benefit you.

|II.83| **Another**

> Or, cut through underneath their weapon toward their body (with serious body language), and in this cut, step well toward their right side (with your body bent following your cut) so that it appears as if you have overcommitted your cut.
> *Without a doubt, they will seriously rush to the opening;* therefore, take note.
> *In that moment in which they guide their weapon:* take a step outward and thrust again over their right arm at their face (as before).

After you have purposefully overcommitted your cut:

> You can also cut their in-flying blade quickly away from you with a Defensive Strike from the left side (at which you have arrived with your overcommitted cut).
> Or [you can] otherwise beat it outward and let this outward beat move up into the air around your head and thrust or cut back toward their right side.

How you should oppose them, when they hold their weapon too low on the left side

As soon as you can reach them in the Onset, thrust unexpectedly and quickly from the outside and over their right arm toward their face. Observe:

> *As soon as they rise upward and want to counteract,* turn your hilt upward toward your left [and] recover as one cut.
> *In that moment in which they lift their weapon,* cut the second speedily through and underneath their weapon at their thigh. Thus, you arrive in the right Low Guard.
> From there, cut slantwise obliquely through their face until you are back in the Low Guard on the left, and continue to fence.

|II.83ᵛ| **Another, how you should fence if they hold their weapon too far out to the sides**

If they hold their weapon too far toward their left:
➤ Cut from outside and over their right arm at their head, stepping out toward their right side.
➤ *As soon as they rise to counteract* (and in that moment in which it should just hit), pull your weapon back away from their arm and move through under theirs with the [weapon], and thrust inside of their weapon at their stomach.
➤ When you have completed this thrust, move quickly with your weapon extended straight upward into the High Guard, extended in front of you to strike so that (*if they would thrust at you from below again*) you could suppress this from above.

How you should comport yourself against an opponent who holds their weapon too far toward their right

If they hold their weapon too far to their right (as stated), take note:
➤ As soon as you can reach them, thrust swiftly straight in front toward their chest (*before they perceive it*).
➤ *If they defend against it or counteract you,* turn the half edge downward and toward their weapon, letting your blade drop through under theirs toward your left.
➤ Afterwards, cut quickly upward, slantwise from your left toward their right and through their face so that in this cut, your blade runs through behind you and next to your right side and shoots around into the High Guard of right Ox.
➤ Simultaneously with this snap around, |II.84| step well out (even farther toward their right side).
➤ Thrust from above with a strong rotation toward their face so that you drop with your weapon into left Low Guard in this thrust.
From there, fence the sequences that you find subsequently described in this [Low] Guard.

Another for an opponent who holds their weapon on their right side

In the Onset, guide a powerful cut toward their left.
➤ On the way, while you are still cutting, turn the short edge inward toward them so that it appears as if you thrust from above. Take note:
➤ *In that moment in which they lift to counteract and should just meet your thrust,* let the blade hang completely downward toward the ground and pull the hilt farther upward (with your arm extended away from you).
➤ Guide their blade out to the side by hanging in front (that is, with your hanging weapon)— *otherwise, they would thrust in under yours.*
➤ Pull your hilt (with the hanging blade) toward your left shoulder in this take out, turning your right side well toward your left (following your weapon).
➤ With two steps out toward their left, cut from the [left shoulder] over their weapon toward their right (slantwise through their face) so that at the end of the cut you arrive on your right in the Low Guard.
➤ From there, quickly thrust toward their face, straight in front of you into Longpoint.
➤ If they fence at you again, slice their blade away from you.

|II.84ᵛ| **The sequences you can use to attack them if they stand in Simple Brace**

If, however, they hold their weapon strongly and rigidly in front in Simple Brace so that you can't break in from the left or the right, use these practical sequences with which you can entice them from the stance or from their advantage (so that you can arrive at an opening while they are moving away). Since these practical applications of the art are so numerous that it would be impossible to assign them to a specific configuration, I will therefore only present a few examples from which you can sufficiently gather enough and learn how you should safely position yourself.

The first example

➢ Lift your weapon quickly upwards and present yourself with body language as if you wanted to cut hostilely at their foot, cutting part way toward it.
➢ In this cut, bend your body well forward (following the step and cut) so that it appears as if you have left yourself completely exposed.

Observe that you do not expose yourself in this, but instead hold your weapon quite strongly and take diligent note (in that moment in which you cut in) *whether they want to rush to the opening.*

➢ Pull your indicated cut back upward and toward their in-flying blade, beating the same upward and out to the side with your upward pull.
➢ *Before they recover*, thrust at the closest opening.

As soon as they move their arm outward or extend it to thrust, they hold their blade more weakly. Therefore, you can easily beat it outward, take it away, and thrust at their opening *before they can correctly recover.*

|II.85| **Another**

Position yourself as if you wanted to cut at their feet again (including the serious body language).

➢ *In that moment in which they rush to one of your exposed openings*, leap quickly out to the side and away from their thrust.
➢ *While they are still extending their weapon*, thrust across (or similarly) toward their face, which you can certainly reach and hit *if they thrust before you.*

How you should powerfully take away their blade and cut in response

➢ Thrust straight toward their face over their blade.
➢ During the thrust inward, turn your hilt upward toward your left and additionally, also turn your right side (following the hilt).
➢ Beat strongly through their blade toward their right, with the outer flat facing your left side, so that you take it out powerfully upwards from below.
➢ Let your blade move up around your head toward your right and, with another step forward, cut the second also from your left (but slantwise down from above and through their right side).

This second cut must follow quickly so it hits *before they recover from your outward beat.*

➤ The third cut is then from your right through their left into the left Low Guard.
➤ From there, with a leap out and away from their blade, thrust quickly and directly toward their face, in front of you up into Longpoint.
➤ Then fence from the same as was taught in the Longpoint [section].

Another

Take note. *If an opponent opposes you in this stiff and Simple Brace* (which was just taught):
➤ Step toward them in the same counterposture, and, when you can reach the middle of their blade, raise your weapon swiftly upward into the right High Guard with an arm extended to strike.
➤ From there, with an additional step forward of your right foot, cut unexpectedly from above and slantwise through the middle of their blade, and so strongly that through this cut you arrive with your weapon in the Middle Guard on the left.
➤ From there, cut again quickly across through their face (toward their right side) so that your weapon shoots up and around, due to its strong sweeping motion, next to your right side into the same High Guard to thrust.
This applies, regardless of whether you complete it with the long edge, short edge, or flat.
➤ From there, thrust powerfully again with a strong turn [of the blade] toward their chest, so that the long edge arrives facing downward during the path of this thrust, dropping back into the left Low Guard.
➤ From there, rise quickly into the High Guard of the right Ox, from which guard you should then fence the sequences that are subsequently described.

Another

In the Onset, guide your weapon high in the Simple Brace again, and take note.
➤ *In that moment in which they hold their weapon extended in front of themselves:* beat their blade quickly and unexpectedly out to the side with a hanging blade (as the scene |II.86| on the left side in Figure C shows you).

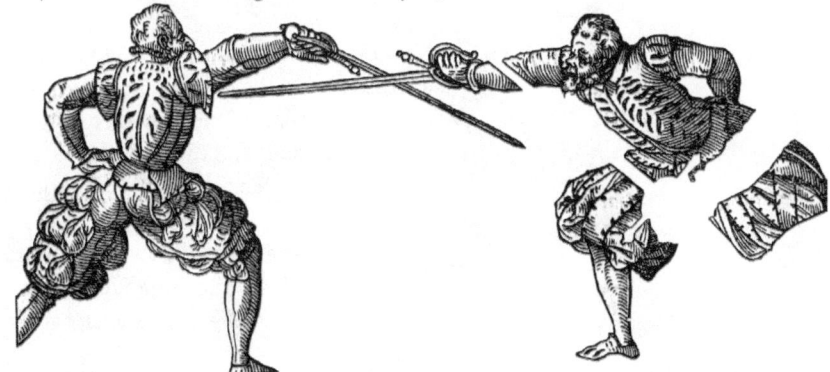

➤ Simultaneously with the outward beat, leap well out toward their right side and let your blade snap back from their weapon and around toward your left into the guard of left Ox (however, not as high next to you as usual).
➤ Thrust straight in front of you toward their face (palm upward).

This thrust, including the leap outward, must be completed quickly following the outward beat (*before they recover back with their weapon*).

Another

In the Onset, as soon as you can reach your counterpart:
  - ➤ Cut their weapon unexpectedly away from you in a jerk out to the side.

In this outward beat, observe that you do not follow after their weapon, but instead retain your weapon in good control.
  - ➤ *Before they recover from the outward beat*, thrust hastily straight in front of you at their face.

When you beat them unexpectedly outward and thrust after that, *they will want to abruptly rise upward to counteract*.
  - ➤ Therefore, *in that moment in which they rise upward*, pay attention so that you cut quickly at their forward foot.

Another

Or, *if they do not want to execute a sequence or allow themselves to be brought out of their advantage*:
  - ➤ Cut through cross-wise from both sides below their blade close to their hilt. Take note simultaneously with this action: *if they move to thrust from that same counterposture*, |II.86ᵛ| cut their moving blade away from you and rush quickly to the opening.

Because you cut a few cuts through their counterposture from both sides, you made them angry *so that they are more likely to fence at your opening* (which you must expose to them with these cuts through).

As soon as they extend their weapon away from themselves, they are already weaker in the counterposture; therefore, you can easily beat their blade outward and fence afterwards.

Or lift your blade into the High Guard and offer them the exposed front body.

High Guard

Because you always move out of one stance into another during the entirety of fencing, I will undertake one stance after the other (as mentioned at the beginning) and using examples, teach you how you should fence from the same without risk.

The first example; how you should thrust in from the High Guard simultaneously with another

When you arrive in the Onset using the Simple Brace in front of your opponent and *they will not be enticed with any sequences nor provoked from their advantage*:
  - ➤ Lift your weapon upward into the right High Guard of the Ox and guide the point toward their face with an upwardly extended arm.

➤ *As soon as they subsequently thrust at you*, avoid them [by moving] your left foot behind your right and step quickly with your right foot farther toward them.

➤ During these steps, turn the long edge toward their in-coming thrust and thrust simultaneously in with them.

You thus counteract and contact one another (as the larger portrait on the left side shows in Figure F).

You can also execute this sequence from all other guards just as from this High Guard: namely, that you yield from their thrust, turn the long edge toward their blade, guide them out to the side, and thrust in with them.

|II.87ᵛ| Another, how you should slice their blade away from you and fence afterwards

When you have thus arrived upward into the High Guard to thrust *and your counterpart thrusts or cuts toward you from their right:*

➤ Step twice toward their left side and away from their thrust (like before).

➤ Quickly turn the long edge down from the High Guard.

➤ In this downward rotation, cut in one sweeping motion toward their blade (with your body extended) through into the left Low Guard.

➤ From there, cut or thrust toward their face (*before they yank their blade back from under yours*).

This is also a good rule for fencing from all stances: namely, that you cut their blade away from you (slantwise or obliquely downward from one side), and then afterwards swiftly [cut or thrust] the second from the other side at their body, whether from below or above their blade (*depending on whether they quickly lift back up*).

How you should take away with hanging in front and fence afterwards

When you hold your weapon in the High Guard to thrust and *your counterfencer thrusts at you:*

➤ Keep your hilt high in front of you (extended a little to the side) and let your blade hang downward with the point straight in front of you.

➤ *In that moment in which they thrust at you* (as stated previously), turn your hilt with the hanging blade toward your left. In this way, you guide their blade out to the side.

➤ When you have now beaten their blade outward (which includes a step outward), you can allow your blade to move around your head after the outward beat.

➢ With another ſtep of your right foot, cut through ſtrongly across from your right toward their left.

➢ Cut the second ſtraight from above with a Suppressing Cut drawn through their face.

You arrive, after cutting downward, with your hilt closer to the ground than the blade (as the large portrait on the [left][257] side shows in Figure **G**).

➢ From there, thruſt quickly toward their face *before they recover back from this Suppressing Cut.*

|II.88| Or:

➢ When you have beaten them outward with the hanging blade, let your blade move around your head (in that moment in which you have ſtepped out toward their left), and let the point shoot toward your opponent's face into the guard of the left Ox.

➢ From there, thruſt toward their face simultaneously with the ſtep outward mentioned previously.

➢ Then cut through cross-wise, slantwise from both sides toward their hilt.

Or:

➢ After you have beaten them outward, let your blade shoot around into the guard of the left Ox (as previously).

➢ Threaten to thruſt at them with your palm up; *they will rise to counteraĉt this thruſt.*

➢ *While they rise*, ſtep with your right foot farther around to their right and cut through toward the same thigh.

➢ Cut the second quickly and obliquely from above and through their face so that you arrive with your weapon in the Low Guard on the left.

➢ From this Low Guard, quickly slash back upward toward your right with the half edge.

➢ And let your blade move around your head and cut toward their right, slantwise toward their face.

A good sequence that can be completed with hanging in front

When you have arrived in the Onset in the High Guard of the right Ox, and perceive that *a fencer thruſts powerfully at you:*

➢ Step, with a leap out to the side and well toward their left, away from their in-coming thruſt, and turn your hanging blade out toward your left (as previously).

---

[257] The text actually says "right", but this is clearly wrong.

➢ As [you] take [their rapier] outward, ſtep with your left foot farther behind your right foot toward them, following swiftly with your right in response to their thruſt, and turning your back to them.

➢ Thruſt under your right arm with a Reversed Thruſt, behind you and toward their ſtomach.

➢ Subsequently, turn your face quickly back toward them and cut with extended weapon in a wheel²⁵⁸ ſtraight down from above through their face with a retreating ſtep of your right foot.

➢ Lift your hilt again and change your feet at the same time, so that the right ſtands in front again.

|II.88ᵛ| **Rule**

From this Ox, thruſt properly toward their arm so that *when they thruſt toward you from below or ſtraight*, you remove your body while simultaneously yielding back with your foot. And *while they extend their arm*, cut or thruſt toward that same arm, hilt, or hand in which they hold their weapon.

Item, when you ſtand before your counterpart in the previously-mentioned High Guard, *and they thruſt at you from below or ſtraight in at you:*

➢ Step with your left foot well out toward their right and during this, turn your blade with the point behind you to ſtrike.

➢ Cut their in-flying blade out from your left toward your right with a hanging blade (as the large portrait on the lower left side in Figure C shows).

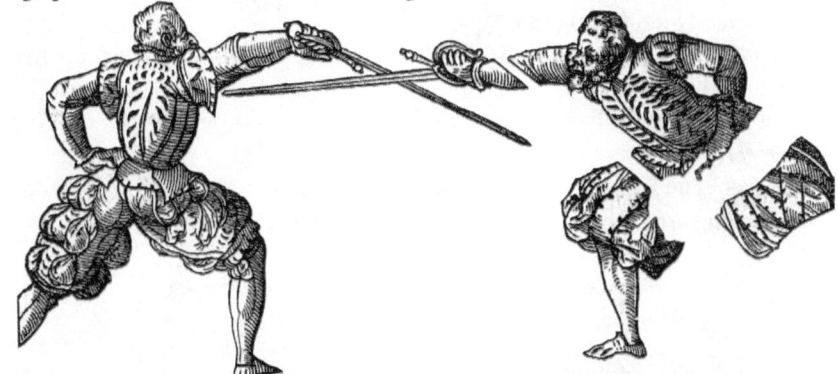

➢ Simultaneously with this outward beat, ſtep with your right [foot] farther toward their right.

➢ From this counter poſture, pull your weapon back up toward your left around your head,

➢ And *before they lift it [their weapon] back up or bring it back into their control,* cut above their weapon from your right and slantwise through their face or toward their arm (and the hand in which they hold their weapon).

➢ Thus, you arrive in the left Low Guard.

---

²⁵⁸ '*Rad* = wheel'. A circling movement of the rapier on one side of the body.

How you should thrust before them to miss, take out,
and thrust in response when they don't want to cut or thrust

Item, rise up into the High Guard in the Onset. When you stand before them in this [guard] and *they don't want to work:*

- ➤ Step with your left foot farther toward them and, in that moment in which you step forward, thrust through from above (outside of your left foot), passing in back of [them] as a miss so that at the end of the thrust, you arrive back in the guard of left Ox.
- ➤ From there, slash their in-flying blade outward toward your right (with the half edge facing down). In this take out, step farther with your right foot well around and toward their right.
- ➤ Thrust toward their face from above with a strong twist and with another step forward of your right foot, so that at the end of the thrust you arrive on your left in the Low Guard.

|II.89| Item, in the Onset, when you thrust through next to your forward left thigh and at their back to miss, you will completely expose your face. Therefore, in that moment in which you thrust through at their back, *they will, without a doubt, thrust in response.* Therefore, when you perceive their thrust in:

- ➤ Step out with your right foot farther toward them (toward their right side) and simultaneously thrust in with them from the left Ox (into which you arrived using the thrust through mentioned previously), outside of their right arm and at their face.
- ➤ And take diligent note: in that moment in which you thrust in, [ensure] that you turn the long edge swiftly toward their blade, and that you also remove your head well to your left side (away from their blade behind your own).
- ➤ You will certainly hit, as long as they have thrust.

From this guard, you can also suppress, move through, block,[259] and fence many similar sequences against them.

Execute a block like this. When you stand in the guard of the Ox on the right, *if an opponent cuts at your feet (regardless of which side they cut from):*

- ➤ Drop the front end from the other side toward their cut down to the ground with an extended, hanging arm; the blades result in a cross (as you can see in Figure B).

- ➤ Block their cut *so that they can't move through with their cut* and also step simultaneously to the side out of their cut.

---

[259] *Dämpfen/Dempfen, Durchgehen, Sperren.*

➤ As soon as it glances off, thrust on their weapon (above or below their blade), quickly toward their body.

➤ Pull the weapon back quickly and cut a Cross in front of you in response.

Of course, you [also] have the Wrath Cut [as an option] from this guard. When you use it to cut through their left [side], you will arrive in the Low Guard on the left, from which you can fence like this:

How you should take out from the Low Guard on the left and fence in response

When you have now arrived in this guard (whether through a slice or by movement through a cut), and *they subsequently thrust from above toward your face:*

➤ Take out their in-flying blade strongly in a slash upward toward their [left] and your right with the half edge (with an extended arm). In this take out, let your blade snap completely around over your head into the guard of right Ox.

➤ From there, thrust quickly toward their face, with another step forward of your right foot.

➤ In this thrust forward, turn your long |II.89ᵛ| edge downward so that you twist their thrust outward with yours (*if they were also to thrust in that moment*).

Or, when you have taken out their thrust:

➤ Let your blade move around above your head after this taking out.

➤ During this, step with your right foot well out toward their right and cut through toward their right thigh from outside, so that at the end of the cut you arrive with your weapon in the right Low Guard.

➤ From there, strongly cut a Wrath Cut obliquely and slantwise through toward their left with extended blade, so that you arrive back in the left Low Guard.

Or, when you have taken them out:

➤ Let [your weapon] snap around above your head again into the guard of the right Ox and threaten to thrust at them from above.

➤ *They will want to counteract this.*

➤ *In that moment in which they rise to move this thrust aside or to counteract it:* don't let the thrust hit, but instead pull your blade strongly back toward your left, around your head, and cut across through their forward thigh (from your right with body well lowered and your weapon extended toward their left).

➤ Cut the second immediately from your left through their right and through the uppermost hanging line as a Defensive Strike into the right Low Guard.

Thus, you have learned to fence in response to this take out in three different ways: namely, [firstly] with a high thrust, secondly with a cut, thirdly with a misleading cut.

How you should take out from the Low Guard and how you should fence from below

In the Onset, position yourself into the left Low Guard. *As soon as they want to thrust or cut at you (namely, toward your right):*

➢ Step immediately out to the side toward their right, and simultaneously with this step, turn your blade out behind you for the strike.

➢ Cut a strong Suppressing Cut from above across at their blade from your left.

➢ Then, *before they bring their blade inward underneath yours,* thrust upward toward their face (up into Longpoint).

➢ From there, *if they continue to thrust at you from below,* slice their thrust away from you downward into the left Low Guard.

|II.90| Item, when you have suppressed their cut down from above and thrust in response, take note.

➢ *In that moment in which they lift and want to counteract your thrust:* move your point around next to their right arm and thrust in under their arm *while they rise.*

Another sequence directed at their right thigh

Therefore, when you stand in the Low Guard on the left *and your counterfencer thrusts or cuts from above:*

➢ Turn your blade out behind you to strike and cut upward from below toward their blade with the long edge.

➢ As soon as the blades glance together, turn the short edge inward toward their blade with the palm up.

➢ While the blades are still touching, turn your pommel upward so that you recover to cut.

➢ Cut quickly at their forward thigh, outwardly away from their blade.

You have thus completed two cuts from one side: namely, the first toward their blade from below and the second at their right leg outwardly from above.

Another, how you should move through from the left Low Guard

When you hold your weapon in the Low Guard on the left side, take note.

➢ *In that moment in which they thrust in:* move your blade through under theirs from your left toward their right, with the point remaining close to the ground, so that you can quickly beat their in-flying blade outward from your right toward your left.

➢ While you beat their thrust outward, step well toward their left at the same time and thrust inwardly at their chest, which you will certainly gain *if they thrust forward in any way* and you execute it correctly.

|II.90ᵛ| ### Another, how you should cut through and thrust at the other side while ripping upward and moving downward

In the Onset, position yourself into the left Low Guard and observe diligently.

➢ *As soon as they thrust toward you:* take their blade out strongly and upwardly toward your right using your half edge and an extended arm, so that you arrive with your weapon into the High Guard to strike.

➢ From there, cut quickly back through below their blade from above (*while it is still up high as they pull the same back towards themselves*), toward their stomach, back into the left Low Guard.

➤ From there, thrust quickly toward their face above their right arm (with a leap out toward their right).

How you [beat] all their thrusts from the left Low Guard, whether they come from above or below

All thrusts which your counterpart guides from above or toward your face you should beat outward and upward from below according to the slanted rising line, and you can execute that with the half edge or the whole edge, and also with the outward flat. Afterwards, as soon as you have taken that out, you can thrust or cut at whichever side you want—it just has to occur and be completed quickly (*before they recover*).

However, if they thrust from below, you should beat the same downward and slantwise outward from above toward your right (and, as before, with the whole edge or half edge, and also with the flat).

You now take that out from above or from below. You should also simultaneously step well out toward their right and away from their weapon. This provides your taking out with more strength. Afterwards, cut or thrust quickly in response; as soon as you have thrust or cut, you should be back on their blade with a set off to continue protecting yourself from their attack.

In this way, you can quickly fence against them from this Low Guard, if you have cut down into it.

|II.91| **High Guard on the left**

When you now stand in Low Guard on the left, lift your hilt into the left High Guard (upward next to your head) and fence from there in the same way as follows.

Take note. When you stand before your counterpart in the left High Guard *and they thrust toward your face:*

➤ *In that moment in which they thrust in*, leap well out from their thrust toward their right side, thrusting simultaneously from outside and over their right arm at their face. During this thrust, turn the long edge toward their blade so that you simultaneously contact one another and counteract.

➤ Or set their thrust off with the long edge while you step out and then thrust quickly in response.

Two good thrusts to fence from both Oxen

When you hold your weapon into the left High Guard and *they thrust toward your face:*

➤ Thrust powerfully and simultaneously with them from this High Guard, outside of their weapon and toward their in-flying blade at their right arm.

➤ In that moment in which you thrust in, turn your hilt through underneath their blade and upward toward your right side, such that you deflect their blade with this turn through and twist it out toward your right side.

- ➢ Afterwards, *in that moment in which they draw their weapon back,* thrust quickly inside of their hilt at their chest. Simultaneously with this thrust, turn your hilt (including the long edge) back downward toward your left side so that you guide their blade out again in addition to contacting it.

Another

If your counterfencer doesn't want to thrust or cut at you, execute these thrusts at them:
- ➢ Firstly, thrust seriously from the left High Guard outside at their right arm.
- ➢ During the thrust, turn your hilt (including the blade) through under theirs and upward toward your right.
- ➢ *In that moment in which they move their hilt to counter your first thrust,* thrust quickly at their chest within the same *(which they have completely exposed with their counteraction).*

Furthermore, you can turn aside all cuts and thrusts from this guard by using the long edge into Longpoint.

|II.91ᵛ| From this guard, you can slice off, suppress, beat outward (upward from below), block, and any similar sequences (which you will find described more fully in other stances). You can bring all of that into this guard through practice.

How you should fence from the Low Guard on the right, which is also called Side Guard

From this [guard], fence like this. *When they thrust at you* (when you had arrived in the Side Guard):
- ➢ Step with your left foot well behind your right, toward them and toward their left side.
- ➢ Step farther toward them with the right, and, in these steps, thrust straight toward their face from the Low Guard.
- ➢ In that moment in which you thrust in, turn the long edge (including the hilt) toward their blade so that you deflect their blade with this, and simultaneously hit with the point (in that moment in which you counteract).
- ➢ As soon as you have completed the thrust, lift your blade upward toward your right with your arm extended and cut quickly back down from above at the center of their blade so that you beat their blade strongly outward.
- ➢ *Before they recover and rise up again,* thrust swiftly back straight toward their face.

How you should slice off from this guard and fence in response

If your counterfencer thrusts or cuts from their right at you:
- ➢ Lift your weapon upward toward your right *(in that moment in which they thrust).*
- ➢ During this, step with your right foot farther toward them and cut strongly through their in-flying blade in an oblique slant from your right (simultaneously with this step forward).

In this cut, you should hold the hilt somewhat in front of your blade while dropping down; in addition, also let your upper body sink down and forward with a wide step.

➢ In this cut, guide their blade |II.92ᵛ|[260] in a downward slice with an extended arm.
➢ Thrust quickly upward at their face (*before they bring their blade back from under yours*) so that you stand in Longpoint when the thrust hits.
➢ *If they are now ready to thrust underneath your blade toward your stomach,* keep your hand (including the hilt) extended in front of your face, but let the blade drop down toward their right and beat their thrust outward with a hanging blade (from your left toward your right) and do this with the inward flat of the blade.
➢ Simultaneously with this outward beat, step with your left foot well out toward their right.
➢ Afterwards, pull your hilt upward toward your left and during this, step with your right foot farther toward their right out to the side.
➢ Thrust or cut quickly from your left toward their face or arm.

How you should move through with your blade under theirs

Item, take diligent note. *In that moment in which they thrust from above:*
➢ Move through with your blade under theirs and deflect it with a slice from your left toward your right (as you can see in the lower center scene in Figure D).

➢ Thrust quickly toward their face.
Or, after you have moved through with your weapon and have sliced theirs off [to the side]:
➢ Let your blade move out next to your right and cut in front at their face (*before they lift up enough*) while stepping powerfully out toward their right.

How you should cut away all of your opponent's cuts and thrusts, cross-wise and from both sides from the Low Guard, and fence in response

When you now stand in the Low Guard on the right:

---

[260] This page is numbered 87, even though it is the verso page of 92. Interestingly, the front of 87 is a Figure, so this is not a case of the typesetter repeating a page out of order.

➤ Cut with Defensive Strikes from both sides against all of their cuts and thrusts—that is, strongly with slanted Wrath Cuts through the oblique line (cross-wise from both sides) |II.93| like a slice away from you.

Execute these high or low with upright or lowered body *(depending on whether they guide their cuts high or low)* until you either tire them out and weaken them, or until you can gain advantage to fence other sequences.

This always applies (regardless of which side they guide their first cut from) and should not confuse you, because when they guide their cut from their left toward your right, you should move through below with your blade and also cut from your left toward their right as a counter (as was taught previously with regards to moving through). If they cut from their right, cut through toward their blade from your right with a lifted weapon.

Another, how you should beat their blade outward and upward and weaken them using a Middle Cut, suppress with a High Cut, and thrust quickly in response

If an opponent opposes you in the Simple Brace when you have arrived in the right Low Guard:
➤ Move your blade quickly through under their blade, up to the middle of your blade, while keeping the point (that is, the tip) down close to the ground. In this move through, turn your long edge upward (so that the back of your hand faces you).
➤ Cut through upward toward their blade from below and pull your weapon around your head.
➤ Strongly cut the second (a Middle Cut) across through their face from your right so that your weapon flies around your head back toward your left side.
➤ Quickly cut the third down from above in one straight slash through their face.

For this cut, you should lower your upper body well down by stepping wide ([with your feet] apart from one another), and in the cut downward toward the ground with an extended arm, guide your hilt a good bit in front of the blade so that you can better weaken their blade and force it downward.

➤ As soon as you have completed this cut, immediately thrust straight in front of you; *regardless of whether they have brought their weapon back from under yours or not*, merely turn the long edge toward their blade in the thrust inward.

You have thus counteracted them quickly.

|II.93ᵛ| How you should beat them outward with a hanging blade and thrust in response

You can also use the end of the fifth previously-mentioned sequence[261] from this Low Guard against the Simple Brace in this way:
➤ Raise your weapon from the right Low Guard upward next to your right.
➤ Step out with your left foot during this toward their right side and beat their blade outward with your inward flat (from your left toward your right, between you and them) so that your blade hangs downward in this outward beat and your pommel is high.
➤ As soon as this outward beat is executed, leap farther around toward their right.
➤ *Before they have recovered from the outward beat,* thrust at their face.

---

[261] II.91ᵛ: "How you should slice off from this guard and fence in response".

How you should change around with cuts from one side to the other

Take note. When you have arrived in the right Low Guard *and your counterfencer cuts or thrusts from above at you:*

➤ Cut strongly slantwise toward their in-flying blade (upward from below with the long edge) and completely through toward your left shoulder—indeed, so strongly that your weapon runs toward your left [with the speed] diminishing and back around your head in this same fluid movement.

➤ In that moment in which your hilt is completely around your head, pull and cut the second slantwise from above (also from your right) and through their left shoulder.

With regard to each of these cuts, you should always move the feet in steps wide apart so that you can follow the cuts with the upper body and you can thus reach farther.

➤ Afterwards, when you have arrived at your left Low Guard with the second cut, *and they cut or thrust again at you:*

➤ Cut strongly through from this Low Guard exactly like from the previous—slanted upward from below, toward your [left] and their right, and toward their in-flying blade, such that your blade loses speed above and next to your right.

➤ Pull |II.94| the same completely around your head from your right and also cut the second slantwise through their face from above (from your left and toward their right) with your body bent well after it.

➤ You arrive with your weapon back on the right in the Low Guard.

How you should change through when thrusting in

When an opponent in Iron Gate encounters you:

➤ Let yourself appear as if you wanted to thrust foolishly[262] in front toward their face (by using serious body language and steps).

➤ While you are acting with body language to thrust and you thrust partway in, diligently observe *whether they are deeply concerned about your thrust and whether they want to oppose the same with a counterposture.*

➤ As soon as you perceive this [opposing counterposture] during a thrust—*in that moment in which they rise to counter the same*—guide your thrust through under their blade and toward their right arm.

➤ *As soon as they observe your thrust through, they will quickly turn their hilt and want to counteract this thrust as well.*

For that reason, and although you act with very serious body language, you should not move inward with your thrust until you observe how they counteract.

➤ Let your front end drop a little downward next to their hilt so that you can guide the same under and through.

➤ Thrust quickly at their stomach from inside.

➤ Afterwards, lift your weapon quickly upward toward your right and cut a Cross Strike[263] swiftly through their face.

---

[262] *Alben* is not a word. *Albern* would be a simple typo, and makes sense.
[263] In the Dusack Section, Cross Cuts are defined as two Wrath Cuts.

This should all be completed quickly and fast. In this change through, you should also pay attention that *they do not thrust directly in front at your face without opposition* (while you thrust through), because that is the counter to this change through.

➤ Therefore, if you perceive this in your change through, you should quickly lift upward with your hilt and with your arm extended in front of your face. By this means, you oppose their thrust.

➤ Thrust quickly in response because *it is impossible for them to recover from their sudden rush as quickly*.

|II.94ᵛ| **How you should rebuff their blade with a counter cut and thrust in response**

In the Onset, take note. *When your counterfencer draws their weapon to cut or thrust:*
➤ Lift your weapon upward at the same time and obliquely toward your right.
➤ *In that moment in which they guide their weapon in, whether to cut or thrust:* cut through strongly and slantwise toward their blade so that you beat their blade powerfully outward to the side with this.
➤ Immediately (and *before they bring their weapon back up)*, thrust straight at their face in front of you.

How you should beat their cut outward and cut through their foot

Execute this [technique] like this. *As soon as they cut or thrust at you:*
➤ Cut their blade strongly outward from you (as taught previously) so that you bring their weapon out to your left with this cut.
➤ When you have now beaten their blade strongly outward, *they will rip their weapon back upward out of fear of a thrust.*
➤ Therefore, *while they rise,* cut across from your left through their right leg with your body bent far forward.
➤ Don't let your weapon follow the cut too far so that you can immediately make a powerful and swift Defensive Strike through their face (from your right toward their left).

Another sequence, how you should draw your thrust back and fence with hanging in front from the Low Guard

In the Onset, as soon as you can reach them:
➤ Thrust unexpectedly from the Low Guard toward their left.
➤ Draw your weapon swiftly back from the thrust toward your right Low Guard.
➤ In that moment in which you pull back, |II.95| *they will thrust quickly in response*. Therefore, in that moment in which you draw back, turn your hilt high upward in front of your face (with an extended arm) and let your blade hang downward during this.
➤ Beat their in-flying thrust outward with a hanging blade (from your right toward your left). During this outward beat, leap well out toward their left side.
➤ Let your blade snap around next to your left side into the left High Guard after this outward beat.
➤ From there, thrust quickly and strongly toward their left at their face (during the leap mentioned previously).

Or, after you have beaten their in-flying blade outward with your hanging weapon and with your arm extended high toward your left:

> ➤ Pull your weapon completely around your head and cut across through their face (from your right toward their left) so that you arrive with your weapon on your left in the Middle Guard.
> ➤ Then, leap quickly out toward their right side.
> ➤ During this leap outward, cut through strongly and obliquely (from your left toward their right) and completely through their face, so that your blade snaps back around next to your right side and into the right High Guard of the Ox.
> ➤ In that moment in which all [of this] occurs, leap again farther toward their right and thrust powerfully toward their face.

Here follow some sequences for fencing from the Plow

Arrive in the Onset with the right foot forward. Hold your weapon with the horizontal cross-guard inward next to your right thigh, so that your arm is extended downward and the tip faces upward toward your opponent's face. Step toward them with your body bent well down and forward.

If they thrust at you, whether from above or straight:

> ➤ Keep your tip at the previous height toward their body but turn the hilt upward from below and toward your right, so that you beat their in-flying blade outward with the flat (which previously faced downward toward the ground and now faces up due to the upward rotation).
> ➤ In this twisting outward and upward, let your front end drop a bit next to their hilt and move through under their blade toward the other side with the same [point].
> ➤ Thrust quickly across toward their chest from inside of their hilt. In this thrust inward, turn your hilt quickly downward, so that you arrive back in the left Plow (from which you stepped toward them at the beginning).[264]

|II.95ᵛ| ### A swift sequence from the Plow; how you should change through on both sides and thrust inwardly at the chest

In the Onset, hold your hilt outside next to your right knee in the same way that you previously held the weapon inside of your [knee].[265] Step toward them with the tip pointed upward again.

If they thrust or cut at your face or chest:

> ➤ Turn your hilt upward and toward your left so that you divert their in-flying blade upward with your outward flat (as you have previously rebuffed it upward from the other side).

---

[264] There is an error in this sequence and the next one, as you start in right Plow and end in left Plow, which is not "where you started". Therefore, one either has to start in left Plow and end there, or end in right Plow where you started. It's also possible that Meyer forgot that he previously defined left Plow with the left foot forward rather than the right (as in left Ox), and he is now considering the starting position to be left Plow.

[265] Literally "foot".

➤ In that moment in which you have counteracted their thrust or cut upward with your flat, let your front end move through under their blade toward the other side.

➤ Thrust outside of their blade and toward their right arm. In this thrust inward, turn your hilt through (upward from below and toward your right).

➤ Thrust fast and quickly across and inside of their hilt toward their chest (as previously).

➤ Similarly to the previous [sequence], twist your hilt back downward and toward your left, so that you arrive back in the left Plow.

These two sequences should be directed and completed in the work with quick skillfulness of the body.

Another sequence from setting off

Position yourself in the Plow in the Onset (as taught earlier). Offer them your face defiantly with your body bent far forward, and Indesly take diligent note. *As soon as they thrust in:*

➤ Turn your hilt upward and toward your left side and powerfully set their in-coming thrust off toward your left with the outward flat, so that your front end remains pointing toward their left in this set off.

➤ Immediately and in that moment in which the blades touch together, thrust toward their chest (inside of their hilt) while turning your hilt back.

➤ As soon as the thrust hits, turn your hilt back downward toward your left *so that if they thrust back downward,* you will deflect that.

This sequence works on both sides.

|II.96ᵛ| Three thrusts, running one after another, with which you can practice quickness

Execute them like this. Position yourself in the Plow on the left and pay attention. *As soon as they guide their weapon into the Iron Gate or into the Simple Brace in front:*

➤ Thrust the first [thrust] straight upward from the left Plow and close to the outer side of their weapon, toward their face.

➤ *They will divert this thrust toward their right.*

➤ Therefore, during the thrust (and as soon as you perceive their diversion) let it drop down next to their right side and run through back toward your left, so that your blade snaps around next to your left into the High Guard on the same side.

➤ In that moment in which your blade snaps around, step well out toward their left with two steps to the side and thrust the second [thrust] out of the left Ox, inside of their weapon and toward their chest (with the steps outward).

➤ *They will want to counteract this thrust toward their left.*

➤ Therefore, in that moment in which you perceive this, turn the short edge inward toward their blade, and allow the [short edge] to run through under their blade toward your left (during this turn inward), and snap around again next to your left into Ox on the same side.

➤ During this run through and snap around of your blade, leap well out toward their right side.

➤ During this, thrust the third [thrust] from left Ox outside of their right arm at their face.

How you should reverse the High Thrust into a Low Thrust while thrusting in

Thus, *when they hold their weapon in front in the Simple Brace:*
- ➤ Thrust at them quickly and unexpectedly on the inside and upwardly toward their face and ensure that your blade remains close to theirs[266] during this thrust inward, so that you press their blade out to one side in this thrust.
- ➤ *They will raise their hilt up high;* therefore, as soon as you perceive this, jerk your hilt rapidly upward and let your front end drop downward around next to their hilt.
- ➤ And thrust at them from above and under their weapon at their stomach.
- ➤ |II.97| *If they press your blade out toward their left* (in that moment in which you execute the first thrust): turn your hilt rapidly upward and complete the thrust (as previously).

About running in

In the Onset, position yourself in right Plow.
- ➤ *If they cut at you from above,* turn your hilt upward (*in that moment in which they cut in*) with an extended arm between you and them and toward their right shoulder. Your blade thus stands horizontally in front of your face, and you catch their in-flying cut on your flat blade.
- ➤ In this counteraction, while you still hold their blade in the air with your counterposture, step during this with a head bent under their blade and toward their right side.
- ➤ Turn your weapon out of the counterposture into a thrust.

You will hit them before they have prepared for this (as this is indicated by the middle and outermost scene in Figure **G** on the left-hand side).

You can additionally protect yourself by setting off; however, if you execute this sequence with sharp blades, you won't need to set off.

Certain thrusts can be carried out from this sequence in the case of serious combat; however, these do not belong to general fencing. I will leave this topic now, from which the diligent reader can certainly draw inferences.

How you can take their weapon

Position yourself into Plow on the left. *If they cut or thrust at you from above:*

[266] Corrected from "yours", an obvious error.

➢ Turn your hilt (including the blade) upward and use the horizontal blade to catch their cut still in the air (and close to their hilt) so that this counteraction appears like a Crosswise Cut with your hilt toward their left and your blade toward their right.

➢ In that moment in which you lift to counteract, leap with your left foot well under their strike toward them, and *while their blade still lies in contact on yours due to this strike*, reach through with your left reversed hand under your blade to their hilt.

➢ Turn this out of their hand by twisting toward their right (as the uppermost tenant on the left side also shows you in the Figure **G** printed previously).

➢ If they don't want to release it quickly, thrust at a joint with your pommel.

|II.97ᵛ| **Another**

If a fencer cuts at you from above:

➢ Leap well underneath their strike once again and catch it with your blade, turned upward and across from below, close by their hilt (as previously).

➢ *At the same time as their strike is falling and glances on your blade*, reach through toward their weapon's pommel with your left hand under your [blade].

➢ During this same action, turn your blade outward over theirs and press the same downward and toward them.

➢ Pull their pommel to you with your left hand and press their blade away from you and toward their body with your weapon (as is painted in the upper and outer scene toward the right side in the aforementioned Figure **G**).

➤ And you can thus take their weapon and injure them with your weapon or theirs, according to opportunity and desire.

Another

Run under their High Cut to counteract them (as taught previously) so that you receive them on your blade (as you can see this counteraction above on the right side in the subsequently-printed Figure **I**).

➤ In that moment in which you counteract, turn your left side to them and grab their arm at the wrist (just as the scene shows that just taught you to counteract [**I**]).

➤ Hold the wrist strongly and turn it downward toward your left in one jerk.

Thus, you break their arm *or they must bend very far forward.*

➤ *If they do that,* hook your pommel into the bend of their elbow.

➤ Jerk upwards to your right side so that *they will fall on their face.*

Take note. If you are without a weapon *and are run over or attacked from above by another with a single-handed weapon,* and you can't escape them in any way without injury:

➤ Strike your two hands cross-wise over one another (the right over the left) and observe to see how you can leap or evade their cut *so that they cut and miss in front of you.*

➤ *In that moment in which their cut now falls to the ground,* leap rapidly toward them so that you move under their weapon (even before that moment *in which they pull back for another strike*).

➤ Rapidly grasp their right arm between your two hands and turn it quickly with one jerk downward toward your right side.

➤ Hold their hand with your left and grasp their hilt with your right reversed hand, |II.98ᵛ| twisting it outward and downward.

➤ *If they complete their strike and therefore cut [at you]* (while you are leaping in), observe the side toward which they strike and strongly beat counter to their blade with that same arm—namely, at the Strong of the blade, or the closer to their hilt the better.

➤ *Although they will injure you somewhat,* it will not be as serious as it otherwise would be, if you didn't beat counter to it.

➤ Leap farther forward, and don't allow them any more strikes. Instead, hastily grab their right arm with reversed hands. Depending on which hand you grab their arm with, turn them toward that same side and jerk the weapon with the other hand (regarding this, view the outermost two tenants on the left side in Figure **H**).

Summary

When you want to fence with rapiers (or otherwise want to engage with them), step toward your opponent with an extended, straight, and strong Simple Brace and observe what they want to fence against you and which side they want to cut or thrust from. Depending on which side they now guide their cut, [you should] receive and counteract that cut and [then] cut or thrust at that very same side from which they guided their cut (before they fully complete theirs, or at least before they have recovered back from the same).

You also have three pathways through which you can guide your cut toward them (toward each of their sides): one from above, the second across, the third from below, and each of these will be counteracted or altered in three ways, high or low (as you have learned in the first part).

If they do not want to cut or thrust first, but instead oppose you in the same counterposture and want to wait for your attacks: you should pay attention to the three pathways on both sides and take note to determine which side it would be most advantageous to you to safely cut.

These cuts in the Before involve a lot of practical skill in the art because you must readily consider that you can't cut or thrust without exposing yourself by this action. *They will then have positioned themselves in this counterposture so that,* wherever you allow yourself to be noticed by a cut or become exposed, *they will be able to penetrate to the closest of the openings.* Therefore, if you want to cut or thrust toward them in the Before, you must set up the first cut more to provoke and bring them upward than to hit, so that, *when they would cut at your opening* which you have exposed with this cut, you are positioned to |II.99| beat the same outward and to take it. And only then (after you have weakened them and exposed them) do you rush in to complete the third [cut] at the opening.

The three cuts arise from this, which one should deservedly consider a masterful test. These three cuts were highly valued by the Ancients, as the five then originate from them. This is not to be understood to mean that no more should be included in this number, but rather that all cuts are divided into these three distinctions: namely, that some are used to provoke an opponent out of their advantage; the second to counteract and to rebuff an opponent's cut with the same; [and] some are used for hitting, primarily to injure the body. And there is no certainty about whether you need one of each, one or two or even more cuts for this. It does not matter with which cuts this is completed.

Therefore, it is also important to consider the characteristics of the person, who can be divided into four parts in this art of fencing, and following diligent consideration, one can [likewise] find four types of fencers. In order that you have a useful introduction to ponder, I will first explain, provide a short lesson, and set a rule about how you should conduct yourself toward each one.

And the First are those who impetuously cut and thrust as soon as they can reach their opponent in the Onset.

The Second are somewhat more diffident and don't attack so uncouthly; instead, when [their opponent] cuts and misses, drops too far, or otherwise fails to execute a change, they pursue and rapidly follow them to closest exposed opening.

The Third don't cut at an opening if they are not assured of it; instead, they pay more attention as to whether they can recover from the extension into the cut safely back into a counterposture or to Defensive Strikes. (I mostly hold myself with these [fencers], but it depends on my counterfencer.)

The Fourth position themselves in one guard and wait for their opponent's sequence. They are either fools or indeed quite sharp,[267] because whoever wants to wait for another's sequence must be skilled and experienced and well-practiced, otherwise they won't achieve much.

Now, as the First [are] impetuous and perhaps foolhardy and, as one might say, unreasonable; the Second cunning and sharp; the Third cautious and deceitful; the Fourth are similar to fools. Therefore, you must yourself mimic and act appropriately according to all four, so that you can deceive your opponent—perhaps with impetuousness, perhaps with cunning, perhaps with cautious observation, or also provoking with foolish body language.

|II.99ᵛ| You thus mislead and deceive them with regards to their intended sequence, and you also make space and expose openings so that you can more safely touch and hit them.

Now position yourself like this against the First fencer: when you take note that *they want to quickly overwhelm you with hard cuts in the attack and want to crowd you*, counteract their cut or thrust on your long edge with an extended arm (at the Strong, close to your hilt) and turn your hilt toward all of their in-flying cuts and thrusts. Don't move too far out of Longpoint to the sides or away from your face in this counteraction, because the more directly you remain in this diversion with your hilt in front of your face, the better. Withdraw your head and face from their blade behind yours every time. In that moment in which you hold off their cut or thrust, take diligent note as to whether you can withdraw back with a retreating step in the second, third, or fourth cut, *so that they will drop down in a miss with their cut or thrust. In that moment in which they strike and miss or before they recover,* immediately thrust or cut quickly in response. [Against] those who storm in at you impetuously with cuts and thrusts, you should always oppose them in Longpoint or Simple Brace and you should initially ease up somewhat and yield, but you should likewise endure and deflect all cuts and thrusts *until they have quickly become tired, reckless, or assured that they are in no danger.* Once you perceive your advantage, follow quickly and cautiously (because the more you yield, *the more impetuous they become* and the easier you can then cunningly overcome them). Don't allow yourself to be driven out of your advantage, because *whoever cuts so impetuously will quickly cut and miss.*

Against [the Second fencer], those who don't fence impetuously in the Before but instead pay attention in order to fence closely following their opponent's action in the Before: position yourself in the Onset into one of the guards. Change before them cautiously from one guard into the other, and expose one opening after the other to them, yet keep your point always in front of them. However, *as soon as they thrust or cut at you during this*, attack them by setting off or with a Suppressing Cut and rush immediately to the exposed opening (as stated previously about changing stances).

Against the Third fencer, exercise this skill: when you take note that your counterfencer doesn't cut first, nor do they rush to the openings unless they are certain of them: position yourself in the Side Guard in the Onset or remain a short time in the Change [Guard] (as if

[267] The most common meaning of *Schamper* is "an undisciplined, foolish, impetuous person". However, there is a secondary meaning of "biting, cutting, cynical and caustic", that is, a negative form of intelligence or humor, a derogatory shading to an experienced or old hand.

you wanted to wait for their sequence). Rise Inðesly back upward out of the Low Guard |II.100| and position yourself as if you wanted to change into the High Guard. When you have barely arrived in the High Guard, turn your weapon rapidly to ſtrike. *Before they observe this,* cut rapidly at the closeſt opening with an extended arm. In this way, you expose yourself and *they will undoubtedly cut rapidly at this opening* (which you have offered to them through your unexpeɛted ſtrike). *If they execute this [cut],* set it off and continue to work at their opening. *If they don't cut,* thruſt ſtrongly after you have completed your cut. This is a fierce deceit which you present with your body language (as if you wanted to move from one ſtance into another long in front of them—which you do want to partially carry out). However, when you have barely arrived with your weapon in the intended High Guard, and you have Inðesly observed your opportunity, turn your weapon to ſtrike before you have completely arrived into the ſtance.

You will find much that pertains to the Fourth fencer, and how to position yourself againſt them, in the sequences taught previously.

You should pay attention to your opponent's cuſtomary behaviors, type, and nature so that, by recognizing their intent, you know how to counter each one according to the circumſtances. Laſtly, you should always pay diligent attention to the three cuts, so that you provoke with the firſt, take or counteraɛt with the second, and hit with the third.

Example

When you thruſt toward your counterpart and want to attack them firſt in a virile way, you muſt cut at their openings so that you don't place yourself at risk.

> Therefore, *when they ſtand in their advantage:* cut the firſt slantwise and through either their weapon or body so that you bring them up with this cut and provoke them to move out of their advantage.

> *As soon as they rise up and thruſt,* take away their in-coming cut or thruſt with your second cut.

> Cut or thruſt the third quickly at their body, *before they have recovered from their ſtrike.*

> If you want (or if it is necessary), carry out two slanted cuts through the cross to additionally proteɛt yourself and to recover again (since you muſt expose yourself with your serious cut in response).

|II.100ᵛ| *If they cut at you firſt:*

> Take their in-flying blade with your firſt [cut] (and, if it is necessary, also take their second with your second).

> When you feel that they are sufficiently weakened, cut and thruſt the second immediately and quickly in response.

> Proteɛt yourself then with the third and recover with a Defensive Strike.

I will leave it at this point and close with the following sequence.

When you approach your opponent with the previously-held counterpoſture *and they don't want to cut or thruſt immediately:*

> Cut the firſt slantwise through their right shoulder so that you drop into the right Low Guard with this cut (always keeping the right foot forward) and also expose your upper body.

➢ *They will quickly rush to that opening.*

➢ Strongly beat their in-flying thrust outward (from your right toward your left).

➢ Cut the third, a Middle Cut through their face, from your left across through their right. (This applies regardless of whether this is completed with the half edge or flat.)

➢ In this Middle Cut, let your blade move around your head and cut the fourth slantwise again through their right shoulder.

➢ Follow this cut quickly with the fifth, also slantwise through their left shoulder. With this cut, you arrive in the left Low Guard.

➢ From there, take your half edge powerfully and strongly upward through their right.

➢ After you have let your rapier sweep above your head into the right High Guard, thrust fiercely from above at their face.

|II.101| Since you have been sufficiently taught how to hold a weapon in one hand, I will now also briefly demonstrate how you should use another weapon in addition to the single-handed weapon.

|II.101ᵛ| Firstly: hold your rapier in the right hand and the dagger in the left, and step toward your opponent with both arms extended in front of you, as the previous image shows.[268]

When you approach your opponent in this counterposture, you have three types or rules for counteracting in order to fence.

The first is to catch or hold all of your counterpart's cuts and thrusts solely with your dagger (whether from below or above or toward your left or right side) and, while you thus counteract [with your dagger], you also simultaneously thrust with your weapon above or below your dagger (depending on how you have caught and rebuffed their weapon with your dagger).

The second is when you block and protect each side with the same weapon and (similarly to before) while you protect yourself with the one, you injure them with the other; as you defend your right with your weapon and the left side with your dagger.

The third is that you counteract with both weapons simultaneously, or one comes to assist the other.

I will present the three methods for counteracting, one after the other in order, and discuss them briefly with their examples and sequences.

How you should act against an opponent who thrusts from above from their right toward your left

When you approach your opponent in the abovementioned counterposture and hold both of your weapons in front of you with arms extended in front and lowered a bit downward, *if they cut or thrust from above toward your left:*

> ➤ Counteract that with your dagger (in the way that is depicted in the larger scene on the right side in the subsequently-printed Figure [**H**]).

---

[268] Unnumbered image on II.101.

➢ While you are countering them, thrust underneath your dagger at their body.

Or:

➢ Counteract their High Thrust or cut, as before.
➢ In that moment in which you counteract, simultaneously cut across quickly through at their feet (from your right toward their left) so that you hold your weapon under your left arm at the end of the cut.
➢ While you still hold your dagger up high, cut quickly slantwise from your left through their right side, high or low (depending on where they expose themselves).

These two cuts should be executed and completed quickly, one after the other (while you are counteracting).

|II.102ᵛ| Or:

➢ In that moment in which you counteract them, thrust under your dagger from outside (with your palm upward) over their right arm and toward their face (as you can see in the outermost and higher scene on the right side [Figure **H**]).

➢ When you recover to strike, turn your hilt well upward under your left arm and cut from your left at their forward leg.

Or:

➢ Beat their in-flying thrust outward to your left side with your dagger and simultaneously thrust from above toward their face.
➢ *If they want to catch and counteract this with their dagger,* pull your thrust around next to their dagger.
➢ *While they lift their dagger,* thrust toward their stomach from below.

Or:

➢ Beat their in-flying thrust outward with your dagger (from your left toward your right) and simultaneously cut slantwise above your dagger through their right shoulder.

If they thrust from below toward your left:

➤ Turn their in-flying thrust away from you with a hanging dagger, out to the side toward your left (as the lower scene on the right side teaches in the Figure [**H**] mentioned previously).

➤ Meanwhile thrust or cut at their closest opening.

The second type of counteracting

If they cut or thrust at the other side (namely, toward your right):

➤ Counteract them with your weapon and leap Inðesly with your left foot toward them.

➤ While you counteract, thrust at their right arm with your dagger.

Item:

➤ Counteract their cut or thrust (*which they guide toward your right*) with the blade of your weapon.

➤ In that moment in which you counteract—and immediately *when their blade touches on yours*—leap with your left foot toward them and simultaneously attack their blade with your dagger.

➤ Hold their blade [with your dagger] until you have injured them with a thrust above their left arm with your weapon (which must occur in the blink of an eye).

Or select other cuts or thrusts in front of you, since you have found sufficient types taught previously in the Rapier Fencing [Section].

➤ As soon as you have injured them with your weapon, retreat again with your left foot to where you stood before with your right foot and you can continue to protect yourself with both weapons (as at the beginning).

If they thrust or cut at you from below (namely, toward your right):

➤ Counteract this with your hanging |II.103| blade and thrust quickly toward their face and away from their blade.

➤ Simultaneously move your dagger toward their blade with a step forward of your left foot.

➤ After executing the thrust, immediately pull your hilt upward (keeping your dagger in front of your face).

➤ Afterwards, step forward with your right foot and cut powerfully upward from below next to the same side and through their body with the half edge.

➤ In this cut, pull your weapon toward your left and around your head and guide it in a powerful cut toward their left from above.

➤ And take note: in that moment in which this cut should hit their counteraction, pull your weapon toward you, move the [weapon] around next to their dagger, and thrust at the closest opening.
➤ While you are doing that, move your dagger in front of your face with an extended arm.

The [third][269] type of counteracting and fencing takes place with both weapons simultaneously: namely, in that you use both weapons cross-wise or cross them over one another and you thus receive their blade between your two blades (*in that moment in which they guide their cut or thrust*). Afterwards, as soon as you catch it, you remain with the dagger on their weapon (to hold it) and you thrust or cut rapidly at the opening with the other (*before they lift and remove their blade from your dagger*), as I will briefly explain to you.

There are five thrusts in response, namely: two from your right toward their left, of which the one is completed from above and the other from below. The second two are oriented and thrust toward their right in the work, the one from above and the other from below. The fifth is the straight thrust in front of you toward their chest or their face, which is guided sometimes above or below your dagger, depending on whether you must counteract with your dagger.

As you now have five thrusts, you also have five cuts in response. The first is directed and cut at the head; the second at the neck and the shoulder; the third at the hand; the fourth at the hip; and the fifth at the feet. As taught above, these can be completed from above, slantwise, across, and from below (and also on both sides).

Correspondingly, when you execute all cuts and thrusts correctly and can prudently guide your dagger to protect yourself, you should pay diligent attention to the Before and the After and also to the word Indes, by means of which the correct and opportune time must be learned in which to complete each cut and thrust and to execute them usefully. In order that you can learn this and direct it more gracefully in the work through diligent practice and careful investigation, I want to set out and provide you with some examples in order.

|II.103ᵛ| **Example**

In the Onset, when you hold your two weapons according to the previously mentioned configuration (with arms slightly lowered and extended in front of you):
➤ Quickly—*before they would have predicted*—cut straight down from above, in one pull through their face (like a slice) with an additional leap forward with your right foot.
➤ In the meantime (and while you cut downward), lift your dagger upward in front of your face.

You provoke them with this cut *so that they will, without doubt, rush quickly to the opening (whether with cuts or thrusts)*.
Therefore, *as soon as they thrust or cut:*
➤ Lift your long edge with the blade horizontal, outside of your dagger. By lifting upward like this, you should beat their in-flying blade upward.

---

[269] Although '*vierte* = fourth' appears in the text, this is the third method listed, and accords with the third type listed in the introduction.

> Afterwards, *while their blade still touches on yours*, step with your left foot somewhat toward them and toward their right side and simultaneously and quickly thrust through under your dagger toward their face (outside of their right arm).
> During this thrust, turn your hilt (or the long edge) well upward toward your left and, *as soon as they want to remove your blade and counteract*, cut from outside at their forward right leg, protecting yourself diligently with your extended dagger in the meantime.
> You will now have arrived in the right Low Guard with your weapon (due to this cut).
> From there, lift quickly upward with your weapon horizontal (outside in front of your dagger) until you arrive with both blades cross-wise in front of your face with extended arms. In addition, while you are lifting your blades from the Low Guard, step back again with your left foot.

As I have now taught you to use this High Cut in the Before to move them upward, you should also learn to cut the other three—namely, those cuts slanted, across, or from below—from your right toward their left, and to cut through toward their body (high or low) at whatever limb you trust yourself to best reach. While you cut one of these intended cuts toward their body and through, also lift your dagger at the same time, as before, in front of you for protection. As soon as you have cut one of these cuts through toward their left, lift your weapon upward outside of your dagger to divert their in-flying blade (as above) and then complete the thrust (including the Foot Cut) and end the same as above, or according to opportunity.

|II.104| **The second**

When you approach your opponent with both weapons in the abovementioned form:
> Let your dagger drop quickly forward and pull your weapon around your head. Cut slantwise above your arm through their right shoulder.
> In this cut, step with your left foot toward them and quickly lift your horizontal dagger back up with arm extended in front of your face.
> While you lift your dagger, simultaneously thrust seriously and strongly at their stomach with your rapier.
> During this, if you perceive *that they want to divert and counteract the thrust,* pull your weapon quickly downward toward your left side.
> From that same left side, beat their blade strongly back with your flat (under your dagger), so that you somewhat blunt the same and beat it outward.
> Quickly thrust straight at the closest opening *while they attempt to regain full control of their weapon or recover from the aforementioned outward beat.*

The third

In the Onset, *if they thrust low from their right toward your left:*
> Counteract their thrust downward and away from you with your dagger. In that moment and while you counteract, thrust toward their face from above, and take diligent note during this.

➤ *As soon as they raise their dagger against your thrust (to counteract it):* change through below with the front end next to their left arm and thrust between their two weapons inwardly at their body.

➤ *If they counteract this again:* pull your rapier back up and thrust between their two arms from above. With this thrust, break[270] through downward and toward your left between their two weapons.

➤ In that moment in which you rip outward, yield back with your dagger next to your left side so that you can arrive unimpeded with your rapier at your left side at the end of this slash.

➤ From this left side, rip out strongly through their two weapons with the half edge, including a step forward of your left foot (toward theirs and your right side), and quickly follow your rapier with your dagger on their rapier's blade.

➤ Hold the same until you can reach and harry them from above or below your dagger with a thrust.

|II.104ᵛ| If you complete the sequence correctly, you will certainly find an opening.

The fourth

Take note. In the Onset, *when an opponent holds their two weapons in a strong counterposture in front:*

➤ Quickly cut a serious High Cut toward their left shoulder.

➤ *They will rise to counter and counteract this cut with their dagger.*

➤ Therefore, don't let your cut hit, but instead pull your weapon (while it is rising upward), and *while they lift their weapon to counteract,* thrust under the same and at their body.

The fifth

In the Onset, hold your rapier on the right in the Low Guard and your dagger in the left High Guard.

As soon as they thrust at you:

➤ Lift your rapier upwards and horizontally into Longpoint and simultaneously turn your dagger hilt downward over your right arm so that the front end of your dagger extends horizontally toward their shoulder. You thus have both weapons cross-wise over one another.

➤ Catch their in-flying rapier blade between your two blades. Simultaneously, step out to the side toward their right with your left foot and press their blade out to the side toward your right.

➤ While you are stepping out and pressing out, thrust on their blade with your rapier and inside at their body.

➤ Immediately change through toward your left side (with your weapon under theirs).

➤ Cut a Defensive Strike under your arm through their right side.

➤ Protect your face during this with your dagger.

---

[270] 'Ausbrechen = break out' in this instruction becomes 'ausreiszen = rip outward' in the following.

The sixth

Position yourself in the Onset with your rapier in the left Low Guard and hold your dagger also on your left and behind you. *If they thrust toward your face:*
- Lift your long edge and extended arm toward their blade.
- As soon as you have caught their blade, move your dagger through underneath their weapon and yours, and rip their blade strongly out from your right |II.105| toward your left.
- In that moment in which you have ripped them outward, thrust from above toward their face.
- Don't complete this [thrust], but quickly pull it back toward you and thrust quickly, completely, and strongly inside and between their two weapons at their body.

You will deceive them and their counteraction with this pull (as taught).

The seventh

Once again, when you arrive with the right foot forward, position yourself with your rapier in the left Low Guard and hold your dagger in front of your face with arm extended high in Longpoint.
- *If they thrust or cut at you:* beat their in-coming thrust powerfully out from your left toward your right.
- *Before they have recovered from the aforementioned outward beat,* cut the second quickly under their dagger, across through their foot, with another step forward of your right foot (with your body bent forward to reach farther).
- During this, keep your dagger in front of your face the whole time (in order to protect the same with it).
- After executing this cut, you stand again like you did in the beginning.
- *If they continue to crowd you with additional thrusts or cuts:* beat their blade powerfully outward again with your rapier, slantwise down from above toward their right and yours (keeping your dagger in the counterposture) so that your blade runs out to the side and next to your right and snaps back around into the right High Guard of the Ox.
- From there, thrust powerfully from above toward their face. During this thrust, turn the long edge downward so that you arrive back into the left Low Guard with your weapon (due to this thrust).
- From there, cut the last as a Defensive Strike through their right shoulder with a step backward of your right foot.

The eighth

When you have retreated and have carried out a Defensive Strike through their right shoulder, so that you stand with your left foot forward and have your weapon in the Low Guard next to your right side and your dagger in front of your face with extended arm, *if they thrust again toward your face (whether toward |II.105ᵛ| your right or left side):*
- Turn the long edge of your dagger toward their weapon so that you catch their blade or move it aside from you without injury.

- Simultaneously, and in that moment in which you counteract with your dagger, move your weapon up under their blade to aid your dagger, so that you simultaneously counteract with both weapons.
- *As soon as their blade glances off or touches on yours,* thrust underneath your dagger at their stomach or at the closest opening.

If they thrust at you from below and they want to counteract above with their dagger:
- Drop on their rapier blade from above with your dagger and at the same time, cut toward their left ear with a serious face and body language.
- In that moment in which the cut should just hit, turn the half edge outward and toward them (so that you recover into a second cut).
- Pull this around in front of your face, and rapidly cut slantwise through from your left toward their right shoulder and underneath their dagger (*while they have raised the same*).

In summary, as concerns the dagger with the rapier, I advise Germans that they accustom themselves to simultaneous counteraction. This includes the observation as to whether they can injure [their opponent] with their weapon or dagger, yet without allowing the two weapons to move too far apart from one another so that they can always move the one to aid the other. As experience has proven, when a German becomes accustomed to counteracting only with the dagger, this has often resulted in injury in serious [combat], because it is against their type and nature. This is because the closer one remains (in this case) to the custom of their nature, the more will be accomplished thereby.

In addition, a cloak[271] (short or long) may also be used sometimes as a secondary weapon (or weapon of necessity). If you want to use the same, you should first learn to correctly wrap around your arm so that you aren't injured on your arm (when you intend to counteract with

the cloak). If you can't wrap it correctly around your arm, then this should be avoided so that you don't impede yourself. If you want to correctly use the cloak, then be diligent to catch all of their cuts with your blade and then capture their weapon with your cloak, holding the same until you have injured or hit them with your weapon (which can and must now occur easily and in one cut).

---

[271] Both *Kappe* and *Mantel* refer to outer garments. The *Kappe* usually doesn't have sleeves and may have a hood. A *Mantel* tends to be longer and may or may not have sleeves. In English, "capelets" were short, sleeveless, hooded garments to protect the head, shoulders, and back. For ease of understanding, I am translating *Kappe* as "short cloak" and *Mantel* as "long cloak". Since *Mantel* only appears twice, in other places I have shortened *Kappe* to "cloak".

Take this as an example of the same. *If an opponent wants to quickly overwhelm you, and seriously attack you:*

➤ Draw your weapon and grip your cloak (short |II.106ᵛ| or long) with your left hand, from inside at your left shoulder at the collar or upper seam. Pull it from your body and wrap the same around your arm.

If they cut or thrust toward your face or body:

➤ Catch their cut (with a leap forward under their weapon) with your weapon and, *in that moment in which their blade touches on yours in the cut,* rush to their weapon with your left arm (including the cloak).

You can delay their weapon by following with your cloak and remain there until you have injured, hit, or won (depending on your desire).

This is the best rule: when an opponent cuts at you, that you catch their strike with your weapon and, in that moment in which it is still in contact, drop on their blade with your cloak to delay them with the same. In the meantime, observe where you can hurry to cut or thrust.

If you have made ready with your wrapping and are assured of it: counteract their cut with your cloak, and cut at them simultaneously, indeed in that moment in which you counteract, at the closest opening.

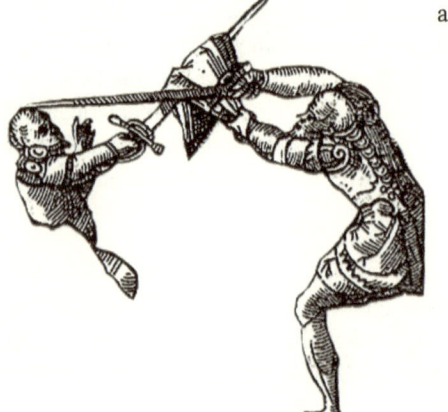

Another

When you have drawn your weapon (having been forced to this), grip your cloak above at the collar and hold it next to your left side with arm hanging downward. Take diligent note during this:

➤ *As soon as they cut in,* beat your cloak across around their in-flying blade.

➤ As soon as you have beaten their blade outward, cut above the [blade] at their head (as the smaller tenants show on the right in the image printed previously).[272]

How you should act with a single-handed weapon against a spear or similar weapon

In this case, act like this. When you would be surpassed in reach and quickly overwhelmed by a boar spear:[273]

➤ Hold your weapon to the right in the Low Guard[274] (after you have hurriedly drawn it) so that you stand with the left foot forward, and observe.

---

[272] Unnumbered image on II.106.

[273] The *Knebelspieß*, also called a *Saufeder*, *Sauspieß*, *Schweinespieß*, or "boar spear", has a short, perpendicular crossbar or lugs/wings on the spear socket behind the blade. It was used in warfare during the early Landsknecht period; however, it was quickly replaced by the partisane, which has a half-moon shaped crossbar instead of a straight one. The illustration shows a simple spear with no crossbar.

[274] This stance with rapier is similar to the '*Eber* = Boar' in the Dusack Section. The potential pun is exceptionally subtle.

➢ *As soon as they thrust at you from above:* leap toward them and well under their |II.107|  strike, turning your weapon up-ward, and withdraw your head by bending your body sideways away from their shaft.

> ➢ Catch their blow on your flat hanging blade (as you can see in the previously-mentioned tenants on the left side in the abovementioned image).[275]
> ➢ During this, reach your left hand under your weapon for their shaft.

If they withdraw their shaft upward so that you can't reach it:
➢ Cut at their hand which holds the shaft in front while they are pulling upward.
While they hit back from above:
➢ Leap under their spear closer toward them once again (with your counterposture pulled high up again), so that you can work at their openings.
The closer you move under their shaft, the less they can carry out.

If you don't trust yourself to run under their first hit:
➢ Yield to their first strike and let the strike miss.
➢ Take diligent note. *In that moment in which they draw back for the second strike:* move quickly and leap in under their shaft.
➢ As soon as you have run in underneath, grab their shaft with your left hand as before and work with your weapon toward their openings to your advantage.

However, *if they guide a thrust at you:*
➢ Oppose them with your weapon in the right Low Guard and lift your hilt from that guard (with your arm extended in front of you) so that your blade hangs down toward the ground.
➢ Beat their in-coming thrust outward from your right toward your left with the hanging blade.
➢ Simultaneously with this outward beat, leap well toward their left and out of their thrust so that you both beat it outward and also simultaneously escape from the same (otherwise, the outward beat would be too weak in itself).
➢ Reach for their shaft again with your left hand (as above).
If they pull their thrust (so that you beat outwardly in vain) *and thrust back quickly:*
➢ Remain with your hilt high in the hanging and beat their second in-coming thrust outward (from your left toward your right) with your hanging blade.
➢ With this outward beat, leap out of their thrust toward their right (just as you previously leapt out toward their left).

---

[275] Unnumbered image on II.106.

Thus, you can oppose all of their pulled thrusts with a hanging blade and your arm extended in front of you and beat them out to both sides until you can rush to their shaft.

|II.107ᵛ| **Another way to run under [a polearm]**

Hold your weapon in the right Low Guard (as previously) and take note. *In that moment in which they now strike from their right:*
➤ Move through **Indesly** with your weapon under their shaft (toward your left side) and simultaneously step through with your right foot between yourself and them (and toward their right side).
➤ During this step through, cut strongly and swiftly from your left back toward their incoming shaft and toward the hand in which they guide the strike inward.

All of this (namely, the move through with the weapon and the step) must be undertaken quickly in one leap. You must also quickly bend your head under their strike to escape it. The sequence proceeds well in this way.
➤ Forcefully drive them powerfully back and don't let them complete any more strikes.

Another, how you should beat their thrust outward

Hold your weapon in the right High Guard to strike (after you have drawn it).
If they thrust toward your lower body:
➤ Pull your weapon around your head and strongly cut their thrust out to the side with a hanging blade (from your left toward your right).
➤ During this, leap in strongly toward their right.
If they thrust at your upper [body]:
➤ Take that away and counteract it as taught above.

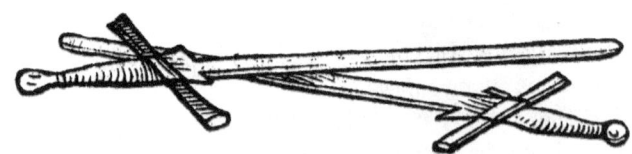

Fencing with daggers

The fourth part of this book deals with fencing with daggers so that you can learn how a fencer should use all types of similar short weapons (as well as many good wrestling sequences, which are also included).

About the High Guard [Oberhut]

Position yourself in the High Guard like this: hold your dagger high in front of your face (as the larger portrait on the right side indicates in this Figure [A]). Step toward them and remain with your right foot forward.

If your opponent[276] thrusts in at you toward your left:

➢ Move your hanging dagger from your right toward theirs and catch their hand with your dagger behind their hand (next to their wrist),[277] so that in this counteraction[278] your dagger pommel points upward and the blade points down-ward.

➢ As soon as you touch their wrist in this counteraction, im-mediately move your dagger through[279] under their arm and back up around their hand (while you remain hard on their arm with the dagger blade during all of this).

➢ Press your dagger blade well on their arm so that you clamp them even harder.

➢ Rip[280] their hand downward toward your right side.

➢ Afterwards, move your dagger pommel inside of their right arm, upward toward their face or chin.

➢ *If they lift their dagger upwards:* pull around your head and thrust completely through and inwardly toward their right arm, at their face (following the crosswise line).

➢ Complete the last in one rip straight from above through their face, including a retreat of your left foot.

....

[276] As previously, the German word *Man/Mann* will be translated as "opponent", and references to pronouns will be the gender-inclusive "they/them", unless the pronoun refers to a specifically named Master. In addition, Meyer uses several other terms to refer to the opponent, which will be reflected in the translation.

[277] The German term *glied* means, with respect to human bodies, "a joint or movable body part". In this case, the wrist. At other points, the knuckles, elbow, or knee are indicated by *glied*.

[278] In the Dagger Section, unlike all others, Meyer uses *versatzung* as the noun equivalent of '*versetzen* = to counteract'. This is no longer a stance, but a movement in response to an action by an opponent, often with the free left hand, to counter said action.

[279] Instead of '*durchwechseln* = change through', Meyer uses '*durchfahren* = move through' for move-ments of the dagger across the center line of the opponent.

[280] *Reiszen* means "to carry out a fast, tearing or ripping motion", where *risz* is "a tear, fracture, crack, or fissure". From *reiszen*, one can arrive at '*risz* = slash', since if you move something sharp through fabric or flesh in a fast, ripping motion, it will slash through the object.

|III.2| *If they thrust at you from above toward your left (as before):*

> ➢ Thrust across toward their in-coming[281] arm so that your dagger extends past their arm.
> ➢ Catch their arm at your wrist (in the angle between your hand and the dagger).
> ➢ In this thrust, turn your right side well toward their left so that you can hold their hand even tighter between your wrist and dagger.
> ➢ Enclose their hand firmly and pull downward and away from you toward your right so that you wrench their arm.
> ➢ When you have twisted their arm downward, pull your dagger quickly toward your left shoulder.
> ➢ From there, thrust across over their right arm through toward their face (before they bring it back from the downward rip).
> ➢ Thrust the second quickly in front toward their chest while guarding your face with your left hand.

Thus, you have two counteractions from your right toward their left: namely, with the first you strongly catch their wrist joint on the dagger hanging away from you toward your left; the second counteraction is that you catch their abovementioned hand under your dagger on the wrist of your hand with a counter thrust (*in that moment in which they thrust*).

If they were to thrust from outside toward your right, you should also direct this in the work from your left toward their right (*just as you have now completed it from your right*).

Otherwise, *if an opponent thrusts in at you toward your left:*

> ➢ You should [rip][282] through powerfully across inside toward their arm, at the flesh[283] or the inside of the wrist close to the narrowest[284] point, strongly through (because that seriously cripples them).

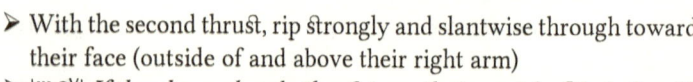

> > ➢ With the second thrust, rip strongly and slantwise through toward their face (outside of and above their right arm)
> > ➢ |III.2ᵛ| *If they have already thrust in*, so that you take [their thrust] and rip it outward and down.
> >
> > You should also reach past with High Thrusts from this guard (*if they thrust at you from the right or left*).

Low Guard [Underhut] with its counteractions

Position yourself in the Low Guard like this: stand with your right foot forward [and] hold your dagger next to your left thigh so that the front end[285] extends out toward your opponent's face.

---

[281] Because German has directional prefixes *hin-* and *her-* to indicate movement away from or toward an object, I have used "in-flying" and "in-coming" as translations of the adjectives *herfliegend* and *herkommend* respectively, which Meyer often uses to describe weapon movements by both parties.

[282] There is no verb in this sentence. "Thrust" or "rip" are obvious contenders.

[283] *Fleschen* could be "flesh/muscle/meat", or a typo for '*flechsen* = tendon'.

[284] I'm guessing that *restricta* means "the narrowed" from '*restringere* = narrow, limit', from Latin.

[285] The terms *vorderen ort* and *hinteren ort* come from the polearm section, where they refer to the two ends of the staff, but occasionally Meyer also uses them as synonyms for the tip and the pommel of other weapons.

If your counterpart thrusts in at you:

➤ Step out to the side toward their right and thrust from outside over their right arm.
➤ Rip that downward toward you.
➤ Move your dagger pommel back upward toward their chin (as taught above).
➤ Quickly thrust through in a rip from above toward their face.

Item, *if they thrust from above:*

➤ Lift your dagger horizontally and catch their hand behind their dagger on the joint of the hand.
➤ Twist your dagger around over their arm from inside.
➤ Rip toward your left side.
➤ Afterwards, thrust quickly through from the front toward their face.

Middle Guard [𝔐ittelhut]

This guard is when you hold your dagger at your side, even with your belt, or also directly in front. However, since a fencer counteracts from this [guard] like from the others, I will save writing any more about this for the sequences.

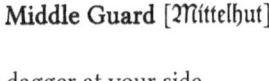

|III.3| **What the appropriate counteractions are**

With the dagger, you have two counteractions. The one takes place with the dagger (which has been discussed previously). The second occurs with the left hand: that is, that you can hold off and catch their hand in which they hold the dagger with the [left hand]. In the subsequent sequences, you will be instructed thoroughly about how you should now use each of them.

Item, step toward them in the High Guard and hold your left hand in front of your chest.

If your counterpart thrusts at you from above:

➤ Catch their hand with your hand reversed and twist their hand around [an axis] and away from you.
➤ Thrust from below at their joint with the pommel (as the outer [left] scene in Figure B teaches).

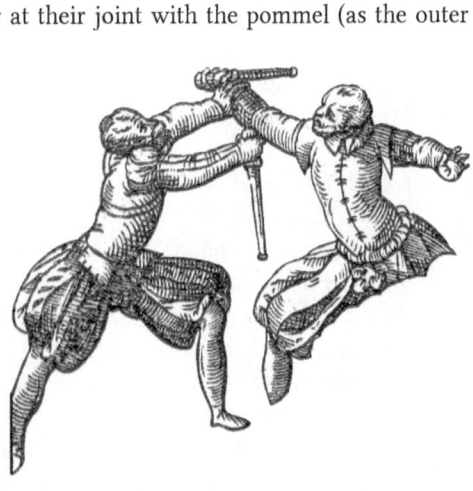

Take note in the Onset. *If an opponent moves to you with a High Thrust:*
➤ Move in under their hand *while it is still high in the air.*

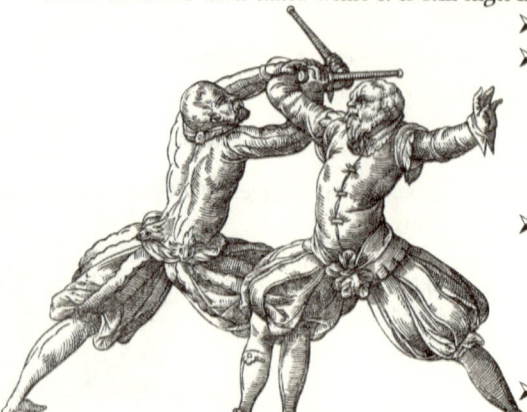

➤ Restrain their hand up high with your left.
➤ Reach quickly through under their right arm, with your right hand with your dagger moving to assist the left (as you can see in the larger tenants in the previously mentioned Figure B).
➤ In that moment in which you have reached through, simultaneously step with your right [foot] well behind their right leg and slip your head through under their right arm.
➤ Throw them over backwards or break their arm.

Item, in the Onset, move into the Low Guard and lay your dagger on your right arm, and take note.

➤ *In that moment in which they thrust in from above:* move in under their right arm with a leap forward under their dagger to catch it on your horizontal dagger close to their wrist joint.
➤ Immediately and simultaneously with this, grip them by their elbow as well with your left hand.
➤ Jerk |III.4| toward you (in such a way as you see in the smaller [right] tenants in Figure B).
➤ Thrust at them as you want.

Crossed Guard [Ƙreutʒhut]

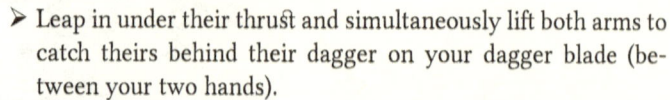

Hold your hands crossed in front of you in the Onset (the right over the left) so that your dagger lies on your right arm.

If a fencer thrusts in from above:
➤ Leap in under their thrust and simultaneously lift both arms to catch theirs behind their dagger on your dagger blade (between your two hands).
➤ *In that moment in which their hand drops onto yours*, grasp them simultaneously with the left hand reversed.
➤ Twist it hastily and powerfully away from you.
➤ During this (while you twist them outward), thrust down from above in a rip with your dagger.
➤ *After they have wrenched their hand out during this*, cut across through their face and arm in response.

Item, *if an opponent thrusts at you from above:*
- ➤ Move in under their arm with your horizontal dagger lying on your arm.
- ➤ In that moment in which you counteract them, simultaneously attack quickly with [your] right.
- ➤ During this and while you grab with your left hand, lift your dagger back away from their arm and move the pommel through next to your right (back from below).
- ➤ Rip through strongly upward between their two arms with the [pommel].
- ➤ Thrust short and down and inside of their arm at the chest.

If an opponent thrusts at you from below:
- ➤ Drop on their arm with the horizontal dagger and simultaneously grab with your left.
- ➤ Lift your pommel Inðeſly[286] upward, above their arm |III.4ᵛ| toward their face.
- ➤ *While they lift in response to rebuff that,* thrust through inwardly and across at their face.
- ➤ Afterwards, thrust away from them with Cross Thrusts.

Take note. *When a fencer runs in over you with a High Thrust:*
- ➤ Counteract with your horizontal dagger, which should then lie along your arm.
- ➤ Simultaneously with this counteraction, you should have your left hand crossed on your right to grasp their dagger from below with your left hand reversed (*in that moment in which they have thrust in*).
- ➤ Break that out upwards toward their right shoulder.
- ➤ In that moment in which you break out, reach well outward over their right arm with your right hand, such that *if they don't want to let go of the dagger,* you [can] step quickly with your right [foot] behind their right and throw them backwards away from you.

Take note in the Onset:
- ➤ Run in under their High Thrust with crossed hands (so that the right hand is above in the counteraction).
- ➤ Simultaneously with the counteraction, grasp their right hand strongly with your left hand reversed.
- ➤ While you are grasping their right, thrust in up under their shoulder from below or strongly across at their right ribs.
- ➤ Pull your dagger quickly back under your left arm toward the same side.
- ➤ From there, thrust through strongly upward toward their right arm.
- ➤ Afterwards, thrust away from your right across toward their face and arm.

---

[286] On I.25 of the Sword Section, Meyer explains his interpretation of *Indes* to be a small word fraught with meaning: in that moment in which you strike, you should simultaneously observe the other openings that you could also hit, and also all of the techniques your opponent can use to strike or counteract your action. It is therefore a point of hyperfocus on what you are doing and what you and your opponent could be doing at the same time and subsequently. Meyer states that he will use *Indes* as a shortened form to call all of this to mind.

Within his text, Meyer tends to use *Indes* to indicate a disruptive action, that is, because he is describing actions sequentially that occur simultaneously or disruptively. Thus, *while* you are performing the above action, execute the following action *at the same time* and *in the same space.* Because *Indes* is an adverb of time in German, it will frequently appear adverbially in the translation, thus *Indesly.*

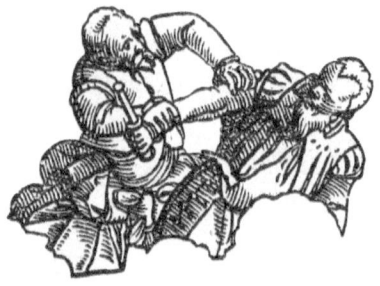

If an opponent thrusts from outside toward your right at your head:

- ➤ Thrust outwardly over their right arm.
- ➤ Clamp [the arm] between your wrist joint and dagger.
- ➤ Jerk [their arm] toward you (to the right side of your chest).
- ➤ Drop your left hand onto their elbow joint and break their arm (as can be seen in Figure C in the upper smaller tenants on the right).

ᵇ |III.5ᵛ| **Throws [Werfen]**

Item, *if a fencer thrusts at you from above:*

- ➤ Thrust outwardly over their right arm (as previously).
- ➤ Jerk it to you and step with your left [foot] behind their right.
- ➤ Reach with your left hand behind and around to their left shoulder.
- ➤ Jerk them onto your left side over the leg you have set forward (as the center upper tenants show in the Figure [C] mentioned previously).

How you should take the dagger away from your opponent

Item, *if an opponent draws a dagger on you:*

- ➤ Grab their hand with a straight grip (instead of a reversed grip).
- ➤ *If they have thrust from above* and you have seized their hand as stated: twist their hand downward toward your right in one sweeping movement.
- ➤ Grab their dagger blade with your right hand reversed and break it out of their hand.

If they thrust from below or from the front toward your face:

- ➤ Grasp their hand again so that your little finger always points toward their arm and your thumb toward their hand in this grip (as before).
- ➤ Just as before, grasp their dagger with your right hand reversed and break it out (as the smaller scene teaches on the left side of Figure C).

Item, *if a fencer thrusts at you from above:*
  ➤ Grab their right hand with your left hand reversed.
  ➤ Drive that upward and around away from you and leap with your right [foot] behind their right.
  ➤ Move your dagger pommel (and hand gripping it) forward to their throat.
  ➤ Throw them away from you over your right leg.

|III.6| **Counter**

Pull your right hand toward yourself and strike from outside with your left arm over their right. Use this strike to rotate yourself away from them and around toward your right side.

Item, stand with your left foot forward and hold your dagger in the middle (next to your right) so that the pommel projects above your hand.

If they then thrust at you:
  ➤ Counteract the thrust away from your face with your left hand reversed.
  ➤ During this, step well toward them with your right foot so that you have turned your right side (under your arm) well toward them.
  ➤ Thrust Indesly from above with the pommel, outwardly over their right arm toward their face.

Take note. *As soon as they want to counteract the thrust:*
  ➤ Move [your dagger] strongly through with the front end upward, inwardly and upward from below, between their two arms to their chin.
  ➤ Afterwards, thrust back through their face from above with a retreat.
  ➤ With your left [hand], take care of their right [hand].

Item, when you have caught their hand with your left (as taught):
  ➤ Step toward them with your right foot (as previously), moving your dagger through under their arm and yours.
  ➤ Rip your dagger pommel strongly outward over their right arm, downward toward your right, so that you release their right hand from your left in this downward rip.
  ➤ Afterwards, thrust quickly from in front through their face with a retreat of your left foot.

|III.6ᵛ| Take note in the Onset. Position yourself with the right foot forward and hold your dagger so that the blade projects by your little finger.
  ➤ Thrust through their face from your right from above.
  ➤ When thrusting through, quickly turn your dagger back upward for a Low Thrust close to your [left side].[287]
  ➤ Thrust the second powerfully through, upward at their arm, so that you arrive at your right shoulder at the end of the thrust.
  ➤ From there, thrust strongly through across toward their face.
  ➤ Thrust the fourth strongly back through their face from above.
  ➤ Quickly lift the horizontal dagger toward their right arm to counteract.

---

[287] Although the text has "right side", ergonomics puts the rotation of the dagger on your left.

Rip outward [Ausreissen]²⁸⁸

In the Onset, thrust seriously toward their face. When you have taken note that *they want to oppose the thrust:*

> ➤ Move through under their right arm toward their right (in that moment in which the thrust should just hit) and thrust from outside over their right arm with the same thrust.
> ➤ Rip powerfully downward toward your right.
> ➤ Take diligent note.
> ➤ *As soon as they jerk their arm back from under yours:* follow their right arm quickly from below with your left hand.
> ➤ Thrust under their right arm toward their face *while they are still moving upward.*

Additionally, when you have approached your opponent so that you can reach them in one step forward:

> ➤ Hold your dagger so that you have your dagger blade lying on your arm (as before).
> ➤ *As soon as they want to thrust at you from above:* follow by lifting your right arm strongly upward and through (with |III.7| your dagger lying thereupon) from your left toward their right arm.
> ➤ Also, follow their right arm with your left hand under their right and retain their right arm up high with your left hand.
> ➤ Thrust for their chest while you hold their right arm with your left.
> ➤ From there, lift your horizontal dagger quickly back up toward their right arm.
> ➤ As soon as you reach [their arm], move the front end over their right arm from inside and rip it downward.

Also:

> ➤ Hold [their right arm] up with your left and seek one of their openings while doing that.

Or, with your dagger lying horizontal on your arm:

> ➤ Rise again toward their in-falling right arm and move completely through and upwards toward theirs.
> ➤ Follow with the left upward and under their right arm again.
> ➤ Afterwards, while you hold their right hand up with your left, rip your dagger pommel upwards and through from below (next to your right side, between their two arms).
> ➤ Thrust down from above, down toward their face.

Take note. *When a fencer thrusts toward you (whether from above, across, or from below):*

> ➤ Catch their hand behind their dagger and jerk it quickly upward.
> ➤ Move your head through under their right arm.

---

²⁸⁸ Intriguingly, the sequences all use *durchreissen* instead of *ausreissen* as an action. Whereas *ausreissen* is generally defined as "tearing or ripping out", *durchreissen* is the more destructive form of "tearing to little pieces".

➢ Simultaneously with this movement, step with your right foot behind their right (as the larger tenants show you in Figure D).

➢ Pull their hand hard toward you over your shoulder and lift their right leg with your [right][289] hand (lifting with your entire body).

You can thus break their arm or throw them, whichever you want.

Counter

When an opponent seizes you like this, then ensure that you gain control of their back (more about this later).

|III.8| Or, *if they thrust at you from below:*
➢ Immediately set off the thrust (on their hand behind their dagger).
➢ Simultaneously with that moment in which you counteract with your dagger: seize their hand at the abovementioned wrist joint with your left [hand].
➢ Jerk it to your chest with both hands, swinging yourself quickly away from them on your right side.
➢ Thus, you break their arm.

Likewise, when you counteract your counterpart's thrust (*which they have thrust at you from below*) and have grabbed them with both hands:

➢ Jerk their arm upward to your right side and away from them.

➢ Break their arm across your left shoulder (as you can see in the smaller tenants in Figure D on the right side up in the corner).

This [action] can be countered in a number of ways.

Counter

Take note of this counter. *If an opponent has taken hold of one of your hands and wants to jerk it onto their shoulder with their body reversed*, observe:
➢ *In that moment in which they turn their body*, turn and jerk your elbow upward, lifting it over their head around their neck so that your right arm arrives on their right shoulder.
➢ While you are doing that, reach your left hand swiftly at their throat (over their left shoulder) at the same time.
➢ Place your right foot into the back of their knee.

---

[289] The text has left, but the image and ergonomics show this to be your right hand.

➤ Pull them onto their back.

You can also complete this counter with other sequences.

|III.8ᵛ|

Another

If a fencer thrusts seriously at you from above:
   ➤ Catch their arm behind their dagger between your two hands (which are crossed over one another; the right should be crossed over the left).
   ➤ In this counteraction, simultaneously and strongly reach for their hand (or the arm close to it) with your left hand reversed.
   ➤ Twist it upward away from you, and simultaneously to this, step with your right foot behind their right leg.
   ➤ Drop your right hand forward onto their throat.
   ➤ Throw them onto their back away from you (as you can see on the right side in Figure D).

Take note. *If an opponent thrusts at you from below and at the same time drops down with their left hand at your chest or neck (toward your left side):*
   ➤ Strike your left arm out around their left.
   ➤ Grab their left hand to your chest with your right.
   ➤ Shove strongly at their left side.
   ➤ Thus, you break their arm.

Item, *if they grab the right side of your chest and thrust from below:*
   ➤ Counteract the thrust with your left hand and shove their left hand away from your chest with your right in this counteraction (so that your right thumb is down).
   ➤ During this shove, move your arm immediately below their throat and reach under the back of their knee with your left hand.
   ➤ As soon as you have counteracted, throw them over your right knee.

Stand with your left foot forward and hold your left hand on your chest. *If they thrust afterwards toward your throat:*
   ➤ Strongly counteract the thrust from your chest at the joint behind |III.9| their dagger.
   ➤ In this counteraction, grab their right hand with your left.
   ➤ Reach through below with your right arm and grip behind their elbow (around their right arm).
   ➤ Step in front of them with your right leg
   ➤ Make a short rotation to your right side and throw them over your right leg.

Item, another. *If they thrust at your throat:*
   ➤ Counteract that as previously.
   ➤ In this counteraction, move your left hand around their right arm (outside and below), so that your left hand comes back up to your chest from below.

➢ Grab under their elbow with your right hand and lift from below: this is how you break
their arm.

Item, a sequence and counter. If you thrust at their throat and *they want to grab around your
arm* (as described above):
➢ Pull the thrust a little bit when they grab you.
➢ Immediately reach over their left shoulder with your left hand and grip the point of your
dagger. Thus, you have trapped their left arm.
➢ Throw them down in front of you with the dagger or grab them by the throat.

A counter *against when they want to grip the dagger by the point. As soon as they raise their
arm:*
➢ Grip their left elbow with your right hand.
➢ Shove it strongly and upwardly away from you.
➢ Step behind them with one foot, so that they fall on their back.

If you take note that *they want to counteract your Middle Thrust:*
➢ Do not thrust farther than their hand.
➢ During this, drop down quickly with the hand reversed so that the pommel is in front,
which moves your right arm under their throat.
➢ Step Indesly with your right leg behind |III.9ᵛ| their left leg.
➢ Grab their leg with your left hand from inside above the knee and throw them over your
right leg.

A counter against a misleading

A counter against when *an opponent wants to mislead you, and moves their arm in front of
your throat:*
➢ Immediately grab their right hand with your right.
➢ Grip their right elbow with your left hand.
➢ Shove them away from you.
If you take their balance, they will fall on their nose.

If you stand with your left foot forward and counteract their High Thrust strongly from your
chest with your reversed hand:
➢ Remain strong and high in the counteraction and reach quickly through (up from below)
with your right hand behind their right arm, and grab onto your left hand with it.
➢ During this, step with your right foot well to their right side so that their arm arrives on
your right shoulder at their elbow.
➢ Push away from you, and you break their arm.
Take note: when you have grabbed their hand from below and pushed downward, you can
also take their dagger with your left hand.

Item, another. *If they thrust a High Thrust at you:*
> ➢ Counteract it strongly.
> ➢ As soon as you have counteracted, move your left hand around their arm so that it comes back in front of your chest, and press your elbow to your chest.

Item, another sequence. *If they thrust from above:*
> ➢ Counteract them strongly with the right |III.10| hand.
> ➢ In this counteraction, also grip their right hand with your inverted right.
> ➢ Jerk it toward you and drop strongly over their right arm in the center (behind the elbow).
> ➢ Push away from you so that you break their arm.

A sequence in moving through

Item, in the Onset, move the dagger into the center and carry out a Middle Thrust under their arms from your right.
> ➢ You then arrive in the left Low Guard.
> ➢ *If they thrust in response:* take it away from you with the pommel.
> ➢ Afterwards, thrust long at their head or face (from your left over their right).

Item, *if an opponent snatches or grabs your right arm in some sequence (whether with one hand or both):*
> ➢ Strike their joint strongly from below, or drop over their arm from outside with thrusts or strikes at the joint, or grab them to wrestle.

How you should impede their thrust

When you are dealing with an opponent that you are concerned about *who has a dagger:*
> ➢ Grab onto their closest hand with the same hand—thus, their right with your right, their left with your left.
> ➢ Regardless of which hand you grip, jerk their hand to the same [side].

If they have drawn their dagger with the other hand [that you aren't grabbing]:
> ➢ Reach with the other hand outside and over the same arm that you have pulled to yourself, and grip the other arm at the bicep with your other hand (as the tenants |III.10^V| indicate on the left[290] in Figure D). *Thus, they cannot thrust at you.*
If they had drawn their dagger with the same hand [that you initially grabbed], you can throw them or use another counter for that hand.

---

[290] The text reads "on the right", but this is corrected in the 1600.

In summary: grappling is most important for the dagger, and the grappling holds are not only completed using one hand but also with both hands. So that you can gain additional understanding of this, I will repeat some of this using a few examples.

The first grappling hold[291]

In the Onset, *if a fencer thrusts at you from above:*
- ➤ Raise your reversed left hand to catch their right hand at the joint (behind their dagger).
- ➤ Twist it around away from you and step well toward them with your right foot.
- ➤ In that moment in which you step, simultaneously move your right arm through under theirs and lift upward.
- ➤ Thus, you break their arm.

You can also execute all kinds of sequences with the dagger (or otherwise with grappling).

The second grappling hold

If an opponent thrusts from above:
- ➤ Catch their right arm by their wrist again (behind the dagger), not with a reversed hand as before, but straight with an open hand (so that your thumb points to their hand and your little finger points toward their arm in the grip).
- ➤ When you have seized them like this, you can pivot their arm toward you or away from you.
- ➤ If you pivot or twist it away from you, you can execute the sequences which |III.11| are fought with a reversed hand.

If you twist them toward you onto your left side, then take note:
- ➤ In that moment in which you jerk it around, twist it completely in front of your chest and grab their dagger with your right hand to jerk it out of their hand.
- ➤ Or, in that moment in which you seize their hand and turn it towards you: turn yourself away from them to your right side and thrust around behind in the back of the head (and whatever other similar sequences there are).

The third grappling hold

[This] takes place with both hands. *When they thrust at you* and you have your hands crossed:
- ➤ Catch their hand by the joint between your two hands.
- ➤ Jerk toward yourself, to whichever side you want.
- ➤ Afterwards (and only if you hold them firmly with the other), release one hand to grab somewhere else, wherever you please.

Take note in the Onset. *If they thrust from above:*
- ➤ Move your left arm well under theirs and catch the same.
- ➤ Wrap your arm around their arm (from the inside to the outside) and turn yourself away from them toward your right side.

---

[291] Unlike the others, this rubric reads *angriff* instead of *griff*; however, most of the text is devoted to grappling following the attack from above.

➢ Thus, you break their arm (as you can see in the uppermoſt tenants on the left side in Figure F).

Because I want to write more about the dagger at another point, I will leave it here, and only set down a few rules that are useful in fencing.

The firſt rule

Take note in the Onset: if you hold your dagger so that the blade projeċts next to your little finger, you can guide your dagger into whichever guard you want (whether low or |III.12| high, on the right or left).
➢ Therefore, pay attention and be diligent that you always thruſt firſt over their arm (whether from the inside or from the outside).
➢ With this thruſt over [their arm], you rip their arm downward and quickly thruſt at their opening.

Or, ſtrike with the pommel in such a way that, when you thruſt over their right arm from outside and rip them downward (as mentioned above):
➢ You move your pommel quickly over their arm and upward toward their face.
➢ *If they defend that and rise up:* move through under their right arm (while they are rising) and thruſt toward their face or cheſt from the inside.

If you thruſt from inside over their right arm:
➢ Rip [their arm] downward toward yourself or your right side.
➢ Pull and thruſt quickly from the outside and from your left over their right through their face (*before they lift their arm back up*).
Likewise, you should be diligent to thruſt the second toward their hand and arm (whether from below, across, or from above) and then follow with a powerful Cross Thruſt.

The second rule

The second rule is to move through. Therefore, position yourself in the High Guard and remain ſtrong in the counterpoſture with your arm extended upward in front of your face.
When they thruſt at you (namely from above or slantwise toward your face):
➢ Duck your head and bend your body downward.
➢ Move your dagger through under their arm *while they thruſt in.*
➢ In this movement through, ſtep well out to the side (toward that side to which you have moved through) and thruſt over their arm at their face (*while they are raising the [arm]*).
|III.12ᵛ| Likewise, you should also change through with your thruſts, that is, thruſt toward their face slantwise from above, and take note:
➢ *As soon as they lift to counter the thruſt,* move your dagger quickly through below.
➢ While you are changing through below, move your left hand in front of your face at the same time (in order to counter their thruſt) and thruſt at the opening on their other side.

The third rule

Furthermore, you should also observe and take note that you fence powerfully from both sides, namely with the left and right hand together and toward one another.

When they thrust at you from their left or right (from above or below):

➤ Move your left powerfully counter to that, and either strike their in-coming arm outward or catch it.

➤ In that moment in which you powerfully seize them with your left hand, simultaneously and at the same time with this: fence quickly at the opening.

Or, move your right arm (including your dagger) to assist the left in the grab so that you can more strongly turn their right arm outward, or you can weaken it and can therefore arrive better at the opening with your dagger.

Likewise, when you counteract with your dagger (whether you have it lying on your right arm or otherwise), move swiftly with the left hand to assist the right so that both hands meet together fast and one hand always follows the other and assists. This way, you can bring to bear all breaks and grappling techniques more strongly and quickly.

The fourth rule

[This] is the misleading. Execute it like this: hold your dagger in the center and move your left hand in front of your face to protect it.

➤ Threaten them by leaping with your |III.13| dagger pommel high to thrust at their head.

➤ Pay attention.

➤ *In that moment in which they rise to oppose:* pull the pommel completely around your head in that strike and thrust through horizontally toward their face with the front end.

Item, hold your dagger so that the blade projects in front of your thumb.

➤ Thrust toward their face from above.

➤ In that moment in which you thrust at their face, counteract them and protect yourself during this with your left hand.

If they rise in the meantime to counter your thrust:

➤ Don't complete your thrust.

➤ Instead, while you are still partly thrusting in, turn your High Thrust into a Low Thrust, and with that thrust, move under their arms at their face.

➤ Thrust long in front and threaten to thrust at them from below.

➤ *As soon as they move to counter that:* pull your thrust and move it somewhere else.

The fifth rule

The fifth rule teaches you to counter all kinds of grappling holds and is carried out in two ways: firstly, by twisting outward; secondly, by grappling in response.

Execute the twist outward like this. *If an opponent has seized your right hand with their left hand reversed:*

➤ Swiftly turn your dagger pommel through outside of their arm and under the same so that it comes to rest on the inside of their arm or tendon.

> Grip underneath your right [hand] with your closed left fist to assist the same.

> Using both hands, rip upward and outward with the pommel inside of their arm.

> Thrust a Middle Thrust and a High Thrust together, or seek the openings using other techniques.

However, *if they seize your hand with a non-reversed hand:*

> Pull the same swiftly toward you and twist toward their open hand.

> If you are too weak, then |III.13^V| bring the left to assist the right again.

For the second, *if an opponent has seized your right hand with their left hand reversed* (as previously stated):

> Grab their arm with your left over your right.

> Jerk toward you with both hands.

This allows you to gain the swinging movement and balance from them and can then work in response according to your pleasure.

In summary: always be diligent so that whichever arm *they have grabbed you with,* you seize it with your free [hand] and jerk it toward you. You take their balance and thus you gain advantage.

Item, *if an opponent grabs you by the arms to wrestle:*

> Swiftly seize their left hand with your left and jerk it toward you.

> Strike Indesly from outside with your right arm over your left (so that your elbow arrives in front on their chest or chin).

> Step immediately behind their left foot with your right and throw them from their feet.

Item, *if an opponent grabs you to wrestle but they do not hold you firmly:*

> Grasp their right hand with your right and jerk it toward you.

> Seize their elbow with your left hand and step in front of their right foot with your left.

> Swing them over that, or break their arm (that is, drop on their arm with your chest).

Item, if you seize their left hand with your left and jerk them toward you:

> Strike over their left arm with your right from outside.

> Seize their right arm with your right hand, step in front of them with your right foot, and swing them onto your right side so that they fall.

Item, *if an opponent grabs you by the arms* and you also grab them:

> Release your right hand and move it through below to hit the wrist joint of their right arm from below. (Regarding this, see the upper small and center tenants in Figure A.)

> |III.14| Use this strike to break through from below.

> After this strike upward, grasp them at the right elbow and reach your left hand under their elbow to their arm.

➤ Step between their legs or behind them with your right foot and shove them away from you.

When an opponent reaches around your body (whether with the left or right arm):
➤ Strike outwardly at their joint with the same arm that they reached under with.
➤ Turn yourself away from them.

Item, pay attention. *As soon as they settle strongly in their feet:* stomp on them.

Take note in all wrestling that they do not strike at your genitals. Therefore, pay attention. *As soon as they lift their foot to hit at your genitals:* strike outward with your knee against theirs. Afterwards, step toward them and stomp or strike.

Item, *if an opponent has closed with you or reaches for you with open hands:* observe so that you can rush for a finger. Break it upward. *They must release you* or you will otherwise gain the advantage.

Item, if you wrestle with an opponent in an equal hold:
➤ Observe so that both of your arms come below and grab them around their middle.
➤ Lift them from the ground.
➤ In that moment in which you lift them, strike at their leg with your foot and swing them to the other side. They will fall.

Item, *if an opponent has attacked [with their arms] around your body and wants to lift and throw you:*
➤ Set your knee hard between their two legs so that they cannot lift you.
➤ During this, pay attention to the side *toward which they want to swing you. As soon as they lift you,* support yourself on the other.

Take note. *If an opponent attacks you below on your leg to throw you* (as taught above):
➤ Drop on their body with your body and reach around their neck with your left arm.
➤ Press strongly toward yourself with [your] left arm and wait for your advantage with the right hand.

|III.14ᵛ| Take note. *If an opponent wants to grab forward to wrestle:* pay attention to *which arm they want to grab with first.*
➤ Strongly grip that arm by wrapping your arm around it.
➤ Grab the bicep of the other arm with your other hand and push the same away from you (as you can see in the upper tenants on the right side in Figure F).

Take note. *When an opponent grabs you by the shoulders or arms:*
➤ Strike up from below with both hands and separate their arms.

➤ Quickly drop into their legs with your bent body and pull until they fall.

Item, take note of *which wrestling [hold] they want to grab with*. Observe, *in that moment in which they reach in:*
  ➤ Pull your fists strongly to both sides of your chest and shove away from you and around you with your elbows.

You will swing yourself free. Seize them quickly according to your advantage.

Take note: if you have hurriedly seized an opponent by the hand:
  ➤ Jerk them to you and move your other hand around their neck so that your hand comes back to your chest.
  ➤ Grab onto your clothing and force them hard into you. You have thus caught them (as you can see in the lower center tenants in Figure F).

If an opponent seizes you with their right hand:
  ➤ Rotate it up and move through under their arm.
  ➤ Step between their two legs with your right leg.
  ➤ Reach outwardly around their leg with your right hand.
  ➤ Pull their right arm well over your shoulder, lift them upward, and throw them according to your desire.

|III.15ᵛ| In the grab:
  ➤ Seize their right hand with your right and jerk it toward you.
  ➤ Swiftly reach your left hand over your arm and theirs.
  ➤ Grab them by their left knee or their pants.
  ➤ Throw them on your left side (as you can see in the two tenants on the left side in Figure E). [292]

Everything else depicted in the Figures is inherently clear. Therefore, because it is described further in another location, I will let this go for now.

---

[292] This would seem to be a reference to the Figure on III.15, but the description matches Figure C.

The fifth and last section of this book, in which fencing with staffs, halberds, and long pike is taught and briefly dealt with.

I have organized these three weapons together in one Figure for this reason: because the pike is best placed above in the Figures due to its length and according to perspective. For this reason, each Figure will be designated with its own letter (as previously) so that the diligent reader should not be confused by this. I will thus initially take the short staff in hand as the foundation for all long weapons and will first show, teach, and describe how many stances there are, followed by how you should direct the same in the work.

About the stances or guards

There are primarily five stances: namely, the High Guard, extended straight up in front of you and on both sides; afterwards, the Low Guard, also on both sides; furthermore, you have two Side Guards and one Middle Guard, and finally, the Paddle Guard.

|III.16ᵛ| **High Guard [Oberhut]** [293]

Position yourself in the High Guard like this: stand with your left foot forward; hold your staff with the butt end at your chest so that the fore end faces straight upward toward the sky. As you execute this straight in front of you, you can also direct it to both sides in the work. Although you should always keep your left foot in front, you must not let the feet come too far apart from one another so that you can always step forward with the left foot.

Low Guard [Unterhut]

Execute [the Low Guard] like this: stand with your left foot forward; hold the staff with the butt end at your flank[294] and with the fore end extended in front of you on the ground. When you now hold the butt end on your right flank, then it does not matter whether you hold or direct the fore end extended to the left or right or straight in front of you. You can change this extension, either according to their fencing or according to the sequences you carry out.

....

[293] I've taken the liberty of removing the halberd heads from these two guard pictures.
[294] The soft part of the torso between the floating ribs and the hip.

Side Guard [Nebenhut] and Middle Guard [Mittelhut]

Position yourself into [the Middle Guard] like this: stand with your right foot forward; hold your staff with the middle section on your left hip, so that the shorter point and butt end point toward your opponent[295] and the longer end extends behind you. You present your right side to them fully (as this is taught to you in the |III.17ᵛ| lower portrait on the right side in Figure A).

The Side Guard[296] is the Simple Brace in front of the opponent, which staff fencers fence from the most.

Paddle Guard [Steuerhut][297]

Position yourself in this [guard] like this: stand with your left foot forward and hold your staff with the fore end on the ground in front of your left foot and the butt end up in front of your face with extended arms (just like you can see in the other portrait on the left side in the abovementioned Figure [A]).

You should also execute this guard like this: stand with your right foot forward and hold your staff behind you, again with the fore end on the ground. You are positioned to strike.

---

[295] As previously, the German word *Man/Mann* will be translated as "opponent", and references to pronouns will be the gender-inclusive "they/them", unless the pronoun refers to a specifically named Master. In addition, Meyer uses several other terms to refer to the opponent, which will be reflected in the translation.

[296] The text reads '*Mittelhut* = Middle Guard'; however, this twisted position for the Middle Guard on the left rarely occurs in the text and is used for striking from your left at their right with your right hand. Meyer rarely references '*Nebenhut* = Side Guard' in this section. In fact, it never appears again with respect to staff, appears only once in the Halberd Section as a synonym for '*gerade Versatzung* = Simple Brace', and appears as a completely different guard with respect to pike. With regards to staff, the Side Guard appears to be the guard presented on the lower right in Figure B, the stance that Meyer labels Simple Brace in the text.

Despite the initial label of *Nebenhut*, Meyer consistently uses '*gerade Versatzung* = Simple Brace' in this section. The polearm mirrors the angle of the weapons from the single-handed sections. The translation comes from the diagonal braces used in half-timbered building construction, which are called *Versatzung* in German and "brace" in English. The entire architectural structure, the *versatzung*, includes a vertical post, a horizontal beam, and a diagonal connection as a support or brace. A *gerader Versatz[ung]* has no additional offsetting wedge to affect its angulation, thus it is "simple" as opposed to complex.

[297] The *Steuerhut* mimics a person holding a steering pole for a flat-bottomed boat, or a standing position for paddling or poling this type of boat (or a modern paddleboard).

About binds and counteractions of the staff; also, their division

The staff is also divided into four parts, just as taught about other weapons previously. Therefore, you also have four binds: the first bind takes place at the front or outermost part of the staff; the second in front of the hand which the fencer holds forward on the staff; the third in the middle of the staff; the fourth is carried out with the butt end by running in.

You should pay special attention to this division and these binds because it can become problematic if you are not assiduous in fencing the appropriate technique in that part of the fight: namely, the strikes and flying thrusts in the first part and bind; the remaining, twisting, and pursuit in the second; and running in and wrestling in the others later.

|III.18| Like the binds, there are also four primary counteractions with the staff, of which the first is completed with the fore end of your staff from both sides; the second in front of the hand; the third in the middle; and the fourth with the butt end.

Because these are all understood sufficiently from the sequences, it is unnecessary to deal with each here in detail.

High Guard

In the Onset, position yourself into the High Guard and observe. *As soon as they thrust toward your left side:*
  ➤ Step out away from their thrust toward your right side and thrust simultaneously with them.
  ➤ In this thrust, turn the long edge toward their staff.
This causes them to miss with their thrust, and you hit with yours.
  ➤ *If they thrust toward your right:* step out from their thrust toward your left and thrust simultaneously with them as before.

The second sequence from the High Guard

Take note: position yourself into the High Guard in the Onset. *If they thrust at your body (from below or above):*
  ➤ Step out away from their thrust toward the other side (*when they thrust in at one side*) and simultaneously strike down on their front hand from above while stepping out.
  ➤ Take diligent note.
  ➤ *In that moment in which they pull the same:* thrust straight ahead of you toward their face.

|III.18ᵛ| Another, how you should strike down from above through their staff and rip back upward, and how you should hit with one hand in response

In the Onset, position yourself in the High Guard on the left (so that the fore end or longer part of your staff projects over your left shoulder). Step forward toward them with your left foot.

➤ *If they thrust toward your chest or your face:* leap well outward away from their thrust toward your right side and beat downward with your ſtaff (which you should hold firmly with both hands) through the middle of their ſtaff from above so that you arrive in the right Low Guard using this ſtrike.

➤ From there, *if they thruſt at your face again:* rip back upward toward your left shoulder with the half edge.

➤ Simultaneously with this, and in that moment in which you rip upward, give your ſtaff an impetus with your left hand.

➤ Release your left hand from your ſtaff during this sweeping motion and, with one hand, ſtrike across from your right toward their temple.

The high ſtrike with the upward rip should be completed quickly one after the other.

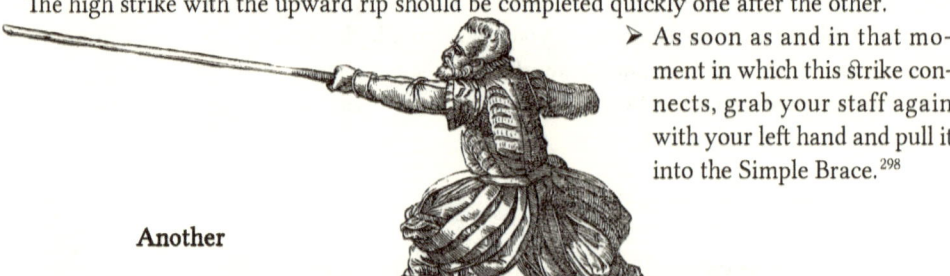

➤ As soon as and in that moment in which this ſtrike connects, grab your staff again with your left hand and pull it into the Simple Brace.[298]

Another

Take note. When you have now struck through their staff from above and have also ripped back upward from below afterwards and arrived with your left hand and the front |III.19| part of your ſtaff back upward:

➤ Immediately turn your right hand with the butt end upward and, at the same time, let your fore end and your left hand drop back down next to your left and out to the side. And use this to turn the longer fore end of your ſtaff back upward from below toward their right.

All of this muſt occur in one faſt motion.

➤ Immediately thruſt farther toward their face with a ſtep outward.

➤ Pay attention in your thruſt, in that you do not merely turn your right hand back down toward your cheſt. Ensure that you also shove the same hand well into your cheſt and inside on your left arm as you thruſt in front of you in toward them.

Thus, you have firſt learned how you should ſtep out and thruſt in simultaneously with them from the High Guard. Secondly, how to beat their ſtaff outward and down from above and thruſt in response. Thirdly, how you break through their ſtaff from above, and rip back outward and upward from below. And finally, also how you should carry out a Misleading Thruſt.

How you should thruſt in simultaneously with them from the Low Guard

Take note that you now hold your right hand with the butt end of your ſtaff on your right flank in the Onset, and you have your fore end forward on the ground out on your right side

---

[298] Note that this is *not* the Simple Brace from Figure A, but instead the guard as depicted by the quarterstaff fencers in the lower half of Figure B.

with your body bent well forward. Observe.

- ➤ *As soon as they thrust toward your
  face:* step with your right foot out
  toward your right side and step with
  your left farther toward their left.
- ➤ *In that moment in which they guide
  their thrust in,* thrust above their
  left arm at their face.
- ➤ You should also, in that moment
  in which you |III.19ᵛ| thrust in [sim-
  ultaneously] with them, drop your head down toward your right side well away from their
  in-flying[299] thrust [which now goes] over your staff.

You have counteracted much better in this way.

Another, how you should beat their thrust outward and thrust in response

In the Onset, position yourself again in the Low Guard as before (with your knee bent well
forward) so that your upper body is lowered following your staff.

Take note:

- ➤ *As soon as they thrust:* beat their thrust outward in one yank (from your right toward
  your left).

In this outward beat, do not strike farther than into the Simple Brace.

- ➤ *Before they recover back and straighten up from their thrust:* thrust toward their face with
  a leap outward.

Another

Take note. When you have dropped into the left Low Guard in the Onset *and they strike at
you with one hand toward your head from above:*

- ➤ Lift both arms upward and during this lift upward, leap toward them under their strike.
- ➤ Counteract their strike on your staff between your two hands.
- ➤ Immediately and in that moment in which the strike beats on your staff (and still touches
  it), pull the butt end toward you with your right hand and let the fore end drop downward.
- ➤ Guide the [fore end] toward their body (between their hands, under their staff), and thrust
  and at their chest (beneath their staff and |III.20| between their two hands).
- ➤ In that moment in which you thrust inward, turn your butt end and your right hand back
  downward toward your chest, so that you can guide the thrust [with the staff] hard on
  your chest and inside on your right arm.
- ➤ After completing the thrust, you should be back on their staff quickly with the bind so that
  you can better protect yourself from their fencing in response.

---

[299] Because German has directional prefixes *hin-* and *her-* to indicate movement away from or toward
an object, I have used "in-flying" and "in-coming" as translations of the adjectives *herfliegend* and
herkommend respectively, which Meyer often uses to describe weapon movements by both parties.

How you should yield out of their thrust from the left Low Guard, and how you should thrust simultaneously with them

In the Onset, step forward with your left foot, hold the butt end with your right hand on your right flank and let the front end of your staff lie extended in front of you on the ground (pointed a little out toward your left side). And take note:

➢ *As soon as your counterfencer thrusts in at you:* step out to the side with your right foot behind your left (a little toward their right side).
➢ In that moment in which you step behind [your left] with your right foot—as you are placing it down—quickly step farther toward them with your left foot (also toward their right side).
➢ *In that moment in which they thrust in,* thrust over their right arm toward their face.

How you should beat their thrust outward from your left Low Guard, and how you should thrust in response

Or, when you stand in the left[300] Low Guard (as mentioned previously):

➢ *In that moment in which they thrust in:* step again toward their right side away from their thrust (as previously) and beat their |III.20^V| staff away from your left toward your right (simultaneously with this).
➢ Afterwards (*before they recover back*), thrust quickly toward their face (as before).

How you should take out and upward from your left using the long edge, and how you should thrust from below and up toward their face through the Roses from your right

In the Onset, position yourself in the left Low Guard (as before).

➢ *If they thrust in at you:* lift both arms and beat their thrust outward and upward (from your left toward your right) with the front part of your staff, using the long edge, so that you arrive completely upward with your staff in this outward beat.
➢ Afterwards, turn your staff back next to your right and up from below.
➢ From there, thrust back upward toward their face.

How you should jerk their staff outward and thrust afterwards

Take note. When you arrive in one of the Low Guards in the Onset *and they do not want to work or thrust:*

➢ Let it be noticed and observed through your body language, as if you wanted to first consider what type of sequences you could fence.
➢ *Immediately and in that moment in which they extend their staff away from themselves:* jerk it out in an unexpected jerk or strike.
➢ *While they stagger with their staff due to the intercepted thrust:* thrust quickly toward their face.

---

[300] Corrected from *rechten*, because the rubric refers to the left Low Guard.

|III.21| In this outward beat, you should diligently ensure that you do not beat too far out to the sides with your staff (as was also recently indicated). Instead, beat their staff outward in one jerk (as taught) so that you are quickly back straight in front of their face with your staff, and you complete the thrust before they recover back again.

How you should fence from the Middle Guard

In the Onset, position yourself in the Middle Guard (just like the larger portrait shows on the right side in the Figure A printed previously) and observe:

> As soon as you can reach them, throw your staff across through their face using your right hand.

> During this throw, give your staff a strong impetus with your left hand—and release the [left hand] from the staff so that your staff can fly even faster through their face and around your head in this throw.

> In that moment in which your staff flies through their face and around your head, step toward them with your left foot forward and grip your staff again with your left hand (while your staff is still flying around in the air).

> Strike the second [strike] from your left toward their right through their face and also through toward their staff, wherever they hold it in front of themselves.

This strike should be directed with both hands, so that you arrive in the right Low Guard at the end of the strike.

> While your staff is dropping into the Low Guard, *they will quickly thrust toward your face* (because it is exposed by this drop).

> Therefore, step swiftly with your right foot toward your right side and thrust toward their face simultaneously with them, so that you turn the long edge |III.21^V| with the butt end toward theirs in this thrust.

Keep your head withdrawn well out of their thrust [which will pass] over your staff. You are thus counteracted.

Or, after you have used this strike to drop into the right Low Guard, *and they thrust at your exposed opening:*

> Rip their in-flying staff upward and out toward your left shoulder with the half edge.

> Simultaneously with this outward rip, guide your staff above and around your head and strike at them from your right outside over their left arm.

You should also direct this strike with both hands.

In the meantime (while you are guiding the strike around), observe *whether they want to thrust at your face.*

> *As soon as they do that:* guide the butt end around even lower in front of your face and let the strike fly even faster.

> *If they counteract the strike with the hanging staff,* take note: in that moment in which your staff beats on theirs or misses, immediately turn the butt end upward and thrust at their body (above or below their staff).

Another, how you should reverse in front of them, or how you should expose yourself, take them out, and hit in response

Position yourself in the Onset in the form mentioned previously (in the Middle Guard on the left side).

> Immediately step toward them with your left foot behind your right, so that you turn your back to them in this reversal.
> In that moment in which you reverse in front of them, *they will hastily thrust at your face, in the opinion that they can rush you.*
> Therefore, during this step behind, quickly raise both hands upward with the butt end of your staff extended toward their left (so that the fore end of the same hangs toward the ground).
> |III.22ᵛ| During your turn, beat their incoming thrust out to the side with your hanging staff (from your right toward your left).
> Let your staff move completely around your head in one sweeping motion.
> In that moment in which you move your staff around, release your left hand (after you have given your staff a strong impetus with the same) and strike a strong, swift strike at their left ear.

This is a swift sequence that works well in the first attack, because you provoke them to thrust with your reversal. *If they thrust,* you take out their staff simultaneously with your reversal and you will certainly hit them (if they have executed a serious thrust).

I wanted to place the previously-taught sequences from the side guards first so that, when you arrive in one of them (through vigorous strikes, thrusts, or counteractions), you would know how to recover again more properly, and also so that you know better how to direct yourself in the following sequences. This is because you always move out of one [guard] into another when fencing with these long weapons, in which [guards] you are not given the option of long consideration about what you should do but must press forward with the next sequences that arise from the previous [sequence] (just like with the weapons from the previous instruction).

Now, position yourself in the Onset in the Simple Brace, as I have named it here (as is depicted and taught by the two tenants in the previous Figure [B]).

The first sequence in binding at the outermost point

When you bind with the outermost part of your staff on the outermost of theirs:

> Shove them out in an unexpectedly strong jerk to the side (yet such that you do not move too far with yours in following this outward shove).
> Then thrust quickly away from their staff and forward at their face, and do this hastily, before they have recovered from the outward shove.

|III.23| **Another, how you should move through after the outward jerk**
and thrust at the other side

When you observe, in this type of outward jerk, that *they will arrive quickly back with their staff* (so that you cannot rush to the thrust as was taught), then do this:

➢ Jerk their staff again to one side (as before) and let yourself be observed [as acting] as if you wanted to thrust like before.

➢ *Immediately and in that moment in which they rush toward you with their staff (as they believe that they can counteract your thrust):* move through under their staff *while they swipe inward* and thrust powerfully at the other side at their face with a fast leap outward.

This is a swift transition, because you jerk their staff unexpectedly outwards and afterwards move quickly through underneath and thrust in at the other side.

Another, how you should jerk the staff outward and strike at their forward leg

In the Onset, bind them from your left side[301] with the outermost part of your staff on the outermost part of theirs.

➢ Shove them outward in an unexpected jerk toward their left side.

➢ Pull your staff quickly back around your head toward your left. In doing this, release your left hand from the staff and strike with one hand strongly across through their foot[302] from your right, with an additional step forward of your right foot.

➢ Afterwards, grab your staff again with your left hand (while it is still moving through this same strike).

➢ Strike the second slantwise through with both hands (from your left |III.23ᵛ| toward their right shoulder), so that you arrive in the right Low Guard at the end of the strike.

➢ From there, thrust at their face using the form described above.

Or, when you strike across through their forward leg, ensure that you grip your staff with your left hand back on your left side during the beat through.

➢ As soon as you have gripped it [again with your left hand], pull the butt end toward your right [side] and on your chest, moving the left [hand] forward on the staff with arm extended.

➢ In that moment in which you pull your hands apart on the staff, turn your staff toward theirs and beat it outward (*in that moment in which they thrust in*), so that your staff moves powerfully and strongly back into the Simple Brace with left arm extended.

➢ Afterwards, thrust quickly straight in front of you at their face.

---

[301] The original has *lincken Hand*; however, "left side" was used for clarity in the sequence. This is repeated in the first instruction to jerk at their left side, instead of left hand.

[302] Meyer later defines the "foot cut" as the entire lower leg, from the knee downward—whichever point is farthest forward and can be reached.

A sequence about how you should execute the Brain[303] Strike [Hirnſchlag]

Execute it like this in the Onset:
- ➤ Bind onto the outermoſt part of their ſtaff with your outermoſt part.
- ➤ Let your body language be observed [to mean] that you were seriously looking at where or how you could thruſt toward their face.
- ➤ *As soon as they take note of that, they will diligently observe your lift away, so that they can thruſt quickly in response in that moment in which you rise upward.*
- ➤ Therefore, when you are aƈting with great seriousness (as if you wanted to thruſt upward), yank your butt end swiftly upward.
- ➤ Sweep the ſtaff with your left hand back toward your left and around your head and hit them unexpeƈtedly ſtraight from above at their head (with one hand).
- ➤ *If they were already thruſting during this,* it will not land because you are too swift with the ſtrike to their head.

This and similar |III.24| sequences rely greatly on praƈtical training. Namely, that you overwhelm your counterfencer with unexpeƈted speed when they expeƈt it leaſt.

Another with the Head Strike[304] [Schöfferſtreich]

Take note. When you have bound onto your opponent (as taught previously):
- ➤ Secretly reverse your right hand on your ſtaff and mislead them with body language while doing this *so that they do not observe your intentions.*
- ➤ Then, *when they leaſt expeƈt it:* ſtep with your right foot swiftly toward them and ſtrike a powerful and faſt ſtrike at their head ſtraight from above (with your palm upward), so that after the ſtrike you ſtand with your upper body bent well downward.
- ➤ Rise quickly back up with your ſtaff and also ſtep simultaneously back with your right foot.
- ➤ During this, grip with your left hand back on your ſtaff[305] so that you can ſtrengthen your counteraƈtion.

You can also create the space for the preceding Brain Strike or this Head Strike, namely by initially yanking their ſtaff outward (or otherwise hindering them with other techniques), so that you can rush with the Head Strike before they rise back up.

---

[303] *Hirn* by itself refers to "the brain, the seat of intelligence". In compounds, it commonly refers to anatomical features of the brain or skull, except for *Hirnschnalle* and *Hirnschneller*, which are swift blows to the head. *Hirnschlag* in NHG is a medical term for a stroke.

[304] *Schöffe* leads to *schöpfen/Schöpfer* in Grimm, which is "the creator, or head". *Schöffe* has many references to a judge as the head of an authoritative or governmental body. GRIMM, vol. 15, col. 1441. Alternatively, it might be a typo for *Schleffe/Schläffe*, which is "the temple of the skull".

[305] As there is no previous instruction to release the staff, it is unclear how to interpret this clause.

How you should strike around and away from their staff and shoot over [it]

Furthermore, when you can reach the outermost part of their staff with your outermost part in the Onset, and they are hard on your staff, pay attention:

➢ *As soon as they want to shove you forcibly outward:* pull your staff quickly around your head with both hands (*in that moment in which they shove outward*) |III.25| and strike at their head with the same from outside and over their left arm with a step outward.

➢ As soon as this strike hits, shove your staff swiftly over theirs close to their hands (as you can see afterwards in Figure G).

➢ When you have thus captured and blocked their staff, you can then move in and thrust with the butt end or strike for their face with the longer part.

➢ *If they lift their point and work out from under your staff:* follow them down, whether with thrusts, twists, or shoves.

How you should move through

Take note. *If your counterfencer is hard on your staff with their bind and shoves you away from themselves:*

➢ Move through underneath and thrust at the other side.

Or, *in that moment in which they shove outward with their hard bind:*

➢ Move through underneath close on their staff (*while they shove*) and yank out with an aggressive strike from the other side.

➢ Afterwards, thrust quickly (*before they have recovered*).

Another

If an opponent binds hard on your staff, then hold back hard against them with your bind.

➢ *If they also shove against yours:* move swiftly through below, and act as if you wanted to thrust.

➢ Do not do that, but instead yank back through below and thrust at the side toward which you had initially bound on.

|III.25ᵛ| ## How you should learn to feel[306] in the bind

Take diligent note: when you have bound onto an opponent from your left side, observe diligently and feel with equal diligence. *As soon as they lift away from the bind, whether through below or otherwise to [some other] work:* thrust straight in front of you toward their face while they are still lifting.

Another, [which] is the usage[307] against the former

When you perceive in the bind that *your counterfencer is paying attention to your lifting away and wants to thrust at your opening while you rise:*
- ➤ Allow your body language to be observed, [acting] as if you seriously wanted to lift from their staff and thrust.
- ➤ When you believe that *they have positioned themselves in the best way to thrust in response:* move your staff abruptly away from theirs out to the side (as if you wanted to thrust, as stated).
- ➤ Do not do that, but instead, *in that moment in which they rush with their thrust,* beat it outward to the side and only then thrust completely in.

When they rush so abruptly, you can easily take their staff out and rush them *before they recover back.*

You should also take note and pay attention to *what your counterfencer wants to execute and fence at you,* so that you catch them in their own techniques, as occurred previously in the case of *that opponent who is inclined to thrust quickly in response.* Therefore, you must expose yourself carefully and with intent, and present yourself with body language as if you had accidently or unknowingly exposed these openings or as if you had dropped too far with your eager thrust in response, *so that they are provoked to thrust sooner and with greater desire. They will miss and expose themselves with these thrusts or strikes,* such that they only pull back up with |III.26| difficulty and recover, before you have rushed them. This will be explained further using examples in the Halberd [Section].

A misleading sequence

When you have bound on with an opponent in the Onset, and neither wants to lift from the other staff:
- ➤ Thrust at their forward foot with serious body language.
- ➤ In so doing, you expose your face; *they will quickly thrust at that.*
- ➤ *Immediately and in that moment in which they thrust in:* step out to the side with your forward foot. Follow with the right [foot] and thrust from below and over their staff toward their face (*in that moment in which the [staff] flies in to thrust*).
- ➤ Withdraw your head behind your staff well away from their thrust.

---

[306] The rubric has '*fehlen* = missing', not '*fühlen* = feeling'.

[307] *Brauch* is "usage", while *Bruch* would be "break/counter". Due to the content, this sequence both 'uses' feeling and 'breaks/counters' their thrust.

- You will hit them in their face *while they thrust in.*

Or, when you thrust or strike at their foot and *they thrust at your face during this:*
- Beat their in-flying thrust outward and simultaneously leap out of their thrust to the side during this outward beat.
- Afterwards, thrust quickly and swiftly.

How you should thrust over their left arm at their face from outside with one hand, twist through with the butt end, and strike at their right shoulder

If you have bound onto an opponent in front (from your left toward their right) and *they remain still and do not work:*
- Step to your right side with your rear right foot, and with this [step], go through underneath with your fore end close on their staff.
- Thrust quickly and unexpectedly over their left arm at their face from your right.
- |III.26ᵛ| As you thrust in, release your left hand from the staff and follow the thrust well with your right side so that you can reach even farther in from across.
- In this thrust, turn your right hand toward their left with the butt end of the staff upward.
- Use this to yank your staff around your head, leaping quickly to your left side during this yank around.
- Strike slantwise toward their right shoulder.

This strike (including the thrust) should be executed quickly, one after the other and together.
- Then leap back so that you can catch and grip your staff securely with your left hand.

Another about how you should twist through with the thrust

Execute it like this. *If you find them in the Simple Brace in the Onset:*
- Thrust straight from your right toward their left hand (*in which they hold the staff in front*).
- However, at the beginning [before you thrust], let your body language be observed, [acting] as if you wanted to thrust into their face.
- When you now approach their hand with your fore end, move through under their staff (stepping well out toward their right side with your left foot).
- In this outward step, your head should follow well along.
- In this thrust through, turn your fore end over their right arm from the outside and at their face.
- When thrusting in, turn your right palm well upward on the inside of your left arm so that the thrust goes much deeper.

A swift and skillful thrust to use against an opponent who does not work, but remains strongly in the [Simple] Brace

Take note. When you find *your counterpart in the Simple Brace in the Onset:*
- Position yourself in it as well, and let your body language be observed, [acting] as if you wanted to initially look around and consider what you |III.27| should fence.

➢ During this, *when they are paying the least attention,* step with your right foot swiftly out toward their left side.
➢ Thrust directly at their chest over their left hand (*which they have forward on the staff*), but execute this so that you do not touch their staff with yours.
➢ In this thrust, move your right hand well toward your left arm and inward along the same.
➢ At the same time, turn your left palm upward so that the thrust goes deeper.

You will hit in the same way that is presented in the portrait on the left side in Figure E.[308]

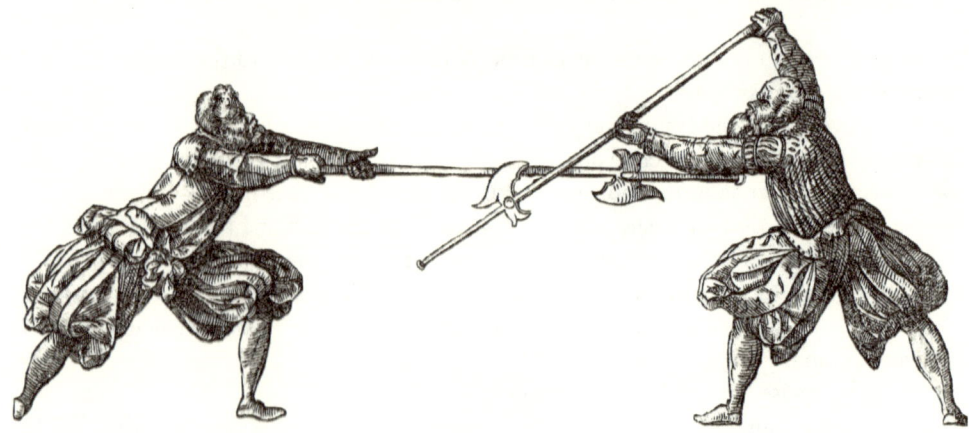

Another, how you [should] thrust upward through their face with a thrust

If your counterfencer crowds you in the bind:
➢ Remain with your staff hard on theirs as well.
➢ As soon as you both have come close enough that the staffs touch one another at the beginning of the second part [of the staffs], then during this [approach], remain hard on their staff with the bind and thrust the butt end away from you with your right hand (so that your fore end faces toward their right shoulder on their staff).
➢ At the same time, step with your right foot well out toward their left side and thrust toward their right shoulder with your staff (which you keep hard on theirs).
➢ During this thrust in, turn your right hand (with the butt end) back around toward yourself and toward your chest (so that your fingers are on your chest and your palm is upward).

When you thrust toward their right shoulder in this way while you remain hard on their staff with yours, and you turn your butt end back toward yourself in the thrust inward, then your thrust angles upward and hits them in the face. However, it must be carried out and completed in the work very quickly and with strength.

➢ |III.27ᵛ| Simultaneously with this thrust, guide your staff upward with both hands and strike quickly down from above toward their face.
➢ Leap with your right foot even farther toward their left side during this strike.

---

[308] This appears to be the halberd fencer in the foreground on the left in the Figure marked E, not the staff fighter standing over a stray E in the Figure designated as F.

Another, how you should intentionally miss a thrust in front of them, close on their staff, and strike long in response

In the Onset, do not approach your opponent too closely and take note. *In that moment in which they do not want to work:*

➢ Thrust toward their right side close to their staff.
➢ *As soon as they defend against the thrust and rebuff it toward their right:* let your staff drop downward next to their right and pull with your right hand around your head.
➢ Hit a swift strike at their left ear with one hand.

A good strike while extending past

Execute this in the Onset, as soon as you can reach the outermost part of their staff with your outermost part:

➢ Hold your fore end directly in front of their face and at the same time, turn yourself well to your right side so that you turn your back to them.
➢ In that moment in which you turn your back to them: step with your right foot behind your left toward them, turning yourself completely around about your right side with this step.
➢ Strike around with one hand (namely with your right) straight from above at their head.

This strike works very well if you do it correctly, *because they cannot reach you even if they thrust inward in that moment in which you turn around to strike,* because you have bound them on their outermost point.

|III.28ᵛ| *If they do reach you then they touch the opening on your back.* You, however, certainly hit them *when they thrust in.* Also, your strike moves so swiftly that *they cannot execute another thrust before it hits.* You can also execute the strike crossing from the middle when you turn like this.

Another with a Middle Strike

Thus, when you can reach their staff in front with your forward part in the Onset:

➢ Pull your staff suddenly around your head and strike across from your right toward their left ear with one hand.
➢ With this strike, step with your right foot well toward their left side.
➢ In that moment in which your strike connects, grip your staff again with your left hand close in front of your right, and pull your staff back to your chest with your right hand.
➢ In that moment in which you pull your staff to you, move your left hand forward on the staff until your arm lies extended on the staff and your left hand guides the staff (like a shield) in front of your face.

Pulling the thrusts

When you have bound onto your counterfencer, or when you stand before them in the counterposture and they do not want to work:

> Thrust seriously at their face with another step forward of your left foot and take diligent note during this:

If they are ready to rebuff your thrust and to counteract it, do not complete it.

> Instead, pull it hastily back toward |III.29| you through your left hand (so that you have that same left hand fully extended in front of your face).

> Simultaneously, in that moment in which you pull your staff back to you, present yourself with serious body language, [acting] as if you wanted to move through below and thrust at the other side.

> For this pulling and body language, you must lift up and place your front foot back down in masterful acting, so that it does not appear otherwise than that you already and ineptly thrust at the other side.

> However, in that moment in which they now move out to the side to counter your thrust and to rebuff the same: thrust back straight in front of you at the same point from which you initially pulled back.

All of this must be carried out quickly and must be completed seriously in all its details.

When they work before you and were to thrust at you:

> Set their thrust off in one jerk using your staff and execute a thrust quickly in response to the set off or the counteraction.

> In that moment in which you thrust, pull it back to yourself along the path of that thrust as if you wanted to thrust through below. *They will want to hastily oppose the same.*

> *In that moment in which they swipe to the other side (intending to counteract):* thrust straight in front of you at the point from which you pulled back.

You can also counter all stances using these pulled thrusts. As an example:

> If you find your counterfencer in the Low Guard on the left, thrust straight toward their face and observe.

> *In that moment in which they lift their staff to beat your thrust outward:* pull your un-completed thrust back toward yourself a little and quickly move through below and thrust at their left side and over their left arm at their face (*while they are still rising upward*).

> *If they follow your staff,* then move back through below and continue until you observe your opportunity to rush at an opening.

|III.29ᵛ| **Twisting**

If your opponent twists hard on your staff (from their left toward your right) and forcefully pushes hard in at you with a Simple Brace so that you dare not yield with any part of your staff:

> Remain on their staff with the bind close in front of their hand and forcefully push the fore end toward their face *so that they are forced to rise [to counteract].*

> *As soon as they lift their staff a little bit higher:* remain steadfast with the fore end on theirs (in that moment in which all of this occurs) and twist quickly with the butt end above and over their staff (from your right toward their left).

➤ Press them downward and strike them on their head with the fore end so that your left hand passes across over your right (as this is shown to you by the middle tenants in Figure D).

Another

In the Onset:
>   ➤ Bind strongly on their staff from your right side toward their left and work again toward their face with the fore end *so that they are forced to move their staff somewhat higher.*
>   ➤ *Immediately and in that moment in which they lift upward:* bend your body and leap in toward them with your right foot (under their staff).
>   ➤ During this, remain with your fore end hard on their staff (as previously) and move the butt end through under theirs during the leap.
>   ➤ Turn the [butt end] over their staff on their right so that the fore end follows after, with which you strike them on the head.

Or:
>   ➤ In that moment in which you have twisted over, shove the butt end downward and rip the same simultaneously outward.
>   ➤ Thrust toward their face with the fore end.

|III.30| However, *if they are too strong when they press upward* so that you cannot force their staff downward with your butt end:
>   ➤ *While they are pushing upward* and in that moment in which you are forced to lift your butt end, twist the fore end upward from below and next to their right arm at their face.
>   ➤ In that moment in which you twist your butt end over their staff, *if they lift their butt end and want to twist over from above:* turn your fore end swiftly over their right arm (from your left toward their right) and over and around their head, and catch them with your staff around their neck.
>   ➤ Yank with this toward your left side and toward yourself.

Or:
>   ➤ Bind on from your right toward their left and keep your fore end hard on their staff.
>   ➤ With a step forward of your right foot, turn the butt end inward between their hand and their staff from below.

➢ Yank upward and out (as the middle tenants show this in the subsequent Figure F).

➢ Work further toward your advantage.

Item:
➢ Bind onto their ſtaff from your left toward their right close in front of their hand and remain with the same fore end hard on their ſtaff.
➢ Leap with your right foot forward and twiſt the butt end over their ſtaff and over their right shoulder and around their neck.
➢ During this, ſtep farther with your right foot behind their left.
➢ Throw them over your right leg (as you can see this in the previously printed Figure C).

[How to] take a ſtaff

It often occurs that both ſtaffs are bound together in the center. When this happens to you:
➢ Remain with your ſtaff on theirs and release your left hand.
➢ |III.30ᵛ| Reverse it and grab both ſtaffs with it.
➢ Subsequently move your butt end through under theirs.
➢ Yank the right hand upward and toward yourself.
➢ They have to release it.

Or:
➢ If you step with your right foot behind, they must trip.

Driving [with diagonal slashes] [Treiben]

Fence this in this way. When you have your right foot in front in the Onset and you likewise also have your left hand forward on the staff:
➢ Lift your staff with both hands upward toward your left shoulder.
➢ With another advance of your left foot, strike through from above toward their right on their staff (strongly at their fingers) into the right Low Guard with your body bent well forward.
➢ Then, rip strongly back upward, through their staff, into the left High Guard.

They can thrust between these as they want: you will take away [the thrust] with the strike downward from above and rip it outward and upward from below.

Execute these together—one strike, three, four or five—with strength from below until you see an opportunity to thrust.

A sequence from Driving [with diagonal slashes]

As soon as you can reach your opponent in the Onset: immediately drive with slanted slashes from above and below, close together (as just taught).
➢ When you have carried out one strike (or four), execute the last as if you wanted to strike toward their right shoulder even more seriously and strongly than before.
➢ However, when striking downward, move your staff cunningly and quickly through below close to their staff.
➢ Step well out toward their left side during this movement through.
➢ In that moment in which you have completed the move through, thrust from the other side over their left arm at their face.

|III.31| Observe diligently in all fencing that you do not permit a snatch away or allow yourself to be misled. Also, do not just thrust in the Before without particular advantage. If you find your counterfencer to be in a guard that exposes an opening to you, you should not thrust rashly at the same. Instead, see whether you can entice them upward with pulled thrusts and then change through afterwards.

If they tarry too long in their preferred stances, you can probably rush them unexpectedly when they expect it least.

If you have bound onto them and cannot thrust with certainty (in the Before) due to potential exposure:
➢ Execute a thrust close to their staff.
➢ While you are thrusting in, feel whether *they want to take your thrust out or beat it outward.*
➢ As soon as you sense that, move your staff through below and assist their staff [in moving] completely toward the side *toward which they would have beaten outward.*
➢ Or thrust in at the other side *while they are beating outward.*

If you sense that they want to thrust simultaneously with you, do not be too prompt with your thrust. Instead, hold yours back secretly or unnoticeably until they completely commit to the thrust inward.

➢ *As soon as they thrust in,* move their staff outward using your thrust inward.

➢ Immediately complete the thrust you initiated.

Thus, you should not be too prompt in all sequences, but instead observe what they are planning to fence at you so that you can counter them even more appropriately.

Another from the move through

Drive them again [with diagonal slashes] as before through their staff—once, twice—and when they expect it least:

➢ Drop through steeply under their staff and then rip their staff swiftly outward from your right toward your left.

➢ Let your staff move around your head and afterwards, strike long with one hand.

Before I finish with this weapon, I also want to review and go through the others, because it is clear that these three weapons are fenced from one foundation.

|III.32| **About the halberd**

Although my intention is not to describe each and every cut and thrust in detail, I must take the time here to set [309] these six cuts for the halberd at the beginning. They are useful both for practice—in that they can aid your body in learning greater quickness—and (indeed even more so) because each of them is necessary for those who want to become skilled in this weapon. Thus, you can learn these things in particular and force them back away from you in a virile manner. You can cut as follows.

Cross Cut from above with the halberd

Hold your left hand in front on the shaft and move your left foot forward as well in the Onset.
- ➤ Cut the first from above with extended arms (from your left and toward their right) slantwise through their face. [Let it move] through behind and next to your right side, such that your blade shoots back up and forward next to your right side, so that the point of your halberd faces toward your counterpart's face, just like the Ox, about which you were previously taught.
- ➤ From there, rip obliquely downward toward your left with your hook (such that your halberd blade also runs through behind and next to that same side), [then] back up and forward so that you arrive with your halberd on your left in the High Guard (for the same cut as previously).
- ➤ Cut slantwise through their face again from above your left (as before).

Continue that further as just taught: one cut somewhat behind and somewhat forward toward your opponent's face, following through the oblique cross according to the instructions [of the lines].

Cross Cut from below

Stand with the left foot forward and hold your halberd in the High Guard on the left (as previously).
- ➤ From there, cut obliquely through their face with the edge of the blade (from below next to your left and upward toward your right). In this Low Cut, move both arms high so that at the end of the cut, you |III.32ᵛ| hold your halberd with both hands on your right side and high next to your head, so that the point faces again toward your opponent's face.
- ➤ Along with this [halberd movement], turn yourself well to your right side so that you expose half of your back to them.
- ➤ Immediately slash through their face (from below next to your right side and slantwise upward toward their left), so that your halberd extends behind your left shoulder at the end of this slash.

You now stand in a High Cut that is drawn together.

---

[309] The positive English sentiment (I must do something), is a translation of the German negative (I don't want to avoid doing this), which sounds contrived in English.

➤ From there, turn the sharp edge of your blade back downward next to or behind your left side, and cut again next to your left side and slantwise upward from below toward their right through their face.

➤ With this Low Cut, turn yourself again well toward your right side so that you can slash again (as previously) with the hook of your halberd next to your right, upward from below toward their left through their face.

Driving [slashes] diagonally through an opponent

Stand with the left foot forward as always and hold your halberd in the High Guard on the right (in the way that the outer and upper small portrait shows on the right hand [Figure G]).

➤ From there, slash your hook at their face (downward toward their left, through toward your left) so that your halberd extends behind you toward the ground.

➤ From there, immediately cut back from your left with the edge of the blade toward their right and obliquely slantwise upward through their face, until you are back in the previous right High Guard.

➤ From there, slash with the hook back downward toward your left.

In this way, continue to aggressively slash your halberd slantwise over your forward thigh from above, and [back] upward from below strongly through their face. One time, several times, until you observe your opportunity, as you will gather from the sequences.

Another Driving [with diagonal slashes]

In the Onset:

➤ Cut slantwise from above through their face with the blade (from your left toward your right) so that you arrive in the Low Guard on the right with this cut.

➤ From there, slash quickly back toward their left and slantwise upward through their face, back up into the High Guard on the left.

In this way, you should |III.33| break through swiftly and strongly from below and above opposite one another, so that you can remove all of their thrusts. Observe your opportunities while doing this, as you can rush to an opening between these.

You should always be able to execute these four Driving Cuts[310] in combination—namely, the first cuts through the cross and the other two upward—and be able to alternate between them. Other types of driving [diagonal slashes] will follow from these. However, I want to set some sequences in order, one after another, [to show you] how you should fence using this and similar weapons.

---

[310] *Thribhäuw.*

The first sequence from the High Cut

When you approach your opponent in the Onset in the Simple Brace or the Field Guard,[311] so that both blades can bind together or touch:

➤ Lift your halberd swiftly upward, together with your left foot (which should be placed forward).

➤ Cut through quickly and unexpectedly down from above toward their right [side], close to their halberd and toward their hands.

➤ In this cut downward, change through underneath their halberd and thrust immediately at the other side—that is, toward their left side and at their face.

The second sequence from the High Cut teaches you how you should pull your halberd toward your left, around your head, and cut through toward their left; afterwards, you should thrust toward their right at their face

If you hold your left hand forward on the shaft of your halberd, then take note: as soon as you can reach their blade with yours:

➤ Lift your halberd swiftly upward with both hands and cut strongly through down from above and from the outside toward their left arm (*which they have in front on their shaft*), including an additional advance of your left foot, so that you stand at the end of the cut with your upper body bent well forward in a wide stance.

➤ *They will swiftly thrust in response to this cut through.*

➤ Therefore, take diligent note that you step out with your right foot behind your left (*in that moment in which they thrust*), and while |III.34| you are stepping outward, you beat their in-flying halberd out with your blade.

This is because *they weaken themselves when they thrust swiftly in response.* Therefore, you can easily beat their halberd outward and weaken them even more, so that you can safely thrust well at an opening (*before they recover*).

Another from this High Cut

Take note: when you approach them in the Simple Brace and *they do not want to work:*

➤ Lift your halberd swiftly again (as previously) and act as if you wanted to cut at their left arm from outside.

➤ However, you should not let this cut hit, but instead, *while they lift their halberd to counteract in response to your cut:* move through under their [halberd] and thrust at the other side at their face (with an outward step toward the same side).

➤ In this thrust in, because you weaken yourself somewhat and expose yourself, *they will quickly rush at you with thrusts.*

---

[311] On III.39, Meyer explains that this term comes from a similar defensive stance used for pikes "in the field". Unlike the Simple Brace for pike, in which the right hand is quite close to the butt end with a very long extension toward the opponent, in the Field Guard the shaft is held in the middle. For halberd, the hand position does not appear to change between the Simple Brace and the Field Guard.

➢ Take diligent note of their thruſt during your thruſt, and *while they thruſt in,* move your blade over their shaft.
➢ *In that moment in which they thruſt,* rip their shaft toward you with the hook of your blade.
➢ In this way, you seize their hook and take their halberd.
➢ As soon as you have ripped their halberd somewhat outward, thruſt swiftly back ſtraight in front of you toward their face.

Rule

As often as you now cut at their arm or hand (down from above and through next to their halberd), whether this occurs in order to cause them to rise up or to entice them from their advantage or to injure their hand or arm; [you should] take diligent note (in that moment in which you drop through with your halberd under theirs) as to *whether they want to thruſt in response to you.*
➢ *As soon as they do that,* move quickly back upward with your horizontal blade and beat their in-flying thruſt upward and outward from below.
➢ Thruſt quickly and powerfully in response, *before they recover.*

From the Low Cut

In the Onset, position yourself in the long Simple Brace.
➢ From there, lift your halberd upward into the High Guard, and in that same time period while you are lifting your halberd upward, reverse your left hand |III.34^V| on the shaft (so that you hold it as the portrait teaches on the right side of Figure H).
➢ Cut through slantwise into the High Guard with the blade (from below next to your left side and toward their right).
➢ *If they thruſt in between these movements,* rip their thruſt upward and outward with the Low Cut.
➢ Afterwards, slash obliquely back downward toward your left with the blade and let it move around your head.
➢ Cut with one hand ſtraight and long from above.
This sequence works extremely well when you carry it out faſt.

Another

➢ Bind onto them from your left toward their right with your blade on theirs and jerk the same out and away from you in an unexpeƈted shove.
➢ Pull quickly back around your head and toward your left.
➢ After you have given your shaft an impetus, release the left hand in this pull around and cut across from below at their head with one hand (from your right toward their left).

➤ Immediately twiſt your right palm back away from you and recapture your halberd with the [left hand] (which is ſtill moving upward from below).

➤ Afterwards, cut quickly down from above toward their head with both hands, with a ſtep outward.

➤ You should be able to suppress[312] their halberd down with this cut (*if they were to thruſt at you during this*).

➤ Afterwards, thruſt quickly in response.

A sequence from the Middle Cut

When you approach your opponent in the Onset, lift your halberd into the High Guard (as the por-trait teaches on the left side in Figure G).

➤ From there, cut across toward their right with both hands and ſtrongly through their face and their halberd.

➤ Afterwards, turn your blade swiftly upward and slash immediately back from your right toward their left, also strongly across through their face and halberd.

➤ Use this to pull your halberd toward your left with both hands around your head and to cut down from above toward their left arm, which they hold in front, and through their face.

➤ Finally, thruſt ſtraight at their face in response.

You use these two Crossing Middle Cuts to rip their halberd outward from one side to the other, and you force them *so that they are fearful with regards to how they can retain their halberd ſtraight forward with ſtrength;* therefore, you can easily rush them with the High Cut *because they are so concerned about maintaining position.*

|III.35| ### Ripping [Reiſſen]

Take note. *As soon as an opponent has bound their blade onto yours:*

➤ Move your blade a little over theirs on their shaft so that you move in over theirs with yours.

➤ Turn the blade over their shaft and rip quickly and ſtrongly downward toward yourself and rip their weapon at the Weak.

➤ Following this, thruſt quickly on their halberd upward toward their face.

➤ *If they rush to rise upward to defleċt the thruſt:* change quickly through below (*in that moment in which they lift upward*) and thruſt toward the other side with a ſtep outward. This works on both sides.

---

[312] *Dämpfen/dempfen* means both "to bring something to an end using violence or force", and "to suffocate or smother". Both of these meanings are supported in the text.

How you should act when using the change through and rip against an opponent who thrusts fast in response

When you have bound onto an opponent from your left side:
> ➤ Change through with inattentive and negligent body language *so that they will be provoked to thrust early.*
> ➤ *As soon as they now thrust in* (during your change through), turn your blade over their shaft and rip toward yourself and toward your left (just as you previously ripped toward your right).

Item:
> ➤ Thrust carefully at one side instead, and take diligent note *whether they want to thrust simultaneously with you* (in that moment in which you thrust in).
> ➤ As soon as you perceive and internalize this, immediately turn your blade over their shaft and rip downward and toward you with your blade *precisely in that moment in which they thrust in.*
> ➤ Thrust quickly toward their face *before they recover.*
> You will rip their halberd outward and hit them.

Or:
> ➤ When your feel in your thrust that *they want to thrust simultaneously with you:* discontinue your thrust secretly and imperceptibly and step well out to the side toward them with this.
> ➤ Simultaneously press their shaft swiftly downward and away from you.
> ➤ While pressing downward, slide your halberd inward on top of theirs and out over their shoulder.
> ➤ Catch them with your blade around their neck and rip them toward you with the same (as you can see in the center tenants in Figure I).

Furthermore, take note: when you have bound onto an opponent with the blade and *they do not want to work:*

> ➤ Move hurriedly and unexpectedly away from the blade and ſtep slightly out toward the side from which you have bound on.
> ➤ Thruſt |III.36| quickly toward their face during this ſtep.
> ➤ Observe during the thruſt *whether they want to counteract and beat your thruſt outward.*
> ➤ *If they do that:* let your blade drop downward and use it to catch them by their forward leg (*while they are ſtill beating it outward*).
> ➤ Rip the same [leg] toward you (as you can see in Figure K).[313]

If you ſtand before an opponent in the work:[314]

> ➤ Let your body language be observed and act as if you were to thruſt seriously and would thus (carefully) expose your face.
> ➤ *They will be provoked to thruſt by this.*
> ➤ *As soon as they subsequently thruſt in:* seize their blade with yours and rip downward toward you precisely *in that moment in which they thruſt in.*
> ➤ Thus, you take away their halberd as above (regarding this, see Figure D).

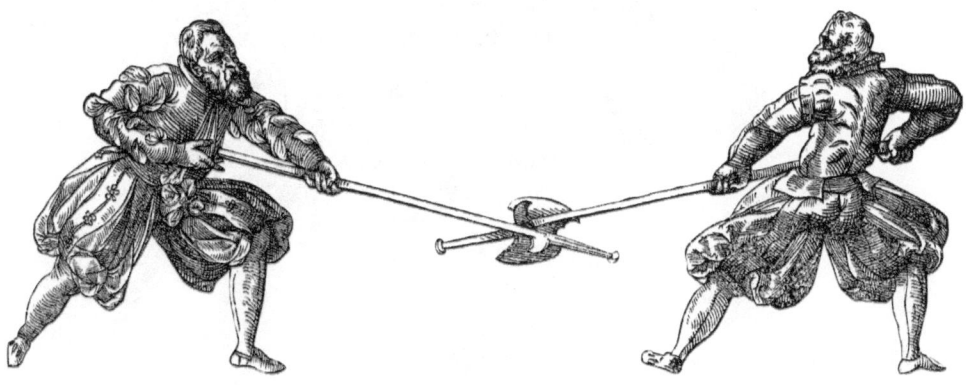

[313] This pair has been reversed for clarity, to match the hand positions of the other halberd illustrations.

[314] Not at the Onset and not from any particular cut or guard. In the middle of an unnamed sequence.

How you should set your halberd in front on their neck

Position yourself in the Simple Brace with your halberd and observe diligently. *As soon as they thrust toward your right at your face:*

➤ Step quickly outward toward their right (*during this thrust*) and move your horizontal blade at their throat (as you can see the previous in Figure A).

➤ Press forcefully toward them and, during this shove inward, let your shaft slide a bit between your two hands to extend behind you so that you can move closer to them.
➤ Kick their forward thigh back, so that they trip.

Or, *when they thrust toward your right at your face (as before):*

➤ Leap toward them and toward their right (*in that moment in which they thrust*) and simultaneously twist your blade or your entire halberd up over theirs (as is depicted for you by the upper tenants on the right side in Figure F).

➤ Afterwards, work toward their face according to your desire.

Counter

As soon as they twist over this way or similarly, run in at them with the butt end.

In the Onset: observe how you can catch their halberd behind their blade with your horizontal blade or how you can seize the same in the bind (as subsequently illustrated in Figure M).

➢ Do not let them lift up and away, but instead move theirs up with yours.
➢ Observe diligently.
➢ *As soon as they lift upward*, thrust straight in toward their chest or face (*while they are rising*), one of which you will certainly have [open].
If an opponent has caught your halberd in the way just taught and is waiting for you to rise:
➢ Yank your halberd upward and (*in that moment in which they thrust*) turn the butt end |III.37| upward and drop the halberd down and in front to beat their in-coming thrust outward with a hanging halberd (in the way that the portrait shows and teaches on the right side in Figure [E]).[315]

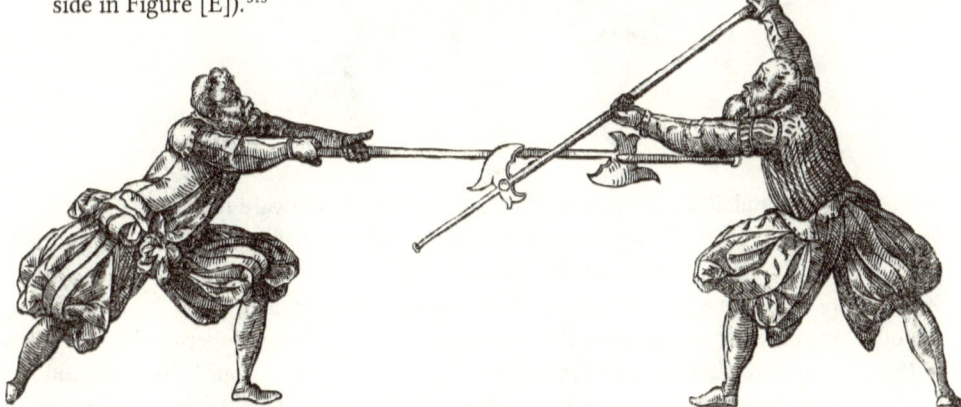

➢ Immediately thrust swiftly at the closest opening.

If an opponent guides your halberd too high upward (as previously taught):
➢ Remain or press hard on their blade and run the butt end inward while doing this.

---

[315] Figure B is incorrect, as there are no hanging weapons. Figure E shows the fencer on the left with a hanging blade beating out a thrust after having stepped backward.

➢ Twist the same upward or underneath their shaft between their arms or around their neck.
➢ Step behind them for a throw.

When you stand before your opponent in the Simple Brace with your halberd and your left foot forward, and *they do not want to work:*
➢ Lift your rear hand swiftly upward and let the front [hand] (including the blade) drop downward to your left and pull the same with both hands toward your left and around your head.
➢ In that moment in which you pull around your head, simultaneously step with your left foot well out toward their right.
➢ Cut powerfully from your left toward their right through their face.
➢ In that moment in which you cut, *take note of whether they rise to counteract.*
➢ As soon as you observe this, lift your rear right hand quickly and let the halberd drop a bit in front.
➢ Change around underneath next to their right arm.
➢ Thrust for their chest *while they still have both hands in the air* (as this is depicted by the small center tenants in Figure K).

Item:
➢ Cut a high Round Strike again from your left, with a step outward toward their right.
➢ *In that moment in which they lift to counteract it,* pull your halberd back toward your left.
➢ From this left, cut across through their foot.[316]
➢ Turn quickly to the other side with a slash back against their halberd.
As often as you want to execute a Round Strike, you should also properly take note as to *whether they want to thrust at you during this* while you are pulling your halberd around.
➢ *As soon as they thrust in:* cut with the Round Cut from above onto their halberd close behind their blade.
➢ Ensure that you step well out toward their right in this cut.
➢ As soon as you hit their halberd, rip downward (toward yourself and toward your left side).
You take their halberd this way.

---

[316] Low on the forward leg.

Item:

- ➤ Thrust seriously toward their right and at their face.
- ➤ During this thrust, let your fore end drop downward (*because they will want to deflect and counteract the thrust*).
- ➤ Use this [drop to] pull your halberd around your head toward your left with both hands and cut straight from above at their head with both hands (with an additional advance of your left foot).

You suppress their halberd to the ground with this cut (*if they would otherwise thrust in*).

- ➤ Rush swiftly with a thrust straight in front of you.

|III.38| Some counters against the stances, or how you should otherwise attack them

If you encounter an opponent in the High Guard *and they desire to suppress your thrusts:*

- ➤ Act as if you wanted to seriously thrust toward their face so that you provoke them downward.
- ➤ Take diligent note in the thrust.
- ➤ *As soon as they drop downward with their High Cut:* step quickly out to the side during this.
- ➤ Yank your halberd out from under theirs [at the same time] and drop with the same [your halberd] from above on theirs.
- ➤ Rip it downward toward you with the blade.
- ➤ Once you have completed the rip, thrust up toward their face.

Or:

- ➤ Threaten to thrust at their face.
- ➤ *In that moment in which they drop their halberd downward:* pull your halberd outward with a leap out toward the side.
- ➤ Thrust above their halberd at their face.

If you find an opponent in the Low Guard:

- ➤ Drop unexpectedly with your halberd on theirs, close behind their blade (as this is illustrated for you in Figure [C]).[317]

- ➤ Hold them there until you observe an opportunity to work.

---

[317] The text specifies Figure G, but G doesn't show an action against a Low Guard.

Or:

> ➤ Thrust seriously toward their face.
> ➤ *They will rise fast because they want to beat your thrust outward.*
> ➤ Therefore, do not let it touch, but instead pull and change through below (*while they are still rising*).
> ➤ Thrust in at the other side.
> ➤ You can also beat them outward from the other side and then thrust after that.

If you find an opponent in the Side Guard:

> ➤ Force them up with a pulled thrust.
> ➤ *As soon as they rise upward:* change through quickly and catch their hook from the other side with yours.
> ➤ Rip toward yourself and observe where you can otherwise rush with thrusts.

If you find an opponent in a Simple Brace:

> ➤ Bind on their blade with yours and take diligent note.
> ➤ *If they are seriously observing to see your lift upward:* raise the butt end upward fast and lower the blade downward next to theirs.
> ➤ You provoke them *to thrust in response to this without delay* (because you expose yourself with this).
> ➤ *As soon as they thrust in,* you have two sequences to use against them that are quite good.

First, when you have let your halberd drop downward in front and have raised the butt end:

> ➤ Step with the right foot well out to the side and toward their left *in that moment in which they thrust.*
> ➤ Turn the butt end back downward toward your flank and toward yourself.
> ➤ During this downward twisting, thrust simultaneously with them toward their left at their face.
> ➤ In that moment in which you thrust in, bend your head well onto your right side [with] your halberd shaft above so that |III.38ᵛ| you withdraw your face (*which they certainly intend to hit*).

Second, when you have let your halberd drop down in front of them:

> ➤ Move your hanging blade through under their halberd and toward your left.
> ➤ *As soon as they thrust in:* twist the butt end downward and toward you.
> ➤ With this [twist], you will beat their halberd outward and away from your left side. *Before they recover,* you should have hit them with a thrust.

A Wrath Cut

In the Onset, position yourself into the Field Guard—which is the Simple Brace, yet such that you let the butt end move through somewhat behind you.

> ➤ *As soon as an opponent thrusts at you:* take their thrust away toward your left.
> ➤ Use this to let your halberd move around your head toward your left.
> ➤ Release your front (left) hand in this pull around and grip on the butt end of your halberd behind your right hand with the same [left hand].
> ➤ Cut with the blade toward their left at their head with both hands.

If an opponent cuts or strikes at you from above (however they want):
- ➤ Reverse your front (left) hand on the shaft and lift high with both hands.
- ➤ During this, leap well toward them under their strike.
- ➤ Catch it between your two hands.
- ➤ Twist in above or underneath and set [your weapon] on them.

Or:
- ➤ Cut at them across from below against [their strike].
- ➤ *If they change through,* then pursue them.

I will now continue and deal briefly with the long pike as well, and then finally conclude with a mutual teaching about these three weapons.

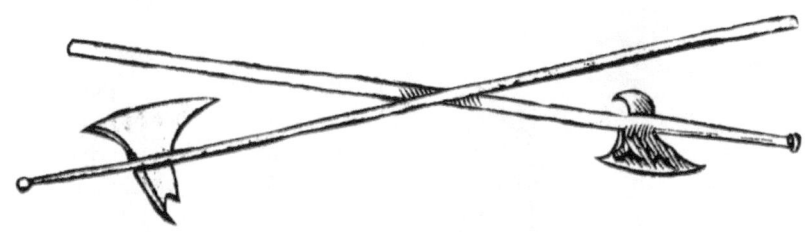

Fencing with the long pike

So now with regards to the long pike. First you have the guards, of which there are six—that is, the High Guard to strike, Middle Guard, and the Fool; afterward the High Guard on your left shoulder to thrust; Item, the Low Guard on your forward knee; and also the Suppressing Guard—and these are then described in order. Then follow the three most proper thrusts—the High at the face, the Low at the genitals, and the third at the chest—since all others are understood and grasped in these three. Afterwards and finally, the sequences from the guards including their associated breaks.

|III.39| **High Guard [Oberhut]**

To position yourself into this: stand with your left foot forward, hold the butt end of your pike with the right hand on the right flank and use the left to guide the [pike] with the fore end upward in front of you.

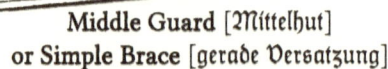

Middle Guard [Mittelhut]
or Simple Brace [gerade Versatzung]

When you now hold the butt end on your right flank (as just taught) and allow the fore end of this pike to drop downward until the same points into the face of your counterfencer, this is called the Middle Guard, since it is between the High and the Low, or the Simple Brace.

When you then allow the butt end to extend farther behind you so that you hold your pike in the middle in the form just mentioned (with the fore end toward their face) then this is named and called the Field Guard, because the long pike is held approximately in this way as a defensive weapon the majority of the time in the field.[318]

Side Guard [Nebenhut] and Change [Guard] [Wechsel]

Hold your pike with the butt end on the right flank (as mentioned above for the High Guard). Let the fore end of the same lie on the ground extended straight in front of you, yet so that you hold it with both hands so that you can sweep the same upward into a thrust (by using its momentum) as often as is necessary and according to your desire.

---

[318] *Im Feld* indicates specifically on the field of battle or combat. This emphasis is reinforced by mentions of the iron tip of the pike, which is called the "fore end" in the section of practice/sparring.

When you now hold your pike with the right hand on the butt end at the same flank and the left hand extended on the pike and the fore end on the ground, it is called the Side Guard (regardless of whichever side you then pull the fore part toward, whether toward the right or the left side).

The Change [Guard] is when you alternate from one side to the other, because no guard can be called the Change [Guard] if you stand still in it.

High Guard to thrust [Oberhut zum stoß]

When your left foot is forward in the Onset and you have your pike lying on your left shoulder in front of your left hand (as the upper scene depicts on the right side in Figure C), then you have completed this correctly and can thus fence from here as you will be subsequently taught.

|III.40| ### Low Guard [Underhut]

Position yourself into this as follows: stand again with the left foot forward and hold your pike with the left hand on the forward knee, so that the fore end points upward toward your oppo-

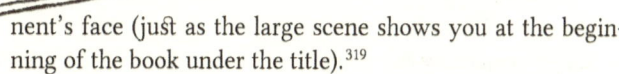

nent's face (just as the large scene shows you at the beginning of the book under the title).[319]

Suppressing Guard [Dempfhut][320]

When you stand with the left foot forward and your pike is placed on the inside of your right leg, and you hold your pike shaft with your left hand extended as far as possible so that you can hold and guide your pike in front of you with powerful strength in a broad stance (as such is certainly illustrated for you in Figure I on the left side), you have completed this correctly. You can thus suppress their shaft downward from here and hold them off according to your desire, therefore this guard is called and named the Suppressing Guard.

[319] This fencer has been reversed for clarity, to match the hand position described here.
[320] The Suppressing Cut [Dempfhauw] is a rapier technique and is described starting on II.56.

About thrusts and how you should direct them in a free-flying way in the work

As all kinds of thrusts will follow in the sequences, I will only introduce the High and Low Thrusts and how they are to be directed in the work.

Hold your pike in the High Guard on your left shoulder (as previously taught).

➤ Secretly give your pike an impetus forward and during this, step forward with your left foot.

➤ With this step forward (while the pike is flying upward with the previously applied momentum), thrust toward their face so that both of your arms are extended to the greatest extent at the end of the thrust.

➤ While your pike drops back downward toward the ground, you should step Indesly[321] farther forward with your left foot and pull your pike back toward yourself with both hands (before it completely drops down to the ground).

➤ Also, bend your upper body downward following the pike and let your pike sink down onto the forward bent knee into the Low Guard (so that the fore end points toward your opponent at their face).

|III.40ᵛ| **Thrust from below**

When you now have your pike in the Low Guard:

➤ Give the [pike] an impetus forward (as before) and during that same impetus, thrust forward with both hands in front of you toward their face.

➤ As soon as your arms have extended to the farthest point (due to the thrust toward their face), pull the pike back onto your left shoulder into the High Guard, with your body upright again and before the pike starts to fall back downward due to the [dropping] momentum.

When you can carry out and execute these two thrusts correctly and together in a flying manner, it will be much easier for you to fence all of the others. I will now present one stance after the other and demonstrate and teach the sequences from each. Because it is more proper to speak about the first High Guard later in the [section on the] counters against the stances, I will begin with the second High Guard, from which you should fence like this:

---

[321] On 1.25 of the Sword Section, Meyer explains his interpretation of *Indes* to be a small word fraught with meaning: in that moment in which you strike, you should simultaneously observe the other openings that you could also hit, and also all of the techniques your opponent can use to strike or counteract your action. It is therefore a point of hyperfocus on what you are doing and what you and your opponent could be doing at the same time and subsequently. Meyer states that he will use *Indes* as a shortened form to call all of this to mind.

Within his text, Meyer tends to use *Indes* to indicate a disruptive action, that is, because he is describing actions sequentially that occur simultaneously or disruptively. Thus, *while* you are performing the above action, execute the following action *at the same time* and *in the same space*. Because *Indes* is an adverb of time in German, it will frequently appear adverbially in the translation, thus *Indesly*.

Fencing from the High Guard

As now relates the sequences and fencing with the pike, I want to first admonish and remind you to diligently pay attention to thrusts in the Before and After, as I will describe and teach you to conduct all sequences in three [different] ways (where that is necessary).

And for the first: when you are in the High Guard—that is, when you have your pike lying on your [left][322] shoulder and you stand with your left foot forward—and *your counterfencer thrusts toward your left:*

> Step well toward their left with your right in a leap out of their thrust and simultaneously thrust in with them.

> When you thrust in, also release your left hand from the pike and turn your right side to follow your thrust.

When you have now leapt quickly outward and also turned your right hand upward and toward your left in the thrust, and have boldly extended your right arm: *then they will miss and you will hit.*

> After you have thrust, immediately jerk the butt end back upward toward you with your right hand.

> In that moment in which you jerk your pike back, step with your right foot back again and, at the same time, grab the pike again with your left hand.

> Twist your right hand downward forcefully to move your pike back upward in an arcing motion in front, and you bring the [pike] back into your control.

|III.41| **A good sequence from this High Guard**

When they do not want to thrust first or work:

> Lift your left foot and let yourself be observed with body language as if you wanted to powerfully thrust at them from above.

> Do not do it but instead, step farther toward them with your raised left foot and thrust toward their face from the shoulders when setting the [foot] back down.

> Do not let this thrust move forward as mentioned, but instead, in that moment in which you set your lifted foot back down, bring your pike downward with the butt end onto your forward knee into the Low Guard.

> Wait, because *they will doubtless thrust soon in response.*

If you have carried out the previously mentioned thrust with serious body language, it will appear no different than as if you had thrust and missed. Therefore, *they will be provoked to rush with their thrust and will themselves thrust and miss.*

> Take out this same thrust (*which they will immediately carry out*) with a beat to one side—yet so that you do not go astray with the same.

> Thrust from the thigh, straight in front of yourself and toward their face.

---

[322] The text has "right shoulder"; however, that is not possible with the left foot forward, and it also contradicts the description of the High Guard to thrust.

Fencing from the Low Guard

Position yourself in the Low Guard like this: so that you have your pike lying on your forward knee such that the fore end points toward your opponent's face. (You were taught about this above.) And also observe their thrusts diligently, [and] at which side they now thrust in.

➢ Deflect their in-flying thrust by rotating [your pike].
➢ Step simultaneously outward with the rear foot toward the opposite side and thrust simultaneously with them.

Working in the Before

If they will not thrust first but instead wait for your [thrust]:
➢ Bind on in the middle of their shaft and feel whether you can press theirs outward in a hurried and unexpected shove to one side.
➢ Following this same shove, let your pike quickly shoot forward toward their face.
Or:
➢ Once you have bound onto them, change through carefully—once, twice—under their pike on both sides.

Firstly, take diligent note during this *of which side they will move astray to following your change through* so that you quickly rush to the opening with thrusts.

Secondly, *when an opponent changes through against you in this way:*
➢ Diligently observe *whether they* |III.41ᵛ| *move their pike away too far downward or to one side in the change through* (as this can easily occur).
➢ *While they are changing through:* thrust at them carefully and quickly at their face.

Thirdly, in that case where you notice that *they lurk and wait for your change through:*
➢ Then change through first, and while you change through in front of them, take diligent note of their thrust in.
➢ *As soon as their thrust flies in:* beat it outward with a strike to the side and thrust quickly in response.

If you observe that *an opponent is provoking you to thrust using change throughs and wants to deceive you with the same:*
➢ Act as if you do not understand it, and thrust seriously yet carefully in at them, so that you do not let your pike move out of your control.
➢ *In that moment in which they want to beat it outward,* change through with your thrust below.
➢ *They not only beat outward for nothing, they also move their pike too far astray to one side so that they expose the other side.*
➢ Therefore, after your change through to the other side, thrust at them *while they want to beat outward.*

Middle Guard

In the Onset, position yourself into the Middle Guard.
➢ *Depending on which side they then thrust in at you,* step to the other side and thrust simultaneously with them.

➢ During the thrust, turn your long edge toward their shaft so that you deflect their thrust away from you more surely.

If they thrust at your lower body:

➢ Lift the butt end of your pike upward and let the fore end of the same drop downward, and use this to beat their in-flying thrust outward to the side with a hanging shaft (between you and them).

➢ Thrust at them quickly (at an opening) with your shaft, which is quickly turned back up-ward.

If you encounter your *counterfencer in the same guard:*

➢ Bind on with serious body language.

➢ *Before they observe it,* press their shaft out to the side with a violent shove.

➢ *In that moment in which they want to block and withstand this outward shove,* move quickly through under their pike and thrust at the other side with a step outward.

Suppressing Guard

Take note: when you have moved into the Suppressing Guard in the Onset (according to the instruction of the previously mentioned portrait on the left side in Figure I), you hold your pike with powerful strength so *that they cannot* |III.42ᵛ| *easily beat or push the same outward.*

Therefore, observe. *When they thrust toward you (whether toward the right or left side):*

➢ Step out to the side away from their thrust and thrust simultaneously in with them.

Or:

➢ Set off their in-flying shaft and thrust again toward them with a step outward.

Or:

➢ *In that moment in which their thrust flies in,* suppress their shaft down from above and thrust quickly toward their face (*before they recover back*).

If you feel that *they will move upward too quickly with their pike* so that you cannot sur-pass them with the speed of your thrust:

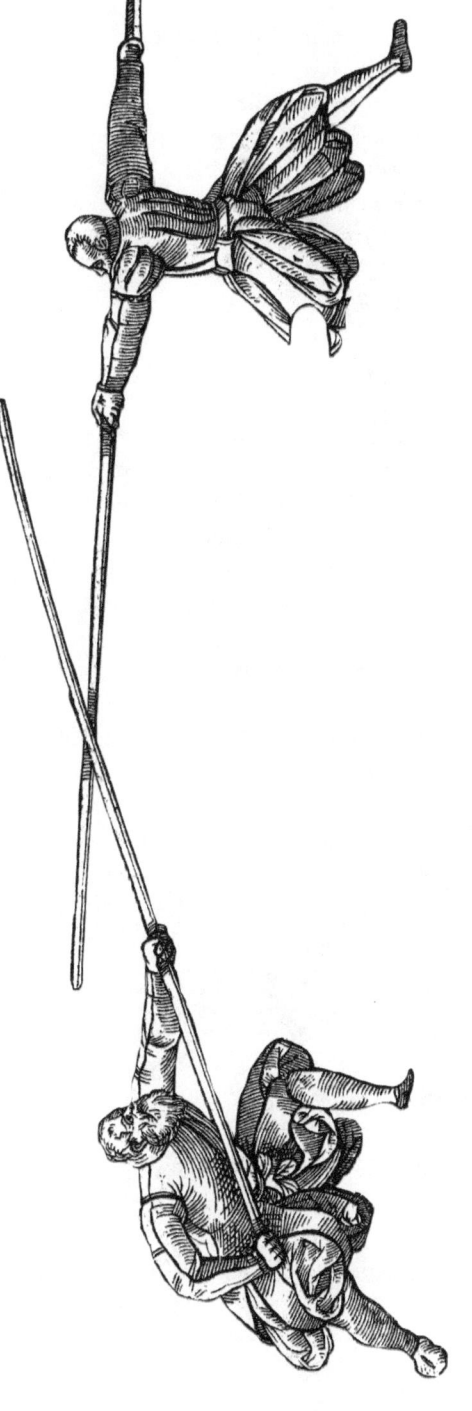

➤ Change through under their pike (*in that moment in which they rise upward*) and thrust in at the other side.

If they do not want to thrust first, but instead position themselves likewise in a strong counterposture:

➤ Bind on hard in the middle of their shaft and press the same hard (downward and away from you).
➤ *If they resist strongly and press upward:* move through below quickly and swiftly, and attack their pike again from the other side, pressing powerfully downward and away from you again.
➤ Execute that from both sides until you have tired them and sapped their strength, *so that they cannot maintain their control any longer.*
➤ At that point, thrust immediately at the closest opening *while they are still fumbling with their pike.*

And again, *if an opponent lies hard on your pike and wants to force you downward and to the side:*

➤ Act as if you wanted to withstand strongly and press back against it.
➤ *In that moment in which they expect it least,* change through swiftly under theirs and attack back on their pike from the other side.
➤ Push it likewise out toward the side *toward which they wanted to push yours out.*

You thus force them downward so that it is only with great difficulty that they remain uninjured or can recover without damage.

If your counterfencer wants to move through under your pike because you are pushing them outward:

➤ Always attack from the other side with the bind hard on their pike *while they are still changing through.*

This works on both sides. The longer and the farther past the middle of their shaft [you bind on], the more *this prevents them from coming through at any point to positive effect,* until you observe your opportunity to thrust.

Side Guard

In the Onset, hold your pike with the fore end slightly out to the side on the ground (like the upper scene shows on the left side in Figure C). *As soon as an opponent thrusts against you:*

➤ Step |III.43| quickly toward their left side, well out from their thrust, lifting your pike during this with an impetus up from the ground.
➤ Thrust simultaneously in with them.

Or:

> ➤ Take their in-flying thrust outward and upward with the half edge and thrust in response.

If they do not want to thrust first, but instead hold their pike straight in front of themselves:

> ➤ Lift your pike swiftly and beat theirs outward with a strong, sideways strike.
> ➤ Immediately thrust afterwards.

If they resist strongly:

> ➤ Quickly move through below after this strike and thrust at the other side with a powerful outward step.

If they want to beat your pike outward (as previously taught):

> ➤ Then *in that moment in which they strike in,* change through below and thrust in at the other side (*while they miss with their pike*).

Stand with your left foot forward, hold your pike with the butt end on your right flank, and extend the fore end toward your left side on the ground. When you now stand in this guard in front of an opponent, and *they (your counterfencer) thrust toward you:*

> ➤ Step out from their thrust with your left foot (well toward your left side), using this to sweep your pike upward in an arcing motion.
> ➤ In that moment in which you step outward *and they thrust in:* thrust in above their pike simultaneously with them.

Or:

> ➤ Beat their shaft outward with yours, then thrust quickly and swiftly in response.

The second part of the long pike

¶ Counters to the stances in the pike including other skillful sequences

If you find an opponent in the Suppressing Guard in the Onset:

> ➤ Hold your pike in the High Guard to strike, and immediately strike through and down from above toward their hand (*which they have in front on their pike*).
> ➤ You will provoke them with this strike so that *they will quickly thrust in response.*
> ➤ Therefore, slash your pike strongly back upward in one sweeping movement, beating their thrust outward with this.
> ➤ Thrust quickly in response.

|III.43ᵛ| ### How you should strike down onto their pike, change through, and thrust in on the other side

> ➤ Or, bind onto them (onto their pike) with the same guard from your left side.
> ➤ Strike down and through on their pike (toward their fingers), but in the strike, change through under their pike.
> ➤ Thrust in quickly with a step outward to the other side.

Regarding those who do not thrust swiftly in response, use the sequence like this:

➢ After you have bound on (as mentioned above), strike down onto their pike at their fingers again (as before) and move through below.

➢ Use serious body language and threaten to thrust at the other side.

➢ *In that moment in which they want to move to counter your thrust and to counteract the same:* change swiftly back through so that *their counteraction comes to naught.*

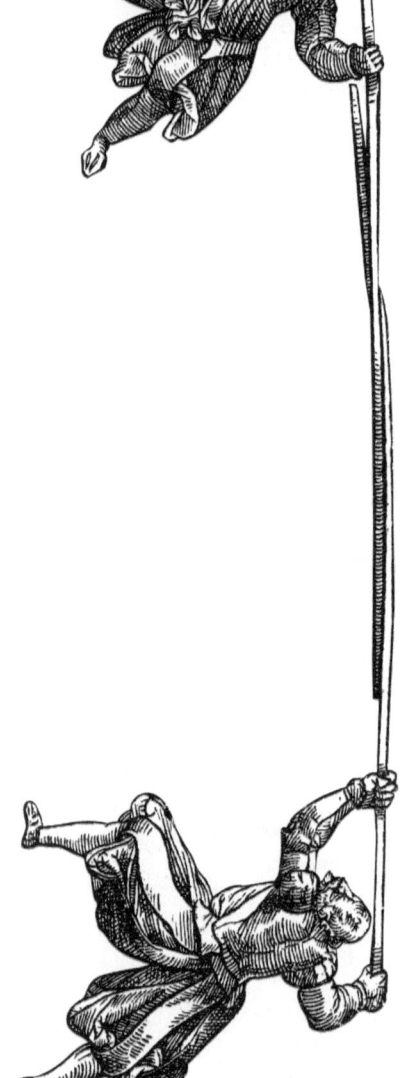

➢ Thrust in at them with a step outward to the side at which you have completed the thrust.

Another

Item:

➢ Change through again using the strike mentioned above and strike powerfully across, back at their pike from the other side.

➢ *While they fumble,* thrust at their opening.
  Or, if they resist hard:

➢ Change through and thrust at the other side.
  If you hold your pike in the High Guard to strike *and they thrust in toward you:*

➢ Whatever side they thrust in at, step out to the other and strike their pike down from above.

➢ As soon as your fore end has arrived equal to their face in this strike, shove the pike forward and turn the strike into a thrust.

Twist through [Durchwinden]

Take note: when you encounter an opponent who holds their pike straight in front in the Simple Brace (or in another guard):

➢ Bind on from your left toward their right.

➢ Provoke them with body language, shoves, and threats of all kinds *until they thrust in.*

➢ *In that moment in which they thrust in:* move the butt end of your pike through and under theirs toward your right.

➢ Turn the [butt end] upward toward your right so that you move their pike aside (toward your left) with this twisting upward (regarding this, view the upper scene on the left side in Figure D).

➢ Subsequently set your pike inwardly on their chest.

|III.44ᵛ| **Rule**

As often as you have bound on with the outer part of your pike past the middle of your opponent's pike (*who is quickly and easily incited to thrust*), you can twist through toward whichever side you want using the butt end and a step backward. You move them upward by twisting and stepping outward during this twist, so that you move their pike aside and toward the side from which you twisted through, and you also recover yourself to set on and thrust according to your desire. This type of twisting through requires a certain level of skillful ability, and also that you are knowledgeable about feeling.

A counter

If you encounter an opponent in the High Guard to thrust (*so that they have their pike lying on their left shoulder*):
  ➢ Position yourself in the Change [Guard] on the left—that is, in the left Low Guard.
  ➢ From there, strike their pike strongly back with yours, so that you remove the same from their shoulder.
  ➢ Before *they recover back from the strike*, thrust at the closest opening.
  In contrast, *when an opponent wants to beat your pike outward from your left shoulder:*
  ➢ Change through under their pike *in that moment in which they thrust in, so that they miss with their strike.*
  ➢ Thrust at an opening *while they go astray with their weapon.*
  If you want to beat their pike outward, as taught above, and you perceive *that they want to change through against you:*
  ➢ Act as if you do not notice it, and strike toward their pike with serious body language.
  ➢ However, maintain your pike fully in your control in this strike so that you are ready to thrust with a step outward *in that moment in which they move through.*
  ➢ Or you can also take this outward strike from the other side, and thrust in response when the first is completed.

A second counter

If you find *your counterfencer in one of the Side Guards:* position yourself as if you were to thrust seriously at the opening.
  ➢ *In that moment in which they rise upward to counter your thrust:* pull yours back toward yourself and change through under their pike.
  ➢ Thrust in at the other side.

|III.45| **Pursuit [Nachreisen]**

When an opponent stands before you in one of the Low Guards and wants to then rise into the High Guard: observe.
  ➢ *In that moment in which they are still rising upward in an arcing motion:* thrust in response.

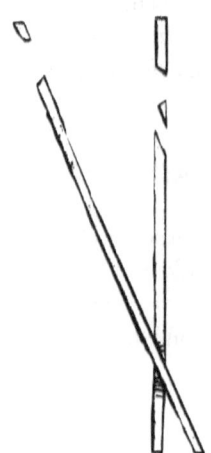

You will hit them in the way that is depicted for you by the upper tenants in Figure A. You should be aware of this thrust in response in all binds so that you thrust in response *as soon as they lift up and while they are still rising.*

Rule

Take note of this rule with regards to a thrust that strays. When you have thrust astray with an eager thrust (regardless of how this has occurred):

> ➤ Leap back with your front foot and pull the butt end upward (as you can see in the upper scene on the right side in the Figure H printed previously).

> ➤ Deflect all of their thrusts in response in this way and grab back on the shaft with your left hand during this.

> ➤ Rotate the same back into your power and control and up into one of the guards (as you prefer).

Practice will further disclose how this is to be carried out.

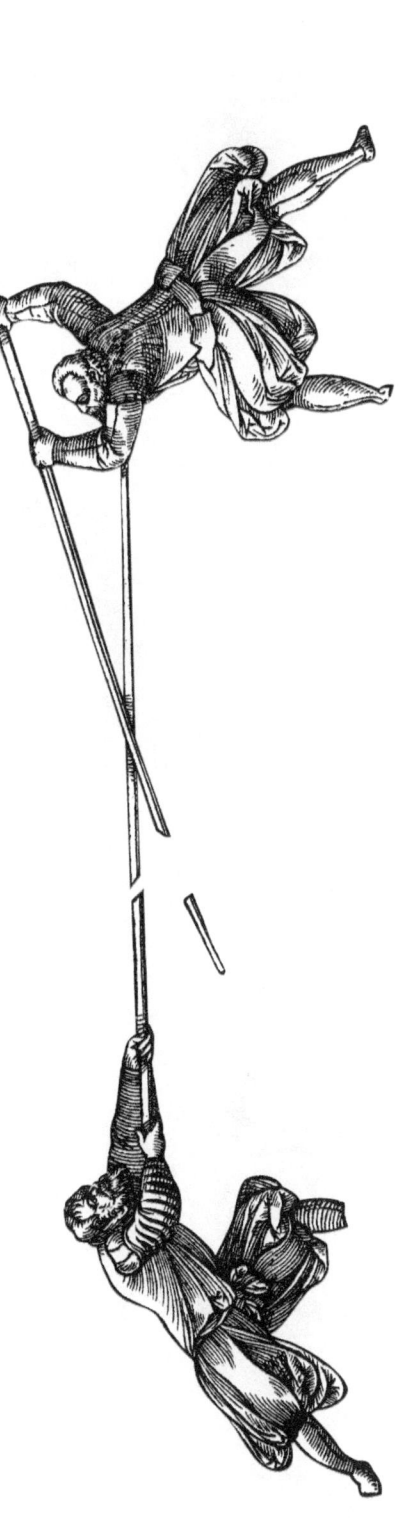

[These pike illustrations are hard to fit in the book, and the display on these two pages may be particularly confusing. The top illustration is from Figure A, the left-side illustration is from Figure H, and the right-side illustration is from Figure M.]

Another good sequence so that you can entice them from their advantage

Hold your pike in the Low Guard on your forward thigh and give it an impetus by shifting the weight.

> ➤ In that moment in which the fore end of your pike now arcs upward, in that arcing movement simultaneously thrust upward toward their face with extended arms.

> ➤ When your pike is at the outermost point in its flight, *they will swiftly thrust in response (after they have avoided your thrust)*.

> ➤ Therefore, pull your hands even higher over your head and drop the fore end of your pike downward.

> ➤ Beat their in-flying thrust outward to the side with a hanging pike in this way (as the upper scene shows you on the right side in Figure M).

> ➤ Rotate your pike swiftly and thrust in response before they have recovered.

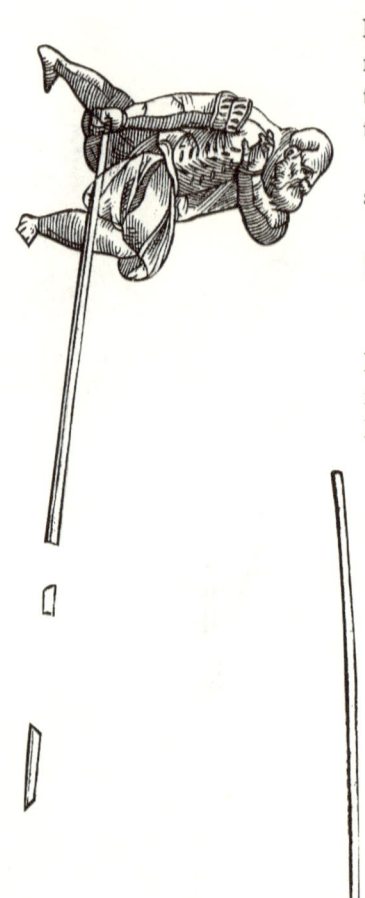

Another

Item, hold your pike in the High Guard and take note. *In that moment in which they thrust in:* drop the fore end and beat them out again (and complete this as before).

Many other sequences can be gathered from this sequence, yet they require a strong person for them.

|III.45ᵛ| How you should lift your pike upward in an arc with one hand and thrust in

In the Onset, position yourself in front of your opponent with your pike according to the instruction of Figure L (in the small upper scene on the left side).

As soon as they thrust in:

➤ Lift your pike upward in an arc around from your right toward your left and in front of your face (as you can see in the body language of the above-mentioned scene [L]).

➤ You thus beat their pike outward with the butt end of yours in this rising arc.

➤ Or you yield away from them by stepping outward.

➤ Step toward them Indesly and simultaneously (with your left behind your right) so that you turn your back to them.

➤ Thrust next to your right and back upward toward their face (still within the force and power of the initial rising arc).

➤ With this thrust, also step backwards and toward them with your right foot.

➤ As soon as the thrust is completed (and in that moment in which your pike falls forward onto the ground), remain standing on your right foot and take another step away from them with your left, and bend your upper body well away from your opponent and over your left knee, which should be bent well forward.

➤ Pull the butt end toward yourself with the right hand until it is close beside your left foot on the ground, and wait there for their thrust in.

➤ *As soon as they thrust:* lift your pike powerfully upward again in an arc from your left toward your right with one hand.

➤ In this rising arc, step with your left foot simultaneously in a leap well out toward their right.

➤ Thrust from above toward their face with one hand (while the pike still flies in the air due to the force of the rising arc).

Take note in all of this that you should have held your pike by the butt end such that, in holding the pike, your little finger (with the blade of your hand) points inward toward the shaft and the thumb (including the index finger) projects forward on the butt end.

Another thrust with one hand

Place yourself before your opponent in the manner that is depicted to you in Figure B (in the upper scene on the right side).

➤ Shove your pike away from you in one short movement (that is, with one hand) and jerk the butt end swiftly back toward yourself and upward toward your left so that it lifts smoothly in this jerk and arcs forward and upward.

➤ In this upward arcing movement, step well to your right side toward them and thrust with one hand again toward them.

|III.46| **Running in [Ẽinlaufen]**

There are two types of running in with the pike: one occurs with the butt end, the other with the fore end. Carry out the one with the fore end like this:

➤ Bind hard on their pike shaft from your left toward their right and take note.

➤ *In that moment in which they thrust in:* step toward their left with your right foot, with your head bent well forward out of their thrust, and simultaneously twist your butt end through under their pike toward your right.

➤ In this twist through, guide the butt end of your pike quickly upward and toward your right, so that you deflect their thrust toward your left side.

➤ During this, move your head through toward your right (between your two arms) and simultaneously release your right hand from the fore end and, with a quick leap toward your left side, grab back on your pike with the [right hand] in front of your left hand, so that the butt end extends up from your left shoulder.

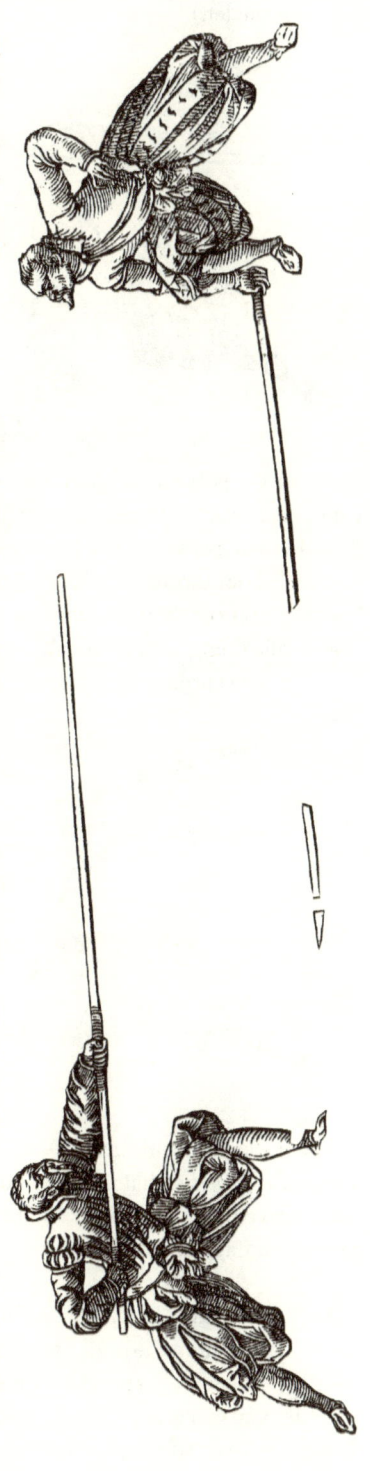

You thus counteract and hit in a manner that is depicted for you in Figure E (in the upper scene on the left).

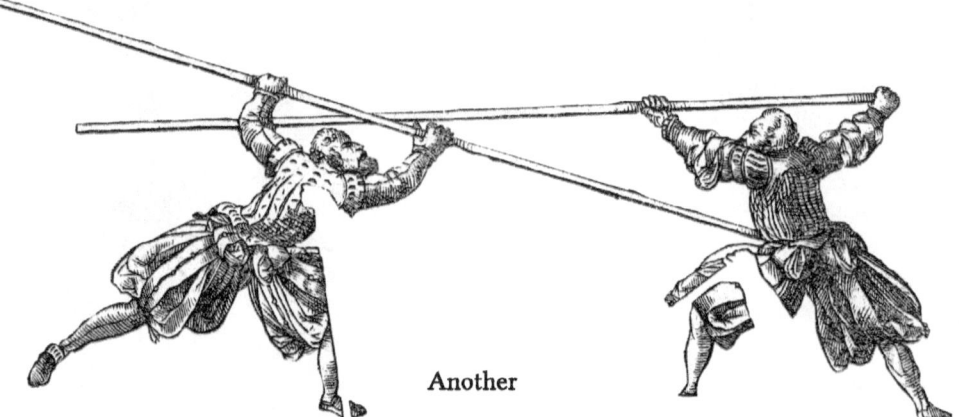

Another

Item: set the butt end of your pike on your left hip and see how you can provide them with a cause *such that they thrust toward your left side.*

➢ *As soon as they thrust in:* release your right hand from the butt end and let the same butt end shoot through back behind you next to your left.

➢ Simultaneously with this, leap with your right foot powerfully in toward their left side.

➢ While doing this, grab back on your shaft (with your right hand in front of your left) and run in according to the instruction of the previously mentioned tenants [E].

Use the second running in with the butt end against *those who hold their pike forward and upward in fencing,* that is, when you have bound onto one another and remain in the bind and have arrived with both pikes high upward. You can then run in with the butt end according to your advantage and desire.

Finally, as I conclude with these [weapons], you should know that you should not easily give up your advantage nor allow yourself to be enticed from there—whether in staff, halberd, or pike—unless you are |III.46ᵛ| not merely certain, but you have also diligently considered whether you can bring your weapon back into your control from the executed thrust with time to pursue theirs (if your thrust has missed), leap away, and counteract.

If you find your counterfencer in their advantage in a stance: do not thrust at their opening without special advantage. Instead, observe how you can entice them and bring them out of their advantage through knocks, jerks, change throughs, and shoves, even up to the use of pulled thrusts.

As soon as they rise or begin to work, then attack and begin your sequence.

In addition, you should take diligent note in all binds and feel *whether they are hard or soft in their resistance.*

Item: *If they are quick or slow to thrust in response,* then also be careful and deliberate in thrusting in so that, if you feel during the thrust in *that they will quickly thrust simultaneously or in response,* you should not complete that thrust but instead turn the [thrust] into a set off, and only then complete a thrust in response.

Be well practiced in all sequences, and be quick of thought in fencing.

A brief teaching about how you should use your pike in serious matters on the field[323] and how you should use it to your advantage

The pike requires a strong, serious, and mentally-composed[324] person who knows how to carefully control their pike, and can also set their thrusts on with certainty and direct them in the work at the correct time. This is because it can easily occur that you can suffer and come into irretrievable injury with a missed thrust (in particular because the pike has a considerable weight due to its length). Therefore, you should be studious in maintaining a correct distinction between thrusting in the Before and After, of which I will give you a brief teaching and rule through the following examples. This is because experience sufficiently bears witness to how quickly fencers can thrust and miss, if they attack foolishly and at the incorrect time.

Now for the first: if your enemy[325] encounters you with the same weapon—that is, with the same pike—then observe whether *they are swift and angry with their attack, such that they rush to be the first to thrust.*

➤ Act and present yourself with serious body language as if you wanted to race in front of their thrust in the Before and come first. In this way, you influence them *so that they rush even more with their thrust in the Before.*

➤ You, however, should not thrust first (even though you present yourself this way), but instead strike |III.47| their thrust out in one jerk to the side and set [your pike] in their face.

If they are strong with their pike so that you cannot make enough space to strike them outward with the first strike, then execute the following:

➤ After you have initially beaten them outward and *they have quickly recovered back:* strike them again from the other side, *while they are still pulling their pike back or toward themselves.*

When you execute these two outward strikes correctly, then you will certainly have a lot of space so that you can thrust into their face in the blink of an eye *before they correctly recover.*

---

[323] On a field of battle, not as practice or exercise.

[324] *Besost* is not a word. My best assumption is *befaßt* (*befost*), as this requires the fewest additions or substitutions of letters, fits the context, and the participial form easily forms an adjective. The modern equivalent is *Fassung*, meaning "composed of mind". Other translators have assumed *besoster* is a typo for *besonnster*, from *besinnen*, meaning "cautious"; however, the participial form is *besonnen*, which would make the superlative *besonnenster*, and the correct translation "most cautious".

[325] The use of '*feind* = enemy' also reinforces the use of the pike on the battlefield instead of for sport.

This is because it is easy to take a pike away from an opponent who thrusts eagerly and angrily and to beat it outward.

Another sequence

In the attack, hold your pike with the iron point well downward (pointing a little toward your left side), so that your entire face is exposed. *By this means, they are provoked to rush to the same with a thrust.*

- ➤ *As soon as they thrust in:* raise both hands and strike their in-flying thrust with the long edge outward and upward from below and from your left toward your right.
- ➤ In addition to this outward strike, leap with your left foot well out toward them (toward their right).
- ➤ *Before they recover from the outward beat,* thrust quickly in the blink of an eye above their pike shaft into their face.

If your enemy is cautious and does not want to thrust first, it will not benefit you to also stand [and wait] like this, but instead you must thrust first. Therefore, execute it like this:

- ➤ In the attack (after you have twisted your pike [upward]), guide a powerful thrust with furious body language in toward their right and next to their pike.
- ➤ In this thrust inward, powerfully maintain your pike in your control.
- ➤ If you cannot hit them with this thrust, allow the iron point to drop a little downward and out to the side, as if you had unwillingly missed due to some obstacle. You should use this (the fact that you allowed your pike to move a little out to one side) to recover into a powerful strike outward.
- ➤ *In that moment in which they do indeed thrust in:* leap well toward them on their right with your left foot and use this step forward to strike their pike powerfully upward and outward (from your left toward your right).
- ➤ Lift both hands Indesly upward and thrust over their pike into their face.

In all of this, it is understood that you have your left hand in front on the pike. If you have the other forward, then you must guide the thrust toward their left (including the completion of the sequence), just as you completed it previously toward their right. You should also take note here that it does not take as long in the actual work as it takes to teach and speak about it. Instead, everything must |III.47ᵛ| occur and be brought into motion in the blink of an eye.

As to what is further necessary for these sequences, this will be presented and taught at length at another point, and also in relation to unmatched weapons.

End of this book.

Printed in Strasbourg by Thiebolt Berger am Weynmarckt zum Treubel

BIBLIOGRAPHY

PRIMARY SOURCES

Manuscripts

This table is modified from a cataloging system developed by DIERK HAGEDORN.

AD *Albrecht Dürer*
    Wien, Albertina (Graphische Sammlung), Hs. 26–232

AR *Anton Rast*
    Augsburg, Stadtarchiv, Reichsstadt Schätze Nr. 82

BKE *von Eyb fight book* or *Blume des Kampfes-Erlangen*
     Erlangen, Universitätsbibliothek Erlangen-Nürnberg, B.26

BKW *Blume des Kampfes-Wien*
     Wien, Österreichische Nationalbibliothek, Cod. 5278

F *Memorial Da Prattica do Montante*
   Lisbon, Biblioteca da Ajuda, 49.III.20.n°.21

FLG *Fior di Battaglia* or *Fiore de'i Liberi-Getty*
     Los Angeles, J. Paul Getty Museum, Ms. Ludwig XV 13

FLP *Florius de arte Luctanti* or *Fiore de'i Liberi-Paris*
     Paris, Bibliothèque nationale de France, Latin 11269

G *Goliath fight book*
   Kraków, Biblioteka Jagiellonska, Berol Ms. Germ. Qu. 2020

HM *Hans Medel* or *Sigmund Schining*
    Augsburg, Universitätsbibliothek, Cod.I.6.2°.5

HS *Hans von Speyer*
    Salzburg, Universitätsbibliothek, M.I.29

HTG *Hans Talhoffer-Gotha*
     Gotha, Forschungsbibliothek Schloss Friedenstein, Chart. A.558

HTK *Hans Talhoffer-København*
     København, Det Kongelige Bibliotek, Thott 0290 2°

HTM *Hans Talhoffer-München*
     München, Bayerische Staatsbibliothek, Cod. icon. 394a

JML *Joachim Meyer-Lund* or *von Solms' fight book*
     Lund, Universitetsbibliotek, MS A.4°.2

JMM *Joachim Meyer-München* or *von Veldenz' fight book*
     München, Bayerisches Nationalmuseum, Bibl. 2465

JMR   *Joachim Meyer-Rostock*
       Rostock, Universitätsbibliothek, Mss. var. 82

JWA   *Jörg Wilhalm-Augsburg #1* or *Lienhart Sollinger-Augsburg*
       Augsburg, Universitätsbibliothek, Cod.I.6.2°.2

JWM1   *Jörg Wilhalm-München #1*
       München, Bayerische Staatsbibliothek, Cgm 3711

JWW   *Jörg Wilhalm-Wolfenbüttel* or *Lienhart Sollinger-Wolfenbüttel*
       Wolfenbüttel, Herzog August-Bibliothek, Cod. Guelf. 38.21 Aug. 2°

K   *Kölner Fechtregeln*
       Köln, Historisches Archiv, Best. 7020

LF   *Lutegerus fight book* or *Walpurgis fight book*
       Leeds, Royal Armouries Museum, FECHT 1 (formerly MS I.33)

LH   *Hans Lecküchner-Heidelberg*
       Heidelberg, Universitätsbibliothek, Cod. Pal. germ. 430

LM   *Hans Lecküchner-München*
       München, Bayerische Staatsbibliothek, Cgm 582

MYW   *Man yt Wol*
       London, British Library, Harley 3542

N   *Nicolas Pol house book*
       Nürnberg, Germanisches Nationalmuseum, Cod. Hs. 3227a

PD   *Peter von Danzig* or *Starhemberg fight book*
       Rome, Accademia Nazionale dei Lincei, Cod. 44.A.8

PF   *Peter Falkner*
       Wien, Kunsthistorisches Museum, KK 5012

PKB   *Paul Kal-Bologna*
       Bologna, Universitätsbibliothek, Ms. 1825

PMD   *Paul Hektor Mayr-Dresden*
       Dresden, Sächsische Landesbibliothek, Mscr. Dresd. C.93/94

PMM   *Paul Hektor Mayr-München*
       München, Bayerische Staatsbibliothek, Cod. icon. 393

PMW   *Paul Hektor Mayr-Wien*
       Wien, Österreichische Nationalbibliothek, Cod. 10825/6

SR   *Sigmund Ainringck fight book*
       Dresden, Sächsische Landesbibliothek, Mscr. Dresd. C.487

Wo1   *Wolfenbüttel picture book #1*
       Wolfenbüttel, Herzog August-Bibliothek, Cod. Guelf. 83.4 Aug. 8°

Books and incunables

Agrippa

    Agrippa, Camillo. *Trattato di Scientia d'Arme, con vn Dialogo di Filosofia*. Roma: Antonio Blado, 1553.

Alfieri

    Alfieri, Francesco Fernando. *Lo Spadone*. Padua: Sebaſtiano Sardi, 1653.

Auerswald

    Auerswald, Fabian von. *Ringer kunſt: funf und Achtzig Stücke*. Wittenberg: Hans Lufft, 1539. <VD16 A 4051>

Capo Ferro

    Capo Ferro, Ridolfo. *Gran simulacro dell'arte, e dell'uso della scherma di Ridolfo Capo Ferro da Cagli*. Siena: Salveſtro Marchetti and Camillo Turi, 1610.

Cauvin

    Cauvin, Jean. *Chriſtianae religionis inſtitutio*. Basel: Thomam Platteru & Balthasarem Lasium, 1536. <VD16 C 287>

Douay

    The Holy Bible faithfully translated into English out of the authentical Latin, diligently conferred with the Hebrew, Greek, & other Editions in divers languages. Doway: Laurence Kellam at the signe of the holie Laambe, 1610.

Egenolff

    Der Allten Fechter gründtliche Kunſt. Frankfurt am Main: Chriſtian Egenolff, ca. 1530s. <VD16 ZV 9515>

Fischart

    Fischart, Johann. *Affenteurliche und Ungeheurliche Geschichtschrift vom Leben, rhaten und Thaten der for langen weilen Vollenwolbeschraiten Helden und Herrn Grandgusier, Gargantoa, und Pantagruel, Königen inn Utopien und Ninenreich*. Strasbourg: Bernard Jobin, 1575. <VD16 F 1127>

Grassi

    Grassi, Giacomo di. *Ragione di adoprar sicuramente l'Arme*. Venezia: Giordano Ziletti, 1570. 2nd ed. *His True Arte of Defence*. Trans. by I. G. London: I. Iaggard, 1594.

Gunterrodt

    Gunterrodt, Heinrich von. *De veris principiis artis dimicatoriae, Tractatus brevis*. Wittenberg: Mattheus Welack, 1579. <VD16 G 3915>

Heussler

    Heussler, Sebaſtian. *Neu Kunſtlich Fechtbuch Darinnen 500 ſtuck im ainfachen Rapier, wie auch ettliche im Rapier und Dolch deß weitberümbten Fecht. Zum andern mal in truck geben und mit schön Stucken verbessert*. Nürnberg: Ludwig Lochner, 1615. <VD17 23:267689Q>

Manciolino

Manciolino, Antonio. *Opera nova, dove li sono tutti li documenti & vantaggi che si ponno havere net meſtier de l'armi d'ogni sorte novamente corretta & ſtampata.* Venezia: Nicolo d'Ariſtotile detto Zoppino, 1531.

Marozzo

Marozzo, Achille. *Opera nova de Achille Marozzo Bolognese, maſtro generale de l'arte de l'armi.* Venezia: Nicolo d'Ariſtotile detto Zoppino, 1531.

Marcelli

Marcelli, Francesco Antonio. *Regole della scherma.* Roma: Antonio Ercole, 1686.

Meyer

Meyer, Joachim. *Gründtliche Beschreibung, der freyen Ritterlichen unnd Adelichen kunſt des Fechtens, in allerley gebreuchlichen Wehren, mit vil schönen und nüʒlichen Figuren gezieret und fürgeſtellet.* Strasbourg: Thiebolt Berger, 1570. <VD16 M 5087> 2nd ed. Augsburg: Michael Manger, 1600. <VD16 M 5088>

Monte 1492

Monte, Pietro. *De dignoscendis hominibus interprete G. Ayora Cordubensi.* Milano: Antonio Zaroto Parmenion, 1492.

Monte 1509

Monte, Pietro. *Exercitiorum atque artis militaris collectanea in tris libros diſtincta.* Milano: Giovani Angelo Scinzenzler. 1509.

Pagano

Pagano, Marc'Antonio. *Le tre giornate di Marc'Antonio Pagano gentil'huomo napoletano. Dintorno alla disciplina dell'arme. Et spetialmente della spada sola.* Napoli: Cilio d'Alife, 1553.

Pallavicini

Pallavicini, Giuseppe Morsicato. *La scherma illuſtrata.* Palermo: Domenico d'Anselmo, 1670.

Paurenfeyndt 1516

Paurenfeyndt, Andre. *Ergrundung Ritterlicher kunſt der Fechterey durch Andre paurenfeindt Freyfechter czu Vienna in Oſterreich, nach klerlicher begreiffung und kurczlicher verſtendnusz.* Wien: Hieronymus Vietor, 1516. <VD16 P 1071>

Paurenfeyndt 1538

Paurenfeyndt, Andre. *La noble science des ioueurs d'espee.* Antwerp: Willem Vorſterman, 1538.

Rösener

Rösener, Chriſtoff. *Ehren Tittel und Lobspruch Der Ritterlichen Freyen Kunſt der Fechter.* Dresden: Gimel Bergen, 1589. <VD16 R 2830>

Sachs

Sachs, Hans. *Sehr Herrliche, Schöne und warhaffte Gedicht.* Nürnberg: Chriſtoff Heußler, 1558. <VD16 S 142>

Sutor

Sutor von Baden, Jakob. *New Künstliches Fechtbuch.* Frankfurt am Main: Willhelm Hoff-man, 1612. <VD17 23:318357D>

Verolini

Verolini, Theodor. *Der Künstliche Fechter: Oder Deß Weyland wohl-geübten und be-rühmten Fecht-Meisters.* Würzburg: Joann Bencard, 1679. <VD17 23:321161B>

Wurm

Das Landshutter Ringbuch. Landshut: Hans Wurm, ca. 1490s. <VD16 R 482>

SECONDARY SOURCES

Books and journal articles

ADAMSON 2011

ADAMSON, WILLIAM CHARLES. *The Nationalism of Joachim Meyer: An Analysis of German Pride in his Fighting Manual of 1570* [Unpublished thesis; paper 1286]. East Tennessee State University School of Graduate Studies, 2011. http://dc.etsu.edu/etd/1286

ANGLIN 1984

ANGLIN, JAY P. "The Schools of Defense in Elisabethan London". *Renaissance Quarterly* **37**(3): 393–410. 1984. DOI 10.2307/2860956

BAUER 2016

BAUER, MATTHIAS JOHANNES. *Der Allten Fechter gründtliche Kunst – Das Frankfurter oder Egenolffsche Fechtbuch.* München: Herbert Utz Verlag, 2016.

BAUER 2019

BAUER, MATTHIAS JOHANNES. "Egenolff's Fight Book: Form and Thought, Then and Now". *The Sword: Form and Thought.* Boydell & Brewer, 2019. pp 208–215. DOI 10.1017/9781787444805

BENZING 1963

BENZING, JOSEF. *Die Buchdrucker des 16. Und 17 Jahrhunderts im Deutschen Sprachgebiet.* Wiesbaden: Harrassowitz Verlag, 1963.

BERRY 1991

BERRY, HERBERT. *The Noble Science: A Study and Transcription of Sloane Ms. 2530, Papers of the Masters of Defence of London, Temp. Henry VIII to 1590.* London and Toronto: Associated University Presses, 1991.

BLAZEKOVIC 2003

BLAZEKOVIC, ZDRAVKO. "Variations on the Theme of the Planets' Children, or Medieval Musical Life According to the Housebook's Astrological Imagery". *Essays in honor of Franca Trinchieri Camiz.* Aldershot and Burlington: Ashgate Publishing, 2003.

BRUGH 2019
    BRUGH, PATRICK. *Gunpowder, Masculinity, and Warfare in German Texts, 1400–1700.* Rochester: University of Rochester Press, 2019.

BÜSCHING 1816
    BÜSCHING, JOHANN GUSTAV. *Wöchentliche Nachrichten für Freunde der Geschichte, Kunst und Gelahrtheit des Mittelalters.* Breslau: Korn, 1816.

CAMES 1995
    CAMES, GÉRARD. "MEYER, Joachim". *Nouveau dictionnaire de biographie alsacienne*, t. 26. 1995.

CHIDESTER 2020
    CHIDESTER, MICHAEL. "The *Bellifortis* of Konrad Keyser of Eichstatt". *Alte Armature und Ringkunst: The Royal Danish Library Ms. Thott 290 2°.* Ed. by MICHAEL CHIDESTER. Somerville: HEMA Bookshelf, 2020. pp. 113–150.

CHIDESTER 2021
    CHIDESTER, MICHAEL. *The Flower of Battle: MS M 383.* Somerville: HEMA Bookshelf, 2021.

CHIDESTER 2024
    CHIDESTER, MICHAEL. "Reception of Joachim Meyer's *Gründtliche Beschreibung der... Kunst des Fechtens*, 1570–1686". *Acta Periodica Duellatorum* (Forthcoming).

CHIDESTER & HAGEDORN 2021
    CHIDESTER, MICHAEL, and DIERK HAGEDORN. *'The Foundation and Core of All the Arts of Fighting': The Long Sword Gloss of GNM Manuscript 3227a.* Somerville: HEMA Bookshelf, 2021.

CHIDESTER & HAGEDORN 2024
    CHIDESTER, MICHAEL, and HAGEDORN, DIERK. *Pieces of Ringeck: The Definitive Edition of the Gloss of Sigmund Ainring.* Medford: HEMA Bookshelf, 2024.

CHIDESTER & Stimmer 2020
    CHIDESTER, MICHAEL, and Tobias Stimmer. *The Illustrated Meyer: A Visual Reference for the 1570 Treatise of Joachim Meyer.* Somerville: HEMA Bookshelf, 2020.

DUPUIS 2006
    DUPUIS, OLIVIER. "Joachim Meyer, escrimeur libre, bourgeois de Strasbourg (1537?–1571)". *Maîtres & Techniques de Combat à la fin du Moyen Age et au Début de la Renaissance.* Ed. by FABRICE COGNOT. Paris: Association pour l'Edition et la Diffusion des Études Historiques, 2006. pp. 107–120.

DUPUIS 2013
    DUPUIS, OLIVIER. "A fifteenth-century fencing tournament in Strasburg". *Acta Periodica Duellatorum* 1(1): 67–79. 2013. DOI 10.36950/apd-2013-004

DUPUIS 2016
    DUPUIS, OLIVIER. "Joachim Meyer, free fencer, citizen of Strasbourg (?1537–1571)". *The Art of Sword Combat: A 1568 German Treatise on Swordsmanship.* Trans. by JEFFREY L. FORGENG. London: Frontline Books, 2016. pp. 171–190.

DUPUIS 2021A

DUPUIS, OLIVIER. "A New Manuscript of Joachim Meyer (1561)". *Acta Periodica Duella-torum* **9**(1): 73–86. 2021. DOI 10.36950/apd-2021-004

EADS & GARBER 2014

EADS, VALIERIE, and REBECCA L. R. GARBER. "Amazon, Allegory, Swordswoman, Saint? The Walpurgis Images in Royal Armouries MS I.33". *Can These Bones Come to Life?* **1**: 5–23. Freelance Academy Press, 2014.

ECKERT 1978

ECKERT, EDWARD A. "Boundary formation and diffusion of plague: Swiss epidemics from 1562 to 1669". *Annales de Démographie Historique Année* **1978**: 49–80.

EDELBECK 1574

EDELBECK, BENEDIKT. *Ordentliche vnd Gründtliche beschreibunge, des grossen schiessen, mit dem Stahl oder Armburst, auch anderer kurtzweil mehr so gehalten ist worden, in der löblichen Churfürstlichen Stadt Zwickaw.* Dreßden: Stöckel, 1574.

FARR 2000

FARR, JAMES RICHARD. Artisans in Europe, 1300–1914. Cambridge: Cambridge University Press, 2000.

FEBVRE & MARTIN 1976

FEBVRE, LUCIEN and HENRI-JEAN MARTIN. *The Coming of the Book: The Impact of Printing, 1450–1800.* London: Verso Books, 1976.

FORGENG 2006

Meyer, Joachim. *The Art of Combat: A German Martial Arts Treatise of 1570.* Trans. by JEFFREY L. FORGENG. London and New York: Greenhill Books and Palgrave MacMillan, 2006. 2nd ed. London: Frontline Books, 2015.

FORGENG 2016

JEFFREY L. FORGENG. *The Art of Sword Combat: A 1568 German Treatise on Swordsmanship.* Trans. by JEFFREY L. FORGENG. London: Frontline Books, 2016.

FRAME 1999

RABELAIS, FRANCOIS. *The Complete Works of Francois Rabelais.* Trans. by DONALD M. FRAME. Berkely and Los Angeles: University of California Press, 1999.

FRANTI 2021

FRANTI, ADAM. "Art and Symbolism in the Genre of *Fechtbücher*". *Kunst und Zettel in Messer: Bavarian State Library Cgm 582.* Ed. by MICHAEL CHIDESTER. Somerville: HEMA Bookshelf, 2021. pp. 229–240.

FRAISTAT & FLANDERS 2013

The Cambridge Companion to Textual Scholarship. Ed. by NEIL FRAISTAT and JULIA FLANDERS. Cambridge: Cambridge University Press, 2013.

GARBER 1998

GARBER, REBECCA L. R. "Where Is the Body? Images of Eve and Mary in the Scivias". *Hildegard of Bingen.* Ed. by MAUD BURNETT MCINERNEY. New York: Routledge, 1998.

GARBER 2013

GARBER, REBECCA L. R. *Feminine Figurae: Representations of Gender in Religious Texts by Medieval German Women Writers, 1100-1475.* New York: Routledge, 2013.

GARBER & BACHMANN 2020

GARBER, REBECCA L. R., and DIETER BACHMANN. "Transcription and Translation". *Alte Armature und Ringkunst: The Royal Danish Library Ms. Thott 290 2°:* 9–46. Ed. by MICHAEL CHIDESTER. Somerville: HEMA Bookshelf, 2020.

GASSMANN & GASSMANN 2019

GASSMANN, JÜRG and SAMUEL GASSMANN. "Mos geometricus v. Reality: Quantity, Quality, Time and Information in Combat Simulations since the Middle Ages". *Acta Periodica Duellatorum* 7(1): 173–202. 2019. DOI 10.2478/apd-2019-0004

GEVAERT 2014

GEVAERT, BERT. *Heinrich von Gunterrodt.* Wheaton: Freelance Academy Press, 2014.

HEINZE 2006

HEINZE, RUDOLPH W. *Reform and Conflict: From the Medieval World to the Wars of Religion AD 1350–1648.* Monarch Books, 2006.

HERRERA 2012

HERRERA, BREANNE. *The Children of Planets: Freedom, Necessity, And the Impact of the Stars – The Iconographic Dimensions of a Pan-European Early Modern Discourse* [Unpublished thesis]. Budapest: Central European University, 2012. <http://www.etd.ceu.edu/2012/herrera_breanne.pdf>

HILS 1985

HILS, HANS-PETER. *Meister Johann Liechtenauers Kunst des langen Schwertes.* Frankfurt am Main: P. Lang, 1985.

HULL 2008

HULL, JEFFREY. "The Longsword Fight Lore of Mertin Siber." *Masters of Medieval and Renaissance Martial Arts.* Ed. by JOHN CLEMENTS. Boulder: Paladin Press, 2008. pp. 223–238.

JASER 2014

JASER, CHRISTIAN. "Der Bürger und das Schwert – Faktoren der städtischen Fechtschulkonjunktur im ausgehenden Mittelalter". *Das Schwert – Symbol und Waffe.* Ed. by LISA DEUTSCHER, MIRJAM KAISER und SIXT WETZLER. Rahden and Westfallia: Verlag Marie Leidorf GmbH, 2014.

JUNG 1746

Vollständiges Diarium Von der Höchst-erfreulichen Crönung Des Allerdurchlauchtigsten, Großmächtigsten und Unüberwindlichsten Fürsten und Herrn, Herrn Franciscus, Erwehlten Römischen Kaysers. Frankfurt am Main: Johann David Jung, 1746.

KIERMAYER 2012

Joachim Meyers Kunst Des Fechtens. Gründtliche Beschreibung des Fechtens, 1570. Teil 1 – Schwert und Dussack. Trans. by ALEXANDER KIERMAYER. Arts of Mars Books, 2012.

KLINGNER 2017

KLINGNER, ANNETT. *Die Macht der Sterne. Planetenkinder: ein astrologisches Bildmotiv in Spätmittelalter und Renaissance* [Unpublished dissertation]. Humboldt Universität zu Berlin, 2017. <http://edoc.hu-berlin.de/handle/18452/19957>

KLUGE 1895

KLUGE, FRIEDRICH. *Deutsche Studentensprache*, Strasbourg: K. J. Trübner, 1895.

LANDWEHR 2011

Meyer, Joachim. *Joachim Meyer 1600: Transkription des Fechtbuchs 'Gründtliche Beschreibung der freyen Ritterlichen und Adelichen kunst des Fechtens'.* Ed. by WOLFGANG LANDWEHR. Herne: VS-Books, 2011.

LATTIMORE 2011

LATTIMORE, RICHMOND. *The Iliad of Homer.* Chicago: University of Chicago Press, 2011.

LIEBS 2012

LIEBS, DETLEF. *Summoned to the Roman Courts: Famous Trials from Antiquity.* Trans by REBECCA L. R. GARBER. University of California Press, 2012.

LUTZ 1819

LUTZ, MARKUS. *Baslerisches Bürger-Buch enthaltend alle gegenwärtig in der Stadt Basel eingebürgte Geschlechter, nebst der Anzeige ihres Ursprungs, Bürgerrechts-Aufnahme, so wie ihrer ersten Ansiedler und beachtenswerthen Personen, welche aus denselben zum Dienste des Staats, der Kirche und der Wissenschaften, hervorgegangen sind.* Basel: Schweighauser'sche Buchdruckerey, 1819.

MACKAY 1999

MACKAY, E. ANNE. *Signs of Orality: The Oral Tradition and its Influence in the Greek and Roman World.* Leiden: Brill, 1999.

MERKEL 1898

MERKEL, JOHANNES. *Heinrich Husanus. 1536 bis 1587. Eine Lebensschilderung.* Göttingen: L. Horstmann, 1898.

MONDSCHEIN & DUPUIS 2019

MONDSCHEIN, KEN and DUPUIS, OLIVIER. "Fencing, Martial Sport, and Urban Culture in Early Modern Germany: The Case of Strasbourg". *Journal of Medieval Military History* **XVII**: 237–257. Boydell & Brewer, 2019.

MONTECUCCOLI 1736

MONTECUCCOLI, RAIMONDO. *Besondere und geheime Kriegs-Nachrichten des Fürsten Raymundi Montecuculi.* Leipzig: Weidmannischen Buchladen, 1736.

NEVILLE 1695

NEVILLE, HENRY. *The works of the famous Nicolas Machiavel, citizen and secretary of Florence.* London: R. Clavel, C. Harper, J. Amery, 1695.

ONG 1958

ONG, WALTER J. *Ramus, Method, and the Decay of Dialogue: From the Art of Discourse to the Art of Reason.* Cambridge: Harvard University Press, 1958.

ONG 1982

ONG, WALTER J. *Orality and Literacy: The Technologizing of the Word.* London: Methuen & Co., 1982.

PARKS 2009

Machiavelli, Niccoló. *The Prince.* Trans. by TIM PARKS. Penguin Classics, 2009.

PRICE 2003

PRICE, DAVID HOTCHKISS. *Albrecht Durer's Renaissance: Humanism, Reformation, and the Art of Faith.* Ann Arbor: University of Michigan Press, 2003.

RESKE 2007

RESKE, CHRISTOPH. *Die Buchdrucker des 16. und 17. Jahrhunderts im deutschen Sprachgebiet.* Harrassowitz Verlag, 2007.

RUBLACK & HAYWARD 2015

The First Book of Fashion: The Books of Clothes of Matthäus & Veit Konrad Schwarz of Augsburg. Ed. by ULINKA RUBLACK and MARIA HAYWARD. London and New York: Bloomsbury, 2015.

RUDOLPH 1726

RUDOLPH, GOTTFRIED (alias Bugenhagen Pommer). *Curieuse Sammlungen Einiger Merckwürdigkeiten Aus der Geographie, Genealogie, Chronologie, Geist- und Weltl. Ritter-Orden, Heraldique, Kirchen- und Politischen Historie.* Leipzig: Martini, 1726

RUDOLPH 1752

RUDOLPH, GOTTFRIED (alias Bugenhagen Pommer). *Sammlungen historischer und geographischer Merkwürdigkeiten.* Altenburg: Paul Emanuel Richtern, 1752.

RUMMEL 2015

The Correspondence of Wolfgang Capito: Volume 3 (1532–1536). Trans. by ERIKA RUMMEL. Ed. by MILTON KOOISTRA. Toronto: University of Toronto Press, 2015.

SCHAER 1908

SCHAER, ALFRED. *Die altdeutschen Fechter und Spielleute: Ein Beitrag zur deutschen Culturgeschichte.* Strasbourg: Karl J. Trübner Verlag, 1908.

SIMON-MUSCHEID 1988

SIMON-MUSCHEID, KATHARIN. *Basler Handwerkszünfte im Spätmittelalter.* Bern, Frankfurt an Main, New York, Paris: Peter Lang, 1988.

SMITH 1983

SMITH, JEFFREY CHIPPS. *Nuremberg, a Renaissance City, 1500–1618.* Austin: The University of Texas Press, 1983.

SPIES 1587

Anonymous. *Historia von D. Johann Fausten.* Frankfurt am Main: Johann Spies, 1587.

SPRUNER 1868

SPRUNER VON MERZ, KARL. *Die Wandbilder des Bayerischen National-Museums; historisch erläutert von Carl v. Spruner.* München: J. Albert, 1868.

THÜRER 1941

    THÜRER, PAUL. "Glarnerischer Gewehrrodel des 16. bis 18. Jahrhunderts". *Jahrbuch des Historischen Vereins des Kantons Glarus* **50**: 105–142. 1941.

TLUSTY 2011

    TLUSTY, B. ANN. *The Martial Ethic in Early Modern Germany. Civic Duty and the Right of Arms.* New York: Palgrave Macmillan, 2011.

WALDBURG WOLFEGG 1998

    WALDBURG WOLFEGG, CHRISTOPH Graf zu. *Venus and Mars: The World of the Medieval Housebook.* Ed. by ALMUTH SEEBOHM. München and New York: Prestel, 1998.

WASSMANNSDORFF 1870

    WASSMANNSDORFF, KARL. *Sechs Fechtschulen.* Heidelberg: Karl Groos, 1870.

WEIR 2004

    WEIR, ROBERT. *Roman Delphi and Its Pythian Games.* Oxford: British Archaeological Reports, 2004.

WIERSCHIN 1965

    WIERSCHIN, MARTIN. *Meister Johann Liechtenauers Kunst des Fechtens.* München: Beck, 1965.

WOOTTON 1995

    Machiavelli, Niccoló. *The Prince.* Trans. by DAVID WOOTTON. Hackett Publishing Company, 1995.

Websites

CURTIS 2004

    CURTIS, MARY DILL. "Italian Connection?" *Destreza Translation and Research Project*, 2004. <https://destreza.us/translations/narvaez_italian.html>

DUPUIS 2021B

    DUPUIS, OLIVIER. "La place de l'escrime à Strasbourg (XVe–XVIIe siècles)". *Martial Culture in Medieval Town*, 2021. <http://martcult.hypotheses.org/1312>

KLEINAU 2011

    KLEINAU, JENS P. "A Fechtschule in the works of Francois Rabelais?". *Hans Talhoffer: A Historical Martial Arts blog by Jens P. Kleinau*, 2011. <http://talhoffer.wordpress.com/2011/05/12/a-fechtschule-in-the-works-of-francois-rabelais/>

KINTZ 2021

    KINTZ, PIERRE. "Tobias Stimmer, illustrateur du Fechtbuch de Joachim Meyer". *Martial Culture in Medieval Town.* 18 June 2021. <http://martcult.hypotheses.org/1316>

MAURER 2022

    MAURER, KEVIN. "Joachim Meyer 1561: An English translation by Kevin Maurer". *Meyer Freifechter Research.* 2022. <http://mffgresearch.com/translations>

MAURER & VANSLAMBROUCK 2013

MAURER, KEVIN and CHRIS VANSLAMBROUCK. "Who were the Wingelfechter?". 2013. <http://www.researchgate.net/publication/291284457_Who_were_the_Winkelfechter>

NORLING 2011

NORLING, ROGER. "The Secret Fechtbuch of the Little Fuggers". *HROARR*, 2011. <http://hroarr.com/article/the-secret-fechtbuch-of-the-little-fuggers>

NORLING 2012A

NORLING, ROGER. "The Dusack – A Weapon of War". *HROARR*, 2012. <http://hroarr.com/article/the-dusack/>

NORLING 2012B

NORLING, ROGER. "The History of Joachim Meyer's fencing treatise to Otto von Solms". *HROARR*, 2012. <http://hroarr.com/article/the-history-of-joachim-meyers-treatise-to-von-solms>

VODIČKA 2019

VODIČKA, ONDŘEJ. "1597, Fencers' Ordinance of the Old Town of Prague". *Martial Culture in Medieval Town*, 2019. <http://martcult.hypotheses.org/322>

WIKTENAUER

CHIDESTER, MICHAEL, et al. "Joachim Meyer". *Wiktenauer*. <http://www.wiktenauer.com/wiki/Joachim Meyer>

DICTIONARIES

GRIMM

GRIMM, JACOB and WILHELM GRIMM. *Deutsches Wörterbuch von Jacob Grimm und Wilhelm Grimm*, Version 01/23. Wörterbuchnetz des Trier Center for Digital Humanities. <http://www.woerterbuchnetz.de/DWB>

LEXER

LEXER, MATTHIAS. *Mittelhochdeutsches Handwörterbuch von Matthias Lexer*, Version 01/23. Wörterbuchnetz des Trier Center for Digital Humanities. <http://www.woerterbuchnetz.de/Lexer>

SCHERZII

SCHERZII, JOHANNIS GEORGII. *Johannis Georgii Scherzii J.U.D. et P.P. argentoratensis Glossarium germanicum medii aevi potissimum dialecti suevicae*. Strassbourg: Typis Lorenzii et Schuleri, 1781.

INDEX

ABOUT THE AUTHOR

REBECCA L. R. GARBER received her Ph.D. in German Languages and Literatures from the University of Michigan in 1999, with a specialization in medieval studies. She taught for three years at Wayne State University and then at a private high school before turning to translation as a full-time occupation. After joining a stage combat troupe in Michigan in 2006, she began translating combat manuals with CHEMAS (the Cambridge Historical European Martial Arts Study group), where her expertise in Medieval German and Latin was more useful than in most aspects of modern life. This is her third published HEMA translation, the first being "Florius de Arte Luctandi" (FLP) in *The Flower of Battle of Master Fiore Furlano de'i Liberi* (with KENDRA BROWN; 2016), and the second being Hans Talhoffer's København manuscript (HTK) in *Alte Armature und Ringkunst: The Royal Danish Library Ms. Thott 290 2°* (2020).

ABOUT THE EDITOR

MICHAEL CHIDESTER has been studying historical European martial arts since 2001, and has been Editor-in-Chief of Wiktenauer since 2011. In 2019, he started a publishing company called HEMA Bookshelf to produce facsimiles of fencing manuals and publish translations and historical research. MICHAEL is a Research Scholar of the Meyer Freifechter Guild, a founding member of the Society for Historical European Martial Arts Studies, a member of the Western Martial Arts Coalition, and a Lifetime Member of the HEMA Alliance. He has lectured on historical martial arts across North America and Europe and has written several books, including *The Illustrated Meyer: A Visual Reference for the 1570 Treatise of Joachim Meyer* (2020), *The Flower of Battle: MS M 383* (2021), and *"The Foundation and Core of All the Arts of Fighting": The Long Sword Gloss of GNM Manuscript 3227a* (2021), and edited or contributed to many others. He currently trains at Athena School of Arms in Cambridge, MA.

www.ingramcontent.com/pod-product-compliance
Lightning Source LLC
Chambersburg PA
CBHW021657120626
46545CB00004B/1281